Amasa Stone and The Ashtabula Bridge Disaster

A Gilded Age Parable of Wealth and Tragedy

Darius L. Salter

EMETH PRESS
www.emethpress.com

Amasa Stone and The Ashtabula Bridge Disaster: *A Gilded Age Parable of Wealth and Tragedy*

Dedication

To David Tobias

The world's foremost excavator and archaeologist of the
Ashtabula Bridge wreck site

Table of Contents

Acknowledgments

I am placing the acknowledgments and sources together because they, if not being one and the same, blend as indistinguishable for resourcing this book. In none of my previous nine books have I relied so much on individuals through face-to-face interviews, emails, and phone calls. Unfortunately, the one person who was the catalyst for many of the articles on which I have relied, especially for the wreck itself, is no longer living. Around the year 2000, Thomas Corts, a native of Ashtabula, Ohio, and then President of Samford University, Birmingham, Alabama, gathered a team of essayists, mostly from the Ashtabula area to investigate aspects of the tragedy. Corts initial attraction to the project was the same as mine, the loss of Philip Paul Bliss. Though the essays are of uneven quality, I have found all of them to be of immense help. Dr. Corts passed away in 2009.

Ashtabula native, David Tobias, a contributor to the Corts' book, has been of invaluable help. David has given his life to excavating, collecting, and organizing artifacts from the wreck. Because of his intrepid dedication to gathering data, books, articles, and countless tidbits of information about the bridge accident, I have unhesitatingly dedicated this book to him. David has combed the location of the wreck more than any other single person. He has repeatedly sent me information and has accompanied me to the wreck site. The foremost fringe benefit or serendipity for a research journey is meeting fellow travelers. David is one of the most delightful individuals I have ever met. Getting to know him has been a true joy.

Barbara Hamilton is the only other individual from the Corts project that I have been able to contact. Her article on "Identifying the Victims" is an example of thorough, and in some ways painful research. Another research scholar who had given a like quantity of time as did Tobias and Hamilton, was Charles Burnham. He founded the Ashtabula Railway Historical Foundation and his article, "The Ashtabula Horror," may be the most recent comprehensive account of the accident. Unfortunately, Burnham passed away in 2018, and his foundation went out of existence.

As cited in this book, Dario Gasparini is the foremost engineering expert on the Ashtabula Bridge. He was one of the first individuals with whom I met in attempting to understand the technicalities of the bridge. He has done his best to step into my knowledge vacuum, and wherever my interpretations have been amiss, the blame is on me. I can hardly think of anything more outside of my knowledge inventory than engineering, though in another lifetime I did manage to finish in the top six on a high school state-wide physics exam. David Simmons, a long-time employee of the Ohio History Center, Columbus, Ohio, who has excelled in studying railroad bridges, and in particular the Ashtabula Bridge, has answered countless questions. He turned over to me a three-inch-thick file of articles and correspondence which he had gathered and collected over the last quarter of a century. Thanks to both Gasparini and Simmons, who have co-authored bridge articles providing essential information.

A more recent arrival to my project is Stephen Ressler, who in the 2022 Kahn Distinguished Lecture Series at Lehigh University gave a presentation, "The Ashtabula Bridge Disaster and the Advent of Civil Engineer Professionalism." He gave convincing reasons for the bridge's failure. Ressler excelled in reconstructing models and experimenting with them in order to give visual and spatial images of design flaws. The lecture as well as follow-up emails and a zoom meeting with Dr. Ressler were critical for understanding Stone's mistakes in designing and erecting the bridge.

Upon arriving in Ashtabula in the fall of 2020 for my initial inquiry into the project, I was informed that a documentary was being done on the Ashtabula Bridge disaster. I normally receive this kind of information with a bit of skepticism, but upon further investigation discovered that Leonard Brown, founder, and president of Beacon Productions had spent the past decade researching and filming *Engineering Tragedy: The Ashtabula Train Disaster*. Thus, another new friend has provided me with much needed information. He had already consulted with several engineers and historians who enabled him to build a scaled model of the bridge. Len has provided me more help than I have been able to reciprocate. I have perhaps fed him some information on Amasa Stone that he did not yet possess. Overall, this book has been a collaborative effort between us, although he started his project some 11 years ago.

I took a week combing Charlton, MA, discovering that the city fathers and librarians had almost completely lost the trail of their native son who became one of the titans of the Gilded Age. Amy Hiatala at the Sturbridge Historical Society and Curtis Meskus at Charlton City Hall furnished me with Charlton historical data and a miniscule, detailed map that could be deciphered with a magnifying glass. Amy also discovered a 1905 picture of the Stone house before it was torn down. Finding the "farm" was especially difficult and required knocking on doors around the area I thought was the approximate vicinity. Brian Mitchell, living almost directly across the road from what had been Amasa Stone's boyhood home, gave Brenda and me a tour on a cold, late, winter evening. He pulled tax receipts that verified the Stone farm location.

Pat Stockwell, more than anyone else in Charlton has researched the Amasa Stone narrative. Upon seeing me interviewed by Leonard Brown, he fortuitously took the initiative to contact me. His research has uncovered obscure information in ancient newspapers and books. He has made critical corrections to this book and may be Charlton's most informed historian and best kept secret.

Two unusual collections have been at my disposal, all of them consisting of newspaper clippings. Gabriel Leverich, a civil engineer, gave a presentation on the Ashtabula Bridge disaster to the American Society of Engineers in 1877. He collected every newspaper article that he could find and placed them in an album. Fortunately, this album came into possession of the Linda Hall Library in Kansas City, MO. The album consisted of some 100 articles. A similar collection was put together by a sister of Edward Trueworthy, who survived the wreck but lost their father and a sibling. Around the year 2000 the scrapbook was placed on eBay, and Thomas Corts purchased the album for $350.00. After his research, he donated the book to the Ashtabula County Historical Society at Geneva on the Lake, Ohio. Volunteers were kind enough to copy the entire scrapbook and mail it to me. Without these two resources, it would have been almost impossible to piece together the testimonies at the Coroner's Jury hearings.

Several libraries had direct bearing on the project. The collection of the Ashtabula County Library was not as extensive as I had hoped. Nonetheless, archivist Carrie Wimer granted me full access to the two large manilla folders. The Bentley Library, adjacent to the University of Michigan, Ann Arbor, possessed the minutes of the Lakeshore and Michigan Southern Railway. These consist of one volume of approximately 500 pages.

The Amasa Stone, Samuel Mather, and John Hay papers are preserved at the Western Reserve Historical Society in Cleveland, OH. Disappointingly, the ink on many of the Amasa Stone letters had bled all over the pages to the extent that the smudged copies were impossible to read. Much of that loss was made up by the personal correspondence between John Hay and Amasa Stone at the John Hay Library of Brown University. The archivists graciously sent me copies of all the letters between Hay and Stone, as well as other correspondence between Stone and the Vanderbilt's. Over 100 in number, these letters are critical for tracking the last two years of Stone's life, and the single most important source for providing a window into his psyche.

Though not as critical as the above libraries and archives, the following have been helpful: the archives of Case Western Reserve University, the Kelvin Smith Library of Case Western Reserve University, and the University of Rochester, which sent me some 15 letters written by Stone. A special thanks goes to Don Guenther, archivist at Old Stone Church, who gave me materials on this architectural wonder, as well as researching newspaper articles. Karen Cohen, facilities manager at Case Western Reserve University, gave Brenda and me a personal tour of this ecclesiastical gem. The Cedar Rapids Library of Cedar Rapids, Iowa, sent me biographical information on Joseph Tomlinson. Zoe Cheek, at the Massachusetts Historical Society in Springfield, MA, furnished materials on Stone's bridge building while he lived in Springfield. For no previous project have I so extensively relied on newspapers. The librarians on the sixth floor of the Cleveland Public Library have never failed to fulfill my requests for articles. The Cleveland City Hall Library was also eager to point me in the right direction. I found valuable help in the Missouri room on the fifth floor of the Kansas City Public Library.

I have relied on several people who wrote works on Amasa Stone, his in-laws, or his environs. They are, not necessarily in the order of importance, Burton Smith Dow, Gladys Haddad, Jan Cigliano, Alan Dutka, Dan Ruminski, Jeanette Tuve, Richard Baznik, Katherine Mackley, Jeannine Love, and John Taliaferro. Of course, Stephen Peet's work on the disaster is indispensable.

As to standard works, I have relied heavily on two of America's foremost biographers, Ron Chernow, and T.J. Stiles. Thanks goes to the many railroad historians from whom I have gleaned for the book, in particular Richard White, John White, and Walter Licht. One of my most valuable finds was *Railroad Leaders, 1845-1890* written by Thomas Cochran.

Without his extensive research on the managerial practices of railroad owners and administrators of the nineteenth century, including Stone, my work would have been greatly impoverished. Albert Chandler's *Visible Hand* continues to be the Bible for analyzing nineteenth century managerial practices.

Of course, there are always dead-end streets on a project of this type. According to Thomas Corts, engineer John Marston Goodwin, 1833-1891, served as a stenographer for the Coroner's Jury in Ashtabula. Because of a lack of funds, the manuscript was never published. The original manuscript possibly exists somewhere in the world, but neither Thomas Corts nor I were able to find it. Being in possession of this valuable source would have made the writing of Chapter nine easier, and I am confident, more accurate. Newspapers for the most part represent second-hand information and someone else's interpretation.

I thank all of the contributors to Wikipedia, a valuable source beyond compare. All historians use it, but I have found no one who gives these researchers the recognition they deserve. Wikipedia has become the single most important resource for tracking down relevant materials. Along with all of my disparaging comments on technology, I bless those who have hung out information in cyber-space or have digitized valuable material.

As to the individuals who have done their best to enable me, I turn to the usual suspects. Debbie Bradshaw, head librarian at Nazarene Theological Seminary has tracked countless resources for me through interlibrary loan. Beth Plank and Kristi Seaton have assisted me in word processing and researching information. Both are married to computer gurus which always helps. A special thanks goes to Bart Nitz, a technological genius who often rescued us from the black hole of cyberspace.

The following people have read either parts, or all of the manuscript, offering insightful corrections and suggestions: William Miller, Dario Gasparini, David Simmons, Leonard Brown, David Tobias, Pat Stockwell, Barbara Hamilton, Stephen Ressler, Jim Stump, Nathan Clark and Stella Harris. For their investment of time and energy, as well as covering territory before I ever explored it, I am grateful. Harris with a professional eye made many editorial corrections.

Appreciation goes to Laurence Wood at Emeth Press and his staff for publishing this work.

As always, my final gratitude goes to my wife Brenda who endures her husband's fascination with events and persons long ago forgotten. She has done much of the manuscript's word processing and tracking relevant resources, as well as e-mailing. We celebrated 50 years together on May 26, 2022. The journey has taken us places we never dreamed of going and doing what we had never foreseen as part of our future. I often believe the roller coaster ride has overwhelmed us with stress for which she did not sign up but has graciously and faithfully endured. I never do anything that I don't have a haunting sentiment that I should be doing something else. I trust that this book has done justice to Amasa Stone, accurately embracing both his triumphs and failures. The value of remembering Philip Paul Bliss as well as the other 97 victims who perished on December 29, 1876, can hardly be exaggerated.

To God be the glory!

Darius L. Salter
August 2022

Introduction

On May 11, 1883 a 65-year-old man stepped into the bathroom of his palatial home in Cleveland, Ohio, and put a bullet through his heart. Less than two weeks later, New York City threw perhaps its greatest party ever. The city exploded "14 tons of rockets and flares," bands played aboard boats in the East River and both employees and students took a holiday.[1] Obviously, the city was not celebrating Amasa Stone's death though all of the major newspapers had reported his suicide, but the two events had more in common than their proximity in time. Both events had to do with a bridge.

Amasa Stone became primarily known for the death of almost 100 people in the collapse of the Ashtabula, Ohio Bridge on December 29, 1876. The event was the deadliest train accident to have occurred in the United States and is still the deadliest train bridge collapse.[2] New York celebrated the completion and opening of the Brooklyn Bridge, up to that time the greatest engineering feat in American history. At its official inauguration on May 24, 1883, 150,000 people walked across the bridge. It is still a favorite promenade for both New Yorkers and tourists. One bridge was a disastrous failure, and the other a total triumph.

But this simplistic contrast deserves a closer look. Both bridges were a first, both demanded geniuses, both builders distrusted anyone's judgement other than his own were dogmatically confident while suffering

---

[1] David McCullough. *The Great Bridge* (New York: Simon and Schuster) 426.

[2] The deadliest train wreck to ever take place was on January 13, 1917, in Ciurea, Romania. Over 1,000 individuals were killed, mostly Russian troops. Christian Wolmar, *A Short History of the Railroad* (London: Dorling Kindersley Unlimited, 2019) 284-285. On November 1, 1918, the Malbone Street Wreck in Bronx, New York, killed 102 people, and on July 9, 1918, two trains collided in Nashville, Tennessee killing 101 people. "Deadliest Train Accidents in American History," enjuris.com.

from digestive disorders, and more critically, the bridges killed the men who conceived and built them. John Roebling, the designer and builder of the Brooklyn Bridge, died of tetanus and lockjaw, having had his toe crushed between a beam and a boat being docked while giving oversight in the early stages of its construction. Even more fascinating was the disability of his son Washington, who assumed the superintendency of his father's wild scheme to bridge the East River from Brooklyn to Manhattan. Hardly anyone other than his immediate employees, a few foremen, and other oversight personnel knew of his whereabouts. Suffering from the "bends"caused by rapid decompression as he ascended from one of the "caissons." Washington became paralyzed, but yet retained sufficient mental acuity to literally oversee the building of one of the most famous bridges in the world from his bedroom window. In Washington Roebling's perception "His nervous system was shattered," almost the exact language Amasa Stone used concerning his own condition.

In short, bridge building was the most challenging engineering feat of the nineteenth century, only to be surpassed by skyscrapers in the early twentieth century. Doing what had never been done before demanded Herculean powers of courageous fortitude exemplified by Amasa Stone and the Roeblings. John Roebling was known to be poised, confident, unyielding, imperious, sincere, all words used of Amasa Stone. Of Roebling it was said he possessed "power of will... tenacity of purpose, and confident reliance upon self... an instinctive faith in the resources of his art that no force of circumstances could divert him from carrying into effect a project once matured in his mind...."[3] This description could have been perfectly applied to Stone. Both men had known hardly anything but hard work and long hours, a genius and toughness exhibited by the few imposed on the many that would bring America to the pinnacle of its industrial and technological triumph over every country of the world.

It is safe to say that not one of the tens of thousands of drivers who daily traverse the Brooklyn Bridge give a passing thought to the caissons that were sunk and the ingenuity of the Roeblings and their employees who suffered from the "bends" while achieving the daunting task of building a bridge over one mile long. The Ashtabula disaster is all but forgotten other than the faint recollection of librarians who can point a researcher to arcane files, a responsibility not to be taken for granted by those of us at their mercy. Specific knowledge of what the "cartons" hold belongs to

---

[3] McCullough, 41.

those of us who naively, if not delusionally believe that whatever facts might be gleaned are still relevant, and possibly of interest to twenty-first century readers.

The motivation that goes into a huge investment of time and energy for a person in their eighth decade is almost as puzzling to the researcher as the subsequent reader. Why bother in a world consumed by pop culture, technological stimulation, and sound bite information? It seems to me (I would like to have explored this with the late David McCullough) there is a possibility to go where no one else has been before by taking a trip backward as much as taking a trip forward. In other words, there has never been a book-length narrative on Amasa Stone and the Ashtabula Bridge train disaster, though I will draw on persons presently who know more about the Ashtabula Bridge failure and its builder Amasa Stone than I do.[4]

Also, I presume there are lessons to be learned that might shed some light on present day issues. Since I have not set out to extol Stone as saint or sinner, it may be disappointing to the reader that there is no smoking gun to be revealed in this book. But yet, there are inescapable moral judgements to be made, and apriori assumptions that some deeds are more salubrious than others, enhancing the betterment of humankind, however differently betterment may be defined. Some of this betterment as opposed to detriment is due to events out of human control, but the historian believes prosperity as opposed to cyclical poverty, safety as opposed to careless endangerment, are due to decisions made in centuries past, or even a few days past.

From the outset I confess I am not a detached observer, as I immerse myself in the life of Amasa Stone, as well as the lives of others who were involved in the failure of the Ashtabula Bridge. Thus, while I do not begin with the assumption that I will excoriate or exonerate any one individual, I nonetheless will exhibit moral judgements which perpetuate a view that in the words of historian Henry Commager, "Moral laws are universal and timeless; murder is always murder and betrayal is always betrayal, cruelty and intolerance are always the same, the historian cannot stand above the moral laws, or stand aside from them, but must acknowledge them and

---

[4] Stephen Peet, a Congregational minister, wrote a book immediately after the wreck which I will later discuss.

participate in them and apply them."[5] And yet these moral laws are not fixed. In the Gilded Age insider trading was a right earned by the financial genius of a Jay Gould, and monopolies by the foresight of a Henry Flagler and should not be judged by judicial regulations that did not yet exist; the same for engineering knowledge that could be attained only by trial and error and would belong to those who gained by the earlier failure of others. I sit from an entirely different perspective than did Amasa Stone and John D. Rockefeller, and confess with Commager that, "The historian, like the judge, the priest, or the statesman is a creature of his race, nationality, religion, class, of his inheritance, and his education, and he can never emancipate himself from these formulative influences and achieve Olympian impartiality."[6]

The following is not a biography of Amasa Stone. I am writing what historians have categorized as microhistory. Rather than my primary purpose being to write a full orbed history of Amasa Stone's life, I am attempting, in historian Jill Lapore's words, to trace my elusive subject, Amasa Stone, through "slender records" and "address a small mystery," in this case the collapse of a bridge and the fairness or unfairness of labeling Stone a "murderer" which many people did. Lapore writes, "If biography is largely founded on a belief in the singularity and significance of an individual's contribution to history, microhistory is founded upon almost the opposite assumption, however singular a person's life may be, the value of examining it lies in how it serves as an allegory for the culture as a whole."[7] Where Lapore uses the word "allegory," I utilize the word "parable" to argue Amasa Stone was an archetypical industrialist of the Gilded Age, a leader in its primary occupation, railroads, and was eventually destroyed by the very forces he helped create. The uniqueness of this book, is though many articles have been written on the "disaster," no one has described the cultural influences that flowed into December 29, 1876 making the bridge failure inevitable. A blizzard which made travel almost impossible on this particular night was bound to happen somewhere and at some time. Or to put it another way, no one has placed the tragedy within

---

[5] Henry Steele Commager. "Should The Historian Make Moral Judgements?" in *A Sense of History: The Best Writing from the Pages of American Heritage* (New York: American Heritage, 1985) 467.

[6] Ibid., 470.

[7] Jill Lepore. "Historians Who Love Too Much: Reflections on Microhistory and Biography," *The Journal of American History* (June 2001) 141.

the larger context of a critical period in American history. For the historian, events do not "just happen." If so, there would be no history to write.

I also work with the assumption that many of the issues facing Amasa Stone in the Gilded Age are still with us today: capitalism within the confines of government regulation; an achieving personality battling the inherent contradictions of his bewildering psychological and cultural complexities in the age of acquisition; a period of time when highly creative individuals attempted to define themselves by industrial magnitude and a variety of inventions that still merit the question, "Did this really make us better?" Of course, indoor plumbing and electricity did make us better, but at what price? Part of a historian's job is to figure out the price tag of the past in order to not pay too much for the future at the cost of health, broken relationships, and the loss of lives.

Amasa Stone was a creative genius, a man of Herculean stamina, and a business entrepreneur unsurpassed in the Gilded Age, an era defined by industrial titans. He deserves far more attention than he has been given. Also, a sesquicentennial trip needs to be taken to Ashtabula where a village of 2,500 in the dark of a cold night witnessed an event that looked as if it had come out of Dante's "Inferno." The Chicago fire had taken place five years before, but until December 29, 1876, no moving vehicle on land had experienced such catastrophic destruction. Unfortunately, tragedy never goes out of style. Thus, we step back into a time and place that produced heroes but also exemplified the futility of circumstances beyond human control.

I am not a train buff, much less someone who understands the complications of train engineering. This project for me has represented a steep learning curve. Thankfully bridge engineers and historians have filled in my knowledge gaps. Some you have already met in the "acknowledgements," and others you will meet in the following pages. The following pages will make an honest and sincere effort to explore exactly what, why, and how the "Ashtabula disaster" happened. I confess the margins of explorations are difficult to establish. Some of my readers may tire of my cultural, philosophical and theological musings. Perhaps I am too sensitive to historians, who have bypassed the Church, when it played such a critical part in the lives of nineteenth-century Americans. Or one might say that the macrocosm is critical for understanding the microcosm. Thus, I have cast a wide net around one of the most fascinating periods of American history, what Mark Twain and Charles Dudley Warner dubbed the "Gilded Age."

Though not as well-known as other "robber barons," Amasa Stone stands as a representative figure. Of course, this perspective begs the question, did he make the acquisitive age or did the acquisitive age make him? There is no definitive answer to that question but simply to confess that I begin with a hypothesis and attempt to demonstrate that Stone was not an anomaly, but very much a product of the nineteenth-century American ethos, shaped by his Puritan ancestors, the uniqueness of his personality evidencing Max Weber's thesis in his *The Protestant Work Ethic and the Spirit of Capitalism.* Our forefathers made sure that America would be founded on the presuppositions of a meritocracy rather than an aristocracy. Benjamin Franklin and Thomas Jefferson were of one mind that government should grant the liberty to pursue happiness. Franklin Foer refers to Franklin and Jefferson as "our great early technologists, profound exponents of liberty. The United States loudly promoted the gospels of technology and individualism, evangelically spreading them over the globe."[8]

But the liberty to define the route to one's happiness proved to be an overwhelming burden for many, including John Roebling and Amasa Stone. Foer further states, "Human nature is malleable. It's not some fixed thing, but has a breaking point, a point at which our nature is no longer really human. We might decide to sail happily past that threshold, but we need to be honest about the costs."[9] The lowly Galilean said something about not building a tower before counting the cost. Before we condemn Amasa Stone, we need to first confess that few if any of us have ever made an exact calculation. We can see into the past much better than Amasa Stone could see into the future.

My initial attraction to this project was a vague awareness that Philip Paul Bliss, a musical genius, had been killed in a train wreck somewhere around Erie, Pennsylvania. Those with an interest in American church history will appreciate reading Chapter 8, "The Loss of Philip Paul and Lucy Bliss." For those not so inclined, the book will retain its narrative if that chapter is entirely skipped. But the reader should be aware Bliss died in attempting to fulfill a request of Dwight L. Moody, the most successful mass evangelist up to that time in America. The chapter explores the irony of religion thriving during an age known for its greed, corruption, and ostentatious materialism. In fact, Moody's success was partially due to the

---

[8] Franklin Foer. *World Without Mind: The Existential Threat of Big Tech* (New York: Penguin Press, 2017) 230.

[9] Ibid., 231.

support of industrial and mercantile titans such as Cornelius Vanderbilt and John Wanamaker.

Amasa is a biblical name, nephew of David, cohort of the usurper Absalom, and killed by David's commander Joab, as narrated in II Samuel. The King James Version places the accent on the second syllable. On one of my first phone calls to the Western Reserve Historical Society in Cleveland, I was informed by someone that the accent falls on the first syllable, that is, the first two letters, "Am." In a letter written to him September 4, 1989, Ohio historian David Simmons learned from a great-granddaughter of Amasa Stone that the accent is placed on the first syllable, and thus, pronounced by the family, AMasa. This is strange in that everyone in Charlton, Massachusetts where Stone was born and raised, still uses the King James pronunciation with the accent on the second syllable. Nonetheless, in the several hundred conversations I have had concerning our subject, I have used the first syllable pronunciation.

Chapter 1

Amasa Stone and His Puritan Heritage

The brothers Simon and Gregory Stone came to America in 1635.[1] Stone roots were deep in the Church of England; the brothers' great-grandfather served as a church warden at Great Bromley, England. He "was a yeoman of substantial estate, and position in the community where he lived. He died at the age of about 50 years."[2] The Stones who migrated to the new world were Puritan non-conformists who disdained the "rule of uniformity" which required strict adherence to *The Book of Common Prayer*. The voyage of 11 weeks included Simon Stone, his wife Julia, along with five children from 16 years to five months. They settled in Watertown, Massachusetts, a hamlet some 15 miles west of Boston.

By 1643 Simon had acquired about 70 acres of land, most of it tillable, in Watertown. Here he built a substantial house with a 20 square-foot dining room and a large kitchen. He was a leading citizen in Watertown and a deacon of the largest Watertown church. Simon died in 1665 and a grandson, Nathaniel Stone, was the first Stone to attain New England prominence. In 1690 he graduated from Harvard and was a pastor for 54 years at a church in Harwick, Massachusetts. He wrote at least one volume in 1731 entitled *The Wretched State of Man by the Fall*.[3] The family bore

---

[1] All information concerning the ancestry of Amasa Stone Jr. is found in J. Gardner Bartlett. *Simon Stone Genealogy: Ancestry and Descendants of Deacon Simon Stone Watertown, Mass., 1320-1926* (Boston: Pinkham Press, 1926) 47. babel.hathitrust.org/cgi/ssd?id=wu.89080563331;seq=13;num=#biblio

[2] Ibid., 33.

[3] Ibid., 67.

another pastor in the person of James Stone who graduated from Harvard in 1754. He filled the pulpit of Holliston, Massachusetts for 13 years until his early death at age 38.[4] His son Nathan graduated from Harvard and pastored a church at Marlborough, Massachusetts for 50 years.

Several of Amasa Stone's ancestors fought in the Revolutionary War, some of them rendering exceptional service. Silas Stone[5] was a Minuteman, and Phillip Stone served with Ethan Allen and his "Green Mountain Boys" when they captured Fort Ticonderoga on May 10, 1775. Major Ambrose Stone (1757-1850) was also at Fort Ticonderoga, Saratoga, and served with Washington at Valley Forge. He founded and operated a "fulling mill" in Goshen, Massachusetts where he became quite wealthy, supplying large amounts of cloth for the Army in the War of 1812.[6] (A fulling mill cleaned wool so that it could be used as fiber for cloth.)

A Farm Boy

The subject of our investigation, Amasa Stone Jr., came into the world on the family Charlton, Mass, farm April 2, 1818, and was named after his father.[7] The father served in the Massachusetts legislature and lived to be 96 years old. "He was a man of energy, became a prosperous farmer, and served several terms as a selectman."[8] Amasa Jr. was the ninth of ten children born to Amasa Stone and Esther Boyden.[9] They were married for 21 years, Esther dying in July 1833, which means Amasa lost his mother at age 15. His father married Rebecca Lamb, wife of his deceased brother Daniel, on January 15, 1834, with whom he had no children. Both of them died in North Brookfield, Massachusetts in 1875; he 96 years old and she 90 years old.

Physicians, lawyers, pastors, teachers, and business owners filled the Stone genealogical tree and until the time of Amasa Jr.'s birth, farming

---

[4] Ibid., 90.

[5] Ibid., 193.

[6] Ibid., 291.

[7] The father did not refer to himself as Sr., but the son at times designated himself as Jr., especially in his later years. This has been confusing to historians.

[8] Bartlett, 226.

[9] Ibid., 473-476. The children were Anna Maria, 1803-1874; Mary Ann, 1803-1822; Lavinia, 1804-1850; Azuba Towne, 1806-1897; Joseph, 1808-?; Daniel, 1810-1863; Liberty, 1812-?; Esther, 1815-?; Amasa, 1818-1883; Andros, 1824-1896.

had been the exception rather than the rule. Amasa Jr.'s grandfather, Lieutenant Daniel Stone, was given a tract of land in Charlton, Massachusetts about 30 miles west of Worcester. Daniel was with Horatio Gates when he defeated "Gentleman" Johnny Burgoyne at Saratoga, New York on October 17, 1777, one of the major battles and turning points in the Revolutionary War. When Daniel Stone died in 1792, the farm was worth 350 pounds.

At the time Daniel Stone died, his son Amasa, Amasa Jr.'s father, was only 13 years old. Daniel's estate was settled in 1801, and at the age of 22 years, Amasa Sr. paid four siblings $266.67 each to take sole possession of the Charlton farm.[10] This expenditure of over $1000 was no small amount for a 22-year-old man. Amasa Stone Sr.'s achievements by way of hard work fits the description of the arch-typical nineteenth century self-made man.

No one ever outworked Amasa Stone Jr. The weariness of clearing land, chopping wood, tilling rocky soil, walking behind mules, milking cows, repairing equipment, mending fences, and the endless array of chores required by the vaguely defined, never quite done, vocation of farming was excellent education for Junior to learn his first important lesson, "I don't want to do this for the rest of my life." Not that the teenager did not gain practical skills that would serve him in his ultimate vocation. The boy was bright, ambitious, self-confident, and blessed with innate mechanical perception and manual dexterity.

There were so many dairy farms in Charlton, the town was given the moniker "cow town"; eighty percent of the total land in Charlton was divided into farms. Amasa Sr. built a large colonial house, one of the finest in the town. The two-story structure had five windows across the upper floor and two windows on each side of the door downstairs. From the hearth rose a large brick chimney, standing some five feet above the ridgeline of the roof. The house was surrounded by trees, and the farm by granite stone fences because rock was in abundant supply, a byproduct of tilling and plowing the earth. Many stone fences are still found in the woods and hills of Charlton, boundaries from farms long ago sold or simply abandoned.

The land today looks much as it did in 1830 when Amasa Sr. and his sons worked the soil. The farm, now owned by developers, but as of this date sitting empty of buildings and other improvements, is located on the Brookfield Road, almost exactly split in half by the Sturbridge/Charlton town line. By some small percentage, most of the farm was located in

---

[10] Ibid., 301.

Charlton, to which taxes on the land were paid. The house burned down shortly after a picture was taken in 1905.[11]

Amasa Stone Sr. is buried in Charlton Bay Path Cemetery next to his first wife, Esther Boyden. Amasa's grave marker is small, measuring approximately 36 inches tall and 18 inches wide. The inscription on the headstone is no longer legible.[12] This simple monument is a curiosity given Amasa Stone Sr.'s stature in the community, and the fact that he had two millionaire sons living in Cleveland. There is no verifiable evidence that Amasa Stone Jr. ever visited Charlton after moving to Cleveland in 1850, or that he attended his father's funeral. His absence may have been due to the marriage of his daughter Clara to John Hay at the almost exact time of his father's death, and his own business ventures consuming all of his time and energy. This still does not explain the lack of finances and expenditure on a more appropriate monument to his father's legacy. One reason may be that the son no longer identified with his father after the death of his mother, and Sr. married another woman. A more plausible explanation is that Amasa Sr. gave little attention to Amasa Jr., expecting full time work, yet giving no more than room and board until his son was 17 years old and left home. Farming was hard work, and little to no nostalgia accumulated in Jr.'s memory bank. The past was the past. Local Charlton historian Pat Stockwell has suggested Rachel Lamb still had four or five children at home when she married Amasa Sr., and Amasa Jr. found himself crowded out of the family matrix.[13]

But in the overall scheme of the stern, corporal punishment-driven ethos of the New England family, it should not be surprising that Amasa Jr. repressed his past including his father. Philip Greven, a scholar of 17th and 18th century "Protestant temperament" in America states, "Their sense of self was systematically depressed in childhood, with life time consequences of feelings, of anxiety, fear, anger, depression, obsessiveness, and paranoia that were directed against the body and the self against other

---

[11] I thank Brian Mitchell who lives close to what was once the Stone farm, and enabled me to locate it as well as sharing local Charlton history. A map of Charlton and a picture of the house are courtesy of Amy Hiatala, Librarian and Archivist at Old Sturbridge Village, Sturbridge, Massachusetts.

[12] From a picture furnished by Pat Stockwell, the inscription seemingly reads, "Amasa Stone, DIED Mar 12, 1875, AE 96 yrs, ESTHER HIS WIFE, Died July 23, 1833, AE 52 yrs.

[13] Email, Pat Stockwell to Darius Salter, 4/7/22.

people."[14] As a whole, 19th Century Americans knew little of the familiarity and communal participation experienced by 21st Century Americans, say in the area of sports. It is well known that Abraham Lincoln never invited his father Thomas to the White House; neither did he attend his father's funeral.

New England soil was not friendly to farming, yet a visitor was impressed with the neatness and richness of its farmland. Francis Asbury, founder of American Methodism, aroused little sympathy in New England for his emotional religion. But as he rode through Massachusetts during the first decade of the nineteenth century, Asbury could not help but admire Yankee habits of economy and industry which "produced rich fields of barley, rye, and potatoes with plenty of cheese, butter, milk, and fish from the millponds."[15] The plenty and prosperity were underscored with New England pride, boasted by America's second President and Braintree, Massachusetts native John Adams, who wrote to his wife Abigail in 1775: "New England has in many respects, the advantage over every other colony in America; and indeed, of every other part of the world that I know anything of…. The people are pure English blood, less mixed with Scotch, Irish, Dutch, French, Danish, Swedish, etc. than any other; and descended from Englishmen, too, who left Europe in purer times than the present, and less tainted with corruption than those they left behind."[16] This Anglo-Saxon ethnocentrism would define Amasa Stone Jr. for the rest of his life.

Working on a farm, Amasa Jr. developed a strong physique with a bronze, ruddy complexion. In his late teens, he grew a neatly trimmed mustache, long sideburns and a shock of dark hair combed straight back which began receding in his early twenties. His manly face, defined chiseled features, straight nose, dark brown eyes, and six foot muscular frame cut an imposing figure. Even as a young man, he developed deep lines in his forehead which gave him a rather stern appearance. As money increased and he moved within "higher" circles, he dressed accordingly with

---

[14] Phillip Greven. "The Self-Shaped and Misshaped: The Protestant Temperament Reconsidered," in *Through a Glass Darkly,* eds. Ronald Hoffman et. al. (Chapel Hill: The University of North Carolina Press, 1997) 354.

[15] Darius Salter. *America's Bishop: The Life of Francis Asbury* (Wilmore, Kentucky: First Fruits Press, 2020) 277.

[16] Richard N. Rosenfield. *American Aurora* (New York: Saint Martin's Griffen, 1997) 262.

a dark suit, white shirt, and dark cravat tie. For an important occasion he wore a cutaway Prince Albert coat with tails.

As he came into full stature as a man, something else about Amasa was immediately apparent: he possessed a wide tall forehead. He was a phrenologist's delight; phrenology was a nineteenth century science correlating head measurements to intelligence. John McClintock, the first President of Drew University in Madison, New Jersey, wrote a friend in 1835 (when Amasa Stone was 17 years old), "Well phrenology must be true for the man gave me a fine head causality, comparison, ideality, etc. in abundance."[17] Phrenology true or not, Amasa Stone Jr. possessed high intelligence, a commanding appearance, and brimmed with self-confidence. He was ready to take on the world.

Amasa Jr. Marries Julia Ann Gleason

Thomas Gleason (1637-1703), was born in England and died in Framingham, Massachusetts, but no record exists of when he actually emigrated to America. Of interest was James Gleason (1721-1801), who served both as a physician and Universalist minister and left behind a treatise on "The First Three Chapters of Genesis."[18] Jonas Gleason was one of the Minutemen at Concord and Lexington, and also fought in the Battle of Ticonderoga. Micajah Gleason was killed at the Battle of White Plains October 28, 1776, at the age of 36.[19] James Gleason [Glezon] (1759-1841), did not lose his life at White Plains, but claimed to have lost: 1 blanket, 1 shirt, 1 knapsack, 1 pr stockings, 1 pr trousers, 1 canteen which he claimed to be worth 4 pounds 15 shillings for which he charged the United States Government.[20] Whether he collected or not is not known.

Julia Ann Gleason was born to John and Cynthia Hamilton Gleason on December 21, 1818 in Warren, Massachusetts.[21] Julia's father John Barnes

---

[17] George Crooks. *Life and Letters of the Rev. John McClintock* (New York, New York: Nelson and Phillips, 1876) 59.

[18] John Bart R. White. *Genealogy of the Descendants of Thomas Gleason of Watertown Massachusetts 1607-1909* (Haverhill, MA: Press of the Nicolas Print, 1909) 80.

[19] Ibid., 107.

[20] Ibid., 111.

[21] Gladys Haddad. *Flora Stone Mather: Daughter of Cleveland's Euclid Avenue & Ohio Western Reserve* (Kent, Ohio: The Kent State University Press, 2007) 3.

Gleason was a sixth-generation descendant of Thomas Gleason. John's father, grandfather and great-grandfather were all named Isaac.[22] Julia's grandfather Isaac had served as a captain in the Revolutionary War.[23]

Cynthia died when Julia was three years old, and Julia then moved in with Cynthia's sister, Susan Hamilton. Her father died when she was seven years old. In 1835, Liberty Stone, Amasa's brother, married Julia Gleason's cousin, Charlotte Hamilton. Another brother, Daniel Stone, married Julia's sister Hulda sometime in 1838. It wasn't difficult for Amasa to meet his future wife who was working as a seamstress in Worcester. Julia and Amasa, each 22 years old, married on January 13, 1842. The couple first lived in Warren, just outside of Boston. They gave birth to Adelbert Barnes July 8, 1844, and Clara Louise on December 28, 1848. When Amasa moved his family to Cleveland, he settled into one of the city's river houses, a Victorian structure, standing at the corner of Superior and Bond Streets. On April 6, 1852, this couple gave birth to the last of their three children, Flora Amelia.

A painted portrait of Julia Ann reveals a handsome face, dark hair parted in the middle, a long straight nose, dark eyes over which were perched prominent eyebrows. Both mouth and chin were gracefully formed, above a somewhat elongated but attractive neck. As a young woman, she was tall and elegant with no excess weight. She had no personal aspirations other than to fulfill the role of a pious wife and mother. Seemingly, the marriage was a perfect match. In spite of ascending into wealth, there was nothing pretentious about Julia Ann Gleason.

When the family took a trip to Europe, June 24, 1868 to July 30, 1869, Julia kept a diary recording all the places the family went which included England, France, Germany, Switzerland, Italy, Spain, and Austria. Julia demonstrated no ability to historically and theologically reflect upon anything she saw. For instance, on February 16, 1869, she recorded. "Drive to the Pantheon, to St. Peters, to one or two shops, purchased some Roman jewels."[24] She made no comment on the British Museum or Westminster Abby. During the whole trip, Amasa kept notes on drawing plans for buildings, calculating costs, keeping track of his stock, bank notes, and what other business occupied him. He commented on nothing in Europe.

---

[22] Ibid., 193.

[23] Ibid., 108.

[24] "Julia Stone Diary," Samuel Mather Family Papers, 1834-1867, Container 23, Western Reserve Historical Society, Cleveland, Ohio.

Opulence and Piety

In 1858, Stone's wealth and status demanded a residence which would make an identity statement. Thus, Stone built perhaps the most elegant and artistically designed mansion located on Cleveland's Euclid Avenue. In this house Stone exhibited both architectural and engineering ingenuity, as well as cosmopolitan taste, though at the age of 40 he had never been out of the United States. His 8,500 square foot Italian style villa, according to Euclid Avenue historian Jan Cigliano, "represented the state of the art in architectural design from its highly stylized romantic details to its structural system and mechanical devices," and "outshining" all other Italianate mansions on Euclid Avenue. The house stood at 1255 Euclid Avenue and was torn down in 1910.[25]

Always ahead of his time, Amasa installed piped water throughout the house as well as central heating. The design consisted of 23-inch-thick walls, constructed from 700,000 bricks, and the interior exhibited the finest of wood and stone work: a grand mahogany hallway, paneled ceilings, molded cornices, rosewood or oak doors, and Vermont statuary fireplace mantels. Stone, understanding the dangers of nature's most particular threat, fire, sheathed his roof with painted tin, constructed his basement with stone and brick, and built a library which recessed his desk, bookcase, and safe with fireproof material. The house was hailed for its "novel and imposing presence."[26]

Flora Stone's biographer, Gladys Haddad, aptly states that after moving into the Euclid Avenue mansion, "the three children Adelbert, then 16, Clara 12, and Flora 8 were reared in the midst of opulence created by Stone's drive for wealth and status."[27] Stone's competitive constant work was off-set by the "simplicity and devotion of a gentle and loving mother."[28] With its heavy dose of Protestant religion, this was a typical nouveaux riches Victorian home.

William Henry Goodrich, a graduate of Yale, pastored the Old Stone Presbyterian Church from 1858 until his death in 1874. His father being a professor at Yale, and his mother, the daughter of *the* Noah Webster, Goodrich was ideal for those who looked for intellectual stimulation and

---

[25] Jan Cigliano. *Showplace of America: Cleveland's Euclid Avenue, 1850-1910* (Kent, Ohio: The Kent State University Press, 1991) 74.

[26] Ibid.

[6] Haddad, 10.

[28] Ibid., 10.

a bit of sophistication in their pastor. It was said of him, "He brought light into every circle and added vigor to every silence, yet with such grace, modesty, and sanctified common sense were his opinions presented and his acts distinguished."[29] It was here that the Stone family attended church. (The current church, built in 1853, is the oldest standing structure in downtown Cleveland.)

The children were further educationally challenged by Linda Thayer Guilford, who taught at the Cleveland Academy where Amasa Stone served as a board member and was a major contributor. Guilford was a Mount Holyoke graduate, and had extensively traveled in Europe. She came as a "missionary" to the frontier town: thus, the Academy had a distinct spiritual tone. Miss Guilford recalled, "…very precious is the memory of the hushed quarter hour of the morning when words of the Holy Writ were dwelt upon for instruction and prayer dedicated all the day. Special to that time is deeply kept in the heart of many a one the prayer meetings led by some older girl, in those recitation-rooms, where nearly all would gather of their own accord after the school."[30]

In 1865 the school was incorporated as The Cleveland Academy, housed in a small brick building with Amasa Stone being one of the 23 original stock holders, and one of eleven people to contribute $1,000. Stone served as the building superintendent at the same time he was building the gargantuan Union Depot. Guilford described the school as a structure of brick, "with a slate roof 60 x 40, two stories above a high basement…. A solid parallelogram with regularly placed windows, it was relieved from positive ugliness by a large two-story portico which covered most of the front."[31] Guilford probably did not exaggerate when she stated that "the overwhelming majority of students were the choicest treasures of many cultivated Cleveland homes, girls of uncommon loveliness, force of character, and great mental promise. There it was easy to bring a high standard of work."[32] When Guilford wrote her history of the Academy she recalled,

> It is not for the purpose of these recollections to dwell on the fundamental religious principles which lay at the base of all plans or efforts for the school. Whatever was sealed with the witness of the Spirit in each soul

---

[29] Ibid., 11.

[30] Linda Thayer Guilford. *The Story of a Cleveland School, from 1848 to 1881* (Cambridge: John Wilson and Son, University Press, 1890) 29.

[31] Ibid., 97.

[32] Ibid., 102.

belongs to its own consciousness, and can neither be added to nor diminished by words from outside. If any were awakened to a solemn sense of duty toward God and their fellow-men, their lives are now testifying to that result. It was surely the aim of all said and done for them from first to last. Believing that the education which is divorced from all religious training is a terrible mistake, sure to be visited upon the communities that sanction it in far-reaching results of evil and misery, the teachers of the Academy in their daily round of influence brought the motives of the gospel to bear with such power they had. More than this we dare not say, leaving it to the inner record of each to manifest how far these motives were effectual. Among the most vivid and precious memories are the voices mingled in the morning hymn, for there was always in the later days someone to play the piano. "Tell Me the Old, Old Story," "I Will Sing for Jesus," "The Lambs of the Upper Fold," were often the prelude to the verses of the Holy Writ, followed by our petition for daily guidance and blessing. Bible lessons were so constant and inseparable from every Monday's work, holding always the place of honor on all public days, that possibly they will be the last trait of the school forgotten by any who learned them.[33]

The parents, the church and the school encouraged the Stone daughters to good works. Old Stone Church had a satellite mission at 46th and Superior Streets, where congregants ministered to immigrants by mending garments and holding "Sunday School festivals." According to Haddad, "They instructed girls in the neighborhood to sew, and while engaged in this activity, society members read to them from selected novels and inspirational materials and conducted discussion of the contents."[34] John Hay biographer John Taliaferro refers to Amasa Stone as "a stern, abstentious Presbyterian, his only indulgence, wife and children and then he indulged lavishly."[35]

Puritan Influence

To the contemporary reader, the word Puritan has little meaning, not even acquaintance with H.L. Mencken's condescending definition of a Puritan as an individual "haunted by the fear that someone somewhere might be

---

[33] Ibid., 215.

[34] Haddad, 31.

[35] John Taliaferro. *All The Great Prizes: The Life of John Hay, from Lincoln to Roosevelt* (New York: Simon & Schuster Paperbacks, 2013) 157.

enjoying himself." Because the fog of confusion is thickened by two centuries of Puritan emaciation between Simon Stone and Amasa Stone, a brief primer of this dilution is in order.

English Puritans were squeezed by a dialectical vice; they, unlike the separatists who boarded the *Mayflower*, believed that they should be faithful to the Church of England, confident that the Church could be "purified" and at the same time not obey the "law of uniformity." First enacted in 1549, the mandate required that England's churches conform to the worship prescribed in *The Book of Common Prayer* and "non-conformity" resulted in fining or some other penalty. Henry VIII pronounced himself "Supreme Head of the Church" in 1532, or to put it in more crass or profane terms, created a schism allowing him to legislate in order to fornicate. To disagree with a 300 lb. egocentric king was a serious liability, as Sir Thomas More discovered when he lost his head at the Tower.

Matters only got worse when Henry's oldest daughter, "Bloody" Mary, took the throne, returning the church to Catholicism and killing hundreds of Protestants. The situation became somewhat better when her younger sister Elizabeth, after Mary's death, returned to Anglicanism. But this only meant reverting to an intolerant state religion. The New World was becoming an increasingly enticing option. By the 1630s, boatloads of Puritans were headed to the "New England" to establish a "city set on a hill," including the Stones, who were on the good ship *Increase*. Prototypical was John Winthrop who as he sat on the *Arabella* in Boston harbor penned his *"A Modell of Christian Charity,"* explicating a covenant between those of the household of faith and God. This covenantal commitment implied a quid-pro-quo theology based more on an Old Testament understanding of Providence than the paradoxical teachings of Christ.

> But if our heartes shall turne away soe that wee will not obey, but shall be seduced and worship [serve *cancelled*] other Gods our pleasures, and proffitts, and serve them; it is propounded unto us this day, wee shall surely perishe out of the good Land whether wee passe over this vast Sea to possesse it;

> Therefore lett us choose life,
> that wee, and our Seede,
> may live; by obeyeing his
> voyce, and cleaveing to him,

for hee is our life, and our prosperity."[36]

Many of us owe much to these paragons of piety, virtue, and perseverance. They were God consumed and ideas about Him were all-important, perhaps too important. The Puritan identity was so rigidly defined that there was no tolerance for anyone who believed differently than the Puritan fathers in the likes of Winthrop, John Cotton, and Increase Mather. Thus, Ann Hutchinson was run out of Salem for extra-Biblical revelation, Roger Williams fled Boston protesting its theocratic government, Quakers were hung in Boston Square for their unorthodoxy, and 19 witches were executed in Salem in 1693 for what was likely little more than collective hysteria.

The Puritans were sincere, but naïve; they cast a vision, but a vision blurred by misperception. Government would consist of a theocracy rather than a democracy and only the "elect" would rule. Thus, two hopeless tasks emerged, reading the mind of God as revealed in Scripture, and demarcating the elect from the non-elect. The autocratic rule, which is recorded in the Old Testament as a form of government, hardly suited a people seeking liberty and freedom of conscience. Thus, there was constant bickering between the ruled and ecclesiastical leaders, the latter believing they had been appointed by God and taking a condescending attitude toward their subjects. Little grace was extended to parishioners, if one believes the following from Pastor Thomas Shepard to his congregation: "You are inherently corrupt, unstable, apt to be led by colours like birds by glasses & larkes by lures & golden pretences which Innouators euer haue."[37]

In other words, the laity had little input into either civil or Church government, which were one and the same. Assuming the Bible could be uniformly interpreted, if the interpreters were simply sincere and honest, led to endless wrangling because perception, often skewed and subjective, lies between the interpreter and the object of his interpretation. Fundamentalist certitude was constantly battered by ancient customs and rules which hardly applied to a people creating a system of government some 3,000 years later than God's chosen people, who were ruled by judges and kings. Governmental rule for the Massachusetts Bay Colony assumed that "Good

---

[36] H. Shelton Smith, et. al. *American Christianity: An Historical Interpretation with Representative Documents,* Vol. I, 1607-1820 (New York: Charles Scribner's Sons, 1960) 102.

[37] Perry Miller. *Orthodoxy in Massachusetts 1630-1650* (New York: Harper Torchbook, 1933) 246-247.

Christians could not demur when the ministers all agreed that the will of Christ had such and such political implications."[38] John Cotton had written, "Democracy I do not ceyde, that ever God did ordeyne as a fitt government, eyther the Church or Commonwealth. If the people be governors, who shall be governed?"[39] Enduring out of the hopeless entanglement of Church and State was a mythic interpretation of America as a chosen nation who dispossessed the original inhabitants just as Israel had conquered Canaan. Charles Sanford in his intriguing book *Quest for Paradise* stated, "More than any other modern nation, the United States is a product of the Protestant Reformation, seeking an earthly paradise in which to perfect a reformation of the Church."[40]

To find out just who the Puritans were, what they believed and what their motivations were, has been the subject of endless historiography. They lived and worked within several dialectics, still challenging the American Church in its myriad expressions today: the tension between this life and the next, an earthly kingdom and a spiritual kingdom, a theocracy and a democracy, Church-regulated behavior and inward spiritual experience, individual freedom and community solidarity, biblical authority and civil authority. For better or worse these dialectics were distilled into a strong individualism, manifest destiny, spiritual superiority or exceptionalism, and above all, the belief that hard work and right living results in material blessing.

By the early nineteenth century, the feared God of the Puritans had vanished, at least a God who condemned individuals in spite of their best effort and most fervent faith. A person was now in control of his destiny and would work out both his salvation and material prosperity. The following, written by Puritan scholar Perry Miller, fully applied to Amasa Stone: "In America, concepts of individualism, free enterprise, self-reliance, the right to make money, of indefinite and inexhaustible prosperity, had become so identified with the eternal law of Almighty God that by no stretch of the historical imagination could these well-intentioned researchers share in the dread that lay at the heart of the Puritan experiment."[41]

---

[38] Ibid., 247.

[39] Ibid., 229.

[40] Charles Sanford. *Quest for Paradise* (Urbana: University of Illinois Press, 1961) 74.

[41] Miller, XII.

The Puritans put a lot of pressure on themselves, always living in the ever-present specter of sin, as taught by Augustine and adopted by both John Calvin and John Knox. They did not rest in the assurance of salvation and the resulting peace and confidence as taught by John Wesley. Within the Scottish Presbyterian theological paradigm from which Amasa Stone descended, one would always have to prove himself. Though the reader may think that Stone's choice of a church was a matter of social status to which I allude in a later chapter, this motive was only partially true. The loyalty to Presbyterianism was due to the Puritan understanding of internal conflict from which the believer shall never be freed as described by Paul in Romans 7. Stone's pastor Hiram Haydn rejected the perfectionism as taught by both John Wesley and Charles Finney, a freedom from sin which could be realized in this life. Charles Cohen, in *God's Caress: The Psychology of Puritan Religious Experience*, quotes Pastor Thomas Shepard: The "body of sinne shal neuer be from vs so long as we liue."[42] For the Puritans, God's caress was not very affectionate.

The last great Puritan was Jonathan Edwards, who experienced revival and growth in his Northampton, Massachusetts church in 1739. He faithfully and forcefully preached the sovereignty of God, original sin, salvation for only the elect, and the fiery damnation of the non-elect. By the time Amasa Stone Jr. arrived on the scene almost a century later, only about fifty miles from Northampton, Edward's original sin, double predestination and eternal damnation had been dismissed by the leading New England theologians Nathaniel Taylor, Horace Bushnell and William Ellery Channing, the last a Unitarian Universalist, believing that no one would be eternally lost.

In 1838, a Universalist Society was officially founded in Charlton, which Amasa Stone Sr. joined, along with over one hundred charter members. The informal gathering of the church began April 1, 1827, pastored by Rev. Messena B. Ballou, who recorded, "I never passed four years of my ministerial life more pleasantly than with the good people of Charlton."[43] Amasa's brother, Liberty, was also a charter member. The church was officially given the name "The First Universalist Society in Charlton."[44] Uni-

---

[42] Charles Lloyd Cohen. *God's Caress: The Psychology of Puritan Religious Experience* (New York: Oxford University Press, 1986) 45.

[43] Holmes Ammidown. *Historical Collections*, Vol. II (New York: published by author, 1874) 189.

[44] Ibid., 188.

versalism was the ultimate rebellion against rigid Calvinism. As to exactly what were Amasa Sr.'s motives and what influence this had on Amasa Jr., we can only speculate. Likely little, because Amasa Jr. had already left home in 1838. We do not know the pattern of church attendance in the Stone family. Spiritual nurture may have been provided by Amasa Jr.'s mother who died when he was 15.

Just because a strict five-point Calvinism no longer ruled New England, does not mean that theological argument expired. That writing of books and pamphlets and a constant exchange of letters and sermons given to theological nuancing and disagreements is confounding to us moderns who have dismissed the "body of divinity" for entertainment and social media. Eminent Church historian Sidney Mead wrote, "Those involved in these arguments had a terrible fascination for their authors because they dealt with the awful things of the sovereign God and that they were vitally important to the life of minister, theologian, and people because that life was lived along an unbroken way in which the New England Street merged imperceptibly with the golden streets of the New Jerusalem."[45] Mead further stated,

> Their language many times was as harsh as the boulder-strewn hills from which they wrested their living, they held their positions as uncompromisingly as the granite rock that everywhere lay just beneath the surface soil of their land, and they had an irritating tendency born of long hours of quiet meditation to spin out a point until the fine line of reasoning became all but invisible. But always they were serious men, never flippant, seriously treating the things they thought were important for eternity.[46]

What this meant for Amasa Stone Jr. was that he was raised in a somber home where there was little place for levity. Heaven and Hell were still serious matters. He would always be a serious man which some interpreted as remote austerity. He did not suffer fools gladly. And as we will later argue, this temperament which was ruled by a demanding conscience separated him from his nemesis, John D. Rockefeller, who tended to laugh off his contradictions or even scoff. He was the Gilded Age champion of compartmentalization, a defense mechanism that Stone never mastered. Puritan scholar Stephen Innes describes the pressure on New England youth instilled in them by their religion. "The ethical tenets imposed on

---

[45] Sidney Earl Mead. *Nathaniel William Taylor, 1786-1858: A Connecticut Liberal* (Chicago: Archon Books, 1967) 96.

[46] Ibid.,97.

New England's young included: diligence in one's calling, strict asceti-cism in the use of material goods or enjoyment of earthly pleasures; and an acute time-consciousness."[47]

As Stone was coming into his adolescent years, Charles Finney was detonating a religious explosion in Rochester, New York. Edwards had described the events that occurred in his pastorate and beyond, as the "Sur-prising Work of God." But there was nothing surprising for Finney; in fact, he believed "the connection between the right use of means for revival and the revival is as philosophically sure as between the right use of means to raise grain and a crop of wheat. I believe in fact, it is more certain in that there are fewer instances of failure."[48] Stone would have fully appreci-ated the analogy, but would have been as confused as the rest of us, as to whether Finney placed his confidence more in his lawyer-like moral suasion arguments or the supernatural power of God. What Stone knew of Finney's evangelistic efforts in Rochester, New York and beyond, we do not know. But we can be certain Stone was aware of Finney's presidency at Oberlin College, 1851-1866, only a few short miles from Stone's home in Cleveland.

Finney and Stone had more in common than a first impression would reveal. They were both Presbyterian, they would both be validated by numbers, Finney the number of souls saved, and Stone the amount of dol-lars earned, as well as the length of both railroads and bridges built. Both were products of the Reformation and were consumed by a rampant prag-matism as creatures of free enterprise. In his seminal work, *The Protestant Work Ethic and the Spirit of Capitalism,* Max Weber argued that the Ref-ormation sanctified work and nothing was more demonstrative of Chris-tian character than conscientiousness, self-denial, frugality, diligence, and honesty in work. The priesthood of all believers accented individualism, and a laissez-faire work ethic, centrist elements of the American ethos.

Unfortunately, though in terms of Reformation doctrine, a person would no longer have to merit his salvation by works, he now faced an additional burden of earning respect and admiration through work. Aris-tocratic bloodlines counted for little. Personal accomplishment was ev-erything. As has been said, "No one in America had to ask, just what is it

---

[47] Stephen Innes. *Creating the Commonwealth: The Economic Culture of Pu-ritan New England* (New York: W. W. Norton & Company, 1995) 125.

[48] Darius Salter. *American Evangelism: Its Theology and Practice* (Grand Rapids: Baker Books, 1996) 95-96.

that a Duke does?" No one *did* more than Amasa Stone Jr. Proving one-self is tough business; it is the most defining activity of the kingdom of meritocracy, as championed not by the lowly Galilean, but by America's leading theologian, Benjamin Franklin, who espoused his formulaic work ethic through his periodical, *Poor Richard's Almanac.* His legacy was far more enduring than that of Jonathan Edwards. Franklin, the quintessen-tial American, wrote the job description for Amasa Stone. Carrying it out killed him.

We can draw a straight line from the Puritan work ethic to the Gilded Age, also referred to as the Acquisitive Age. This self-defeating correla-tion American Christianity has never solved. Innes states, "The very suc-cess of the Protestant ethic in fostering time-consciousness and striving behavior in producing wealth as well as fostering literacy and strong civic institutions raised a larger moral problem, one that New England minis-ters' sermons and published tracts returned to indirectly again and again: how to draw the line between industrious enterprise and self-regarding acquisitiveness."[49]

As I will later argue, it is not simply coincidental that most of the "rob-ber barons" were quite "pious."

---

[49] Innes, 159.

Chapter 2

The Bridge Builder

Early American Bridge Building and the "Truss"

In a Ph.D. dissertation done for the University of Pennsylvania, George Danko effectively argued the truss bridge went through an evolution from empirical/experimental design to a scientific/rational design. This paradigm shift created an occupation for which there was a new designation, "the civil engineer." Because of the demands for transporting military artillery and supplies, building bridges and roads, as well as constructing anything that would be of advantage in a battle such as breastworks and siegeworks, the Army Academy at West Point, New York, would produce the most antebellum civil engineers. They were not unaware that they were using civil means for uncivil purposes, but only history can note that these skills of mathematical formulas and exact measurements enabled the North to win the "Civil War," the most profound oxymoron in American history.

In the beginning of bridge building, as well as other construction projects, the designer literally built a model and jumped up and down on it, or placed heavy weights on it and implemented other modes of destruction. Timothy Dwight, President of Yale, recalled such a demonstration:

> Six Gentlemen placed themselves together upon an exact model of one of the arches, ten feet in length, in which the largest pieces of timber were half an inch square, and the rest smaller in proportion. Yet not the least injury was done to the model. Of this fact I was a witness; and was informed by the gentleman present, that eleven persons had, a few days before, stood together upon the same model, with no other effect, than compacting

it more firmly together. The eleven were supposed to weigh at least sixteen hundred pounds. [1]

Obviously, this primitive method for calculating the strength of a bridge had its limitations. Amasa Stone may not have been a trained civil engineer, but his intuitive abilities to calculate, measure, and construct by minute preparation accumulating over long years of experience, eliminated such arcane procedures.

The first bridges in America were straight beam bridges over narrow spans and for wider crossings, arch bridges, which would hold up any structure by compressing throughout the arch. Medieval churches are still standing today because of using the stone arch. If stones of sufficient strength and size were used, it was almost foolproof for wagon loads, stage coaches, and all other modes of animal drawn transportation. Of course, the longer the span, the greater the cost. Builders discovered they could save money and give even more stability to the bridge by underpinning it with a "truss". Webster defines a truss as "Any of various structural frames based on the geometrical rigidity of the triangle and composed of straight members subject only to longitudinal compression tension or both, functions as a beam or a cantilever to support bridges, roofs etc."[2]

Thus, a truss would consist of squares or parallelograms, reinforced by diagonal bracing which crossed in the exact middle of the square, creating triangles.[3] If a person stands some distance from a truss bridge, he will observe hundreds if not thousands of these triangles. For these early truss bridges, the rudiments of geometry and trigonometry, along with the trial-and-error method of carpentry were sufficient. Danko states, "The education available to the skilled builder in the late 18ᵗʰ century was overwhelmingly utilitarian in nature. The basic operation of arithmetic, and the

---

[1] George Danko. *The Evolution of the Simple Truss Bridge, 1790-1850: From Empiricism to Scientific Construction* (unpublished Ph.D. dissertation, University of Pennsylvania, 1979) 47.

[2] *Webster's Encyclopedic Unabridged Dictionary of the English Language* (San Diego: Thunder Bay Press, 2001) 2030.

[3] Almost all large bridges today are suspension bridges. For instance, the bridge over the Mississippi River in St. Louis, connects Interstate 64 with Interstate 70. There are also many truss bridges in the U. S. For example, the bridge over the Missouri River on Route 291 south of Liberty, Missouri, and the bridge on Route 41 over the Ohio River south of Evansville, Indiana. The Golden State Bridge in California is a suspension bridge with a truss under the roadbed.

axioms of descriptive geometry were used to calculate amounts of material and render the design of his projects."[4]

Railroad Transformation

Then along came the railroad, the transporting of tonnage that had never been conceived, much less prepared for in the history of humankind. To this day, railroads transformed America perhaps more than any other technical achievement. They stimulated and created the growth of cities, speed of communication, transportation of goods, a huge labor market, vocations for the uneducated lower class, and necessitated formal schools of education for the expanding role and number of civil engineers. In short, it was a cultural revolution. Eminent historian Daniel Walker Howe claims, "The sheer size of the railroad companies altered the American economy. The major railroads came to dwarf the antebellum manufacturing concerns, even the Lowell Mills. Railroads became the largest corporations since the demise of the BUS (Bank of the United States), and the first nationwide secular enterprises under entirely private control."[5] John H. White, in an impressive piece of scholarship, *The American Railroad Passenger Car*, begins by reminding us, "Americans once spent a considerable portion of their time in railroad cars. From approximately 1860 to 1930, railroads were the most common means of inland transportation, and at their peak just before the first world war, 98% of all intercity travel was by rail. No other means of transportation, not even the automobile has achieved such a monopoly."[6]

"Such a monopoly" necessitated all kinds of inventions, innovations, and ancillary industries from food services to metal fabrication, from gigantic railroad sheds to miniscule toilets. None were more important than a safe reliable bridge to span chasms and rivers for routes over ever challenging topography.

A Changing Educational Paradigm

The immensity of technical problems imposed by the railroad called for a new philosophy of education. The curriculums consisting of philoso-

---

[4] Danko, 65.

[5] Daniel Walker Howe. *What Hath God Wrought: The Transformation of America, 1815-1848* (New York: Oxford University Press, 2007) 566.

[6] John H. White. *The American Railroad Passenger Car* (Baltimore: The Johns Hopkins University Press, 1978) XI.

phy, theology, history, rhetoric, and a particular emphasis on the ancient languages was no longer equal to the task required for the new world of technology and mass production. Education became far more utilitarian. For instance, student Dennison Holmstead, for his course in natural philosophy at Yale College in 1838, "learned to dimension the breadth, depth, and length of a beam to sustain a given concentrated load, or experiment with the most economic shape for a simple beam designed to carry a moving load."[7] Union College in Schenectady, New York, was one of the first schools where a graduating senior was to have completed "two terms of trigonometry, descriptive and analytical geometry, natural philosophy, and (including statics, dynamics, and hydrostatics) an additional year of mechanics."[8]

The first school to offer an almost completely pragmatic and technological curriculum was founded in Troy, New York, in 1847. Rensselaer Polytechnic Institute's courses included "Theory of Bridge Construction, Truss and Arch Construction in Wood and Iron, The Suspension Bridge, and Principles of Iron Tubular Bridges."[9] It is not coincidental that Charles Collins, civil engineer of the Cleveland, Painesville, and Ashtabula Railroad when the Ashtabula Bridge collapsed, was a graduate of this school; he had trained to be a civil engineer. The question deserves to be asked, did he have the personality and temperament to accept the responsibilities which the title "civil engineer" demanded? We will address this question later.

Of course, there was push back to such rampant pragmatism, what was considered a radical reductionism of sacred institutions, which at one time had turned out only ministers, doctors, lawyers, and teachers. If one was really educated, he would travel to Europe, study at Tubingen, Germany, the Sorbonne in Paris, or the Hague in Holland. The second level of rejection, often in the form of scoff and sarcasm, came from a generation of builders who had been successful without the advantages of higher education. The only school that really mattered was one of hard knocks, perseverance through failure, until finding what actually worked. As late as 1869, Lewis Wernac, an employee of the Baltimore and Ohio Railroad, wrote to its President Phillip E. Thomas:

All the knowledge and experience that I am at this time in possession of I

---

[7] Danko, 79.

[8] Ibid., 81.

[9] Ibid.

have not got from theory but from seven knocks on my knuckles. I have handled all them tools for many years from a crow bar to a moulding plane. To be at this eleventh hour of my days under the control of engineers as they are called I would make myself an unhappy being. . . I have forgot they are before me in calculating. They may do it quicker to know how many cubic yards it would take out of a hill to fill a hole but not in anything else.[10]

This same letter could have been written by Amasa Stone when he arrived in Cleveland some two decades earlier. He may have been aware that in 1832 James Rennich in his *The Elements of Mechanics* had developed an exact formula for measuring stress loads.[11] It seems incredible to us that someone could make such exact measurements with the wide variation in the strength of wood, even if it was the same kind of wood. It is not that we need to completely understand Rennich's formula (for which I am grateful), but Amasa Stone did need to understand not only that formula, but the many other advancements that were made between then and 1865 when he built the Ashtabula Bridge. Because he had little formal education Stone's exact knowledge of engineering formulas is left open to speculation. He was not able to produce the "strain sheet" when asked for it by the Ohio Legislature.

Iron and the Howe Truss

Wood had to go; it cracked, it frayed at the ends, it compressed which loosened the bolts, it rotted, and most critically it often caught fire from the cinder spewing locomotives for which the bridge was built. More importantly, iron became more economically feasible. Laminating arch beams was a time-consuming process that after all of the work was not as precise as iron castings. And the rods, beams, etc. could be mass produced at a rolling mill while castings were made in a foundry and shipped by rail to the building site.

Enter William Howe, Stone's brother-in-law, who along with Stone, built a truss bridge over the Connecticut River at Springfield, Massachusetts. Howe advanced a truss design made popular by Stephen Long in 1830 by using iron bracing, cross ties, and utilizing 2 inch-diameter iron rods for the tension members which he patented. According to Danko, this was the "First time in a railroad structure, a major member was constructed

---

[10] Ibid., 167.
[11] Ibid., 162-163.

in metal."[12] By using the threaded ends, Howe achieved a tighter connection, with less vibration than could be achieved with wood. He did not do this without the insight and suggestions of his brother-in-law. Danko gives Stone due credit: "Amasa Stone, Howe's brother-in-law and co-contractor on the Connecticut River Bridge acquired the patent rights, formed the firm of Boody, Stone, and Company, and with excellent business acumen, pushed the popularity of the design."[13]

The iron rods for the tension members could be pre-stressed, that is tightened with nuts on the top of the upper chord and the bottom of the lower chord. This process was more exact than driving wooden wedges into the bottom and tops of wooden tension members. The Howe truss became the dominant design for railroad bridges because they could be pre-stressed, pre-fabricated, and more easily repaired. Replacing iron members was more exact than replacing wood members. The latter varied in strength, even though they might be the same size. According to Dario Gasparini and David Simmons, "As railroads began to experiment with all iron bridges in the 1840s, the Howe truss with cast iron for the compression members and wrought iron for those in tension, became the natural choice of some of the first designs."[14]

Even though iron was more exact in its strength and uniformity, perfection was elusive, because of variations in carbon content, the presence of air bubbles, and differences in cooling rates. At the time of the building of the Ashtabula Bridge, there was no way to detect an air bubble on the inside of iron which had been cast. Also, to roll hundreds of I -beams in the 1860s of the same strength and size was simply not possible. Iron and steel historian Emory L. Kemp concluded, "Thus despite the various books and articles published in the 19th century that give typical values for wrought and cast iron, such values cannot necessarily be associated with a particular structural component. Depending on the exact composition and production processes, there is a wide range of properties found in iron members from the 19th century. Manufacturers were not required to meet

---

[12] Ibid.,195.

[13] Ibid.,194-195.

[14] Dario Gasparini and David Simmons. "American Truss Bridge Connections in the 19th Century, 1829-1850," *Journal of Performance of Constructed Facilities,* Vol. II, Number 3 (August 1997) 125-126.

any national standards, nor to produce stress strain curves or chemical analysis of their products."[15]

Finding a Vocation

As an early teen, Amasa assisted an uncle in cabinet making, learning to build furniture and craft almost anything necessary for both the interior and exterior of a house. At age 17, along with his older brothers Daniel and Joseph, Amasa began building churches and houses. One of the first which they built is still standing in East Brookfield, Massachusetts on Route 9. The totality of his formal education consisted of a couple of months a year in the local Charlton school, and one year at a Worcester, Massachusetts academy paid for by a small sum which he had saved. In 1837, The Baptist Work Manual Labor School under the direction of Silas Bailey consisted of a heavy dose of mathematics including algebra and geometry which would serve Stone well for his life's vocation. There was little cost for attending the school. A year after its founding in 1834, Bailey wrote the trustees, "The Manual Labor Department has been more efficient than any time previously. Many students have been able to pay their tuition by labor (at eight cents per hour,) and some have paid their board also by like work; and it is but justice to state that those who have been shown greatest improvement are the young gentlemen who have spent a portion of each day in manual labor."[16]

Bailey had studied under the famed "mental and moral" philosopher, Francis Wayland, at Brown University. There was a high moral and spiritual tone to the school. Bailey added a by-law "that if any student after having been kindly admonished for the offense, shall persist in the use of profane language in any of its forms of irreverence, it shall be the duty of the principal immediately to dismiss him from the Institution."[17] As we will later see, both of Amasa Stone's sons-in-law, John Hay and Samuel Mather, referred to Amasa Stone's purity of speech. The school may have had as much spiritual influence on Amasa as vocational.

---

[15] Emory L. Kemp. "The Introduction of Cast and Wrought Iron in Bridge Building," *The Journal of the Society for Industrial Archeology* Vol. 19, Number 2, (1993) 14.

[16] George Otis Ward. *The Worcester Academy: Its Locations and Its Principals: 1834-1882* (Worcester, Massachusetts: The Davis Press, 1948) 21.

[17] Ibid., 23.

Daniel Stone

The story of Daniel Stone is even sadder than that of his brother. Daniel, older than Amasa by nine years, and who had been the most influential person by enabling him in the art and business of construction, took his life at age 54. He had continued to live in Springfield after Amasa left in 1850, but at some point moved to Philadelphia because much of his work was in that area. Four years before his death, he had fallen from a bridge on which he was working and incurred an injury from which he never recovered. On Thanksgiving Day of 1862 while the family was in church, he threw himself into the Schuylkill River, and drowned before anyone could reach him. The Worcester newspaper, *The Palladium* reported that,

> Mr. Stone, like the late Hugh Miller, the famous Scottish geologist, was the victim of an over-wrought brain, laboring as he did in the various public enterprises and works of internal improvement of the day, with a degree of enthusiasm in his occupation such as is so rarely found, he became at times partially deranged. The peculiar nervous excitability incident to his nature, together with the effects of an injury of the head produced by a fall some four years since, had warned his friends and family that if they would save his health the cares of business must be laid aside.[18]

Because the two brothers Daniel and Amasa committed suicide, some have suggested a strain of mental illness in the family. No genealogical records support this thesis. The family life of the ten children in the Amasa and Esther Boyden Stone home seems to have been relatively stable. However, as we have argued in Chapter 1, the Puritan work ethic and the Victorian understanding of success produced what today would be referred to as "workaholism". Visible production and accumulation of wealth, even for the "pious," and maybe especially for the "pious," were the two standards for a meaningful life. That the brothers, all living elsewhere, other than Liberty who stayed back on the family farm, never gave any attention to the death and interment of their father, suggests a situation that the brothers wanted to forever leave behind. We will never know.

What we do know is that Amasa adopted Daniel's two youngest children when his wife died in 1869. (The oldest son, Augustine, 29 years old, stayed behind in Philadelphia, and would later move to Cleveland, and work for Andros Stone.) Amasa, whom others considered to be gruff

---

[18] "Melancholy Suicide," *The Worcester Palladium* (Worcester, Massachusetts: December, 1863) 3.

and aloof, always had a soft spot in his heart for children, a characteristic which we will later illustrate. In 1870, Emma and Eddie Stone came to live with their Uncle Andros in Brooklyn, New York. For the children, it was a stroke of fortune as both of them would be included in Amasa's will.[19]

Early Bridge Success

Amasa Jr. was never confronted by a job which he believed to be beyond his abilities. At age 19 he framed a house for a Colonel Temple in Worcester. He was paid with a $130 "note of a manufacturing firm," but the firm failed before he could cash it, and the nineteen-year-old lost all of his money.[20] For a year he helped his brother-in-law build two churches and several houses in Warren, Massachusetts. It would seem that at age 20 Stone was headed into a career of ecclesiastical architecture. Not that God did not appreciate his houses of worship, but William Howe decided to go another direction, providing a vocation which would occupy Stone for the next quarter of a century, make him a millionaire, and ironically lead to the disaster which would be his primary legacy.

Inventive genius ran in the Howe family; his brother Tyler introduced the first spring mattress and his nephew Elias Howe invented the sewing machine. William Howe, a millwright from Spencer, Massachusetts, schooled his younger brother-in-law in the art of spanning a wide river. By the use of diagonal braces and vertical iron rods, Howe had developed a workable solution for the railroad's most challenging obstacle, any topography demanding a bridge: gorges, valleys and in particular, rivers. The Howe truss provided sufficient strength for supporting a railroad bed that was less expensive, and more quickly built, than stone arches or any kind of suspension apparatus. As historian David Simmons states, "Time is money in America. A Howe truss was cheaper in large part because it required less labor than a stone arch. Stone arch construction was more common in Europe because labor was cheaper on the Continent."[21]

Dario Gasparini, the foremost authority on the bridge building of Amasa Stone, notes that Stone and Howe built their first bridge in 1839 in

[19] "Laws of the General Assembly of Pennsylvania," 1870. No. 201, "An Act." Courtesy of Pat Stockwell.

[20] Maurice Joblin. *Cleveland Past and Present: Its Representative Men* (Whitefish, Montana: Kessinger Publishing, LLC, 2010) 458. Retrieved from https://www.gutenberg.org/files/9328/9328-h/9328-h.htm. on July 15, 2021.

[21] Email from David Simmons to Darius Salter, January 31, 2022.

Warren, Massachusetts. By 1840, Howe introduced threaded iron rods to connect the top and bottom cords of the bridge and pre-stressed the trusses by tightening the nuts on the threaded iron. Because of these innovations, the chief engineer of the Western Railroad, G. W. Whistler, awarded the contract for a bridge over the Connecticut River at Springfield, Massachusetts to Howe and Stone. At 22 years old, Stone assisted Howe in building a 1,260-foot bridge (about a quarter of a mile) across the Connecticut River. The bridge was so strong that when a locomotive passed over a span, the sag was only a quarter of an inch. The bridge required 28,636 feet of board. Each span necessitated the use of 21,810 lbs. of wrought iron, a total for the seven spans of 152,670 lbs. The spans were 180 feet, perhaps the longest spaces that had ever been transversed, 30 feet longer than the entire Ashtabula Bridge later built.[22]

Howe and Stone prefabricated the bridge on ground before erecting it over the river, not requiring master timber framers climbing to great heights. Thus, much of the danger of building massive bridges was eliminated. Gasparini states, "All the bridges west of Springfield used Howe trusses; approximately 40 bridges with a combined length of 3,900 ft. were built in less than two years. Many of these bridges were no doubt built by Amasa Stone." Gasparini quotes early Cleveland historian Maurice Joblin, claiming that by 1869 Stone had built over ten miles of Howe bridges. Gasparini further claims, "The Howe Truss Bridge became the dominant form for railroad bridges for approximately 30 years until all – iron and all – steel bridges were built." He argues that Stone was so successful with the Howe truss that the form was adopted around the world being used throughout Europe, Russia, Australia, and possibly Asia.[23]

Stone was a quick study, and it wasn't long before he knew everything about bridge building that his brother-in-law knew. The novice quickly got the attention of the New England railroad community. He purchased from his brother-in-law the William Howe truss patent for $40,000.

---

[22] Lewis M. Prevost, Jr. "Description of Howe's Patent Truss Bridge Carrying The Western Railroad Over The Connecticut River at Springfield, Massachusetts," *Journal of the Franklin Institute of the State of Pennsylvania, and Mechanics' Register* (May 1842) 289.

[23] From a non-published lecture given by Dario Gasparini at Case Western Reserve University, September 2017, I am indebted to Dr. Gasparini for providing me his notes from the lecture. He is an M.I.T. Ph.D., and taught 39 years at Case Western Reserve University.

Now on his own, Stone took a construction crew to the Green Mountains of Vermont in the winter of 1841 and built a span of similar length to the aforementioned Connecticut River bridge.[24] He was a leader by example, not afraid to get his hands dirty, a demanding foreman who barked orders to men who toiled in sub-zero temperatures; sawing, nailing, climbing, carrying lumber and whatever other tasks required by a particular project. This was no Sunday School picnic, but a construction crew, consisting of 20 or so men, housed in shanties, and working from sun up to sun down in brutal weather. Stone was relentless with both himself and his men. Even with the difficulty of crossing a river, which looked to most people to be an almost impossible task, it was not nearly as difficult as the cultural chasm he would later face with his nouveaux riches in Cleveland, Ohio, a transition Stone never completely made. Building bridges was a lot less formidable than bridging cultures.

Stone transcended the reputation of every bridge builder in New England, if not all the United States. He formed a corporation with Azariah Boody, "Boody, Stone, and Company," and according to Gasparini, he set up the first pre-fabrication bridge shop in America. It was a fortuitous decision in that Providence and American ingenuity furnished the two men with an industry that would become America's predominant business in the nineteenth century, railroads. Railroads would learn as they went, adapting rapidly changing technology to endless obstructions and challenges. No one was more representative of Richard Hofstadter's description of Gilded Age industrialists than Amasa Stone as men of "heroic audacity and magnificent exploitive talents – shrewd, energetic, aggressive, rapacious, domineering, insatiable."[25]

A Colossal Accomplishment

At the age of 27 while maintaining his relationship to Boody, the New Haven and Springfield Railroad hired Stone as its Superintendent. The Railroad was immediately confronted with a crisis. The Railroad's quarter-mile bridge over the Connecticut River at Warehouse Point, Connecticut, was destroyed by a hurricane on October 14, 1846. The Railroad commissioned Stone to rebuild it, and when finished he watched as railroad cars

---

[24] Joblin. 458.

[25] Richard Hofstadter. *The American Political Tradition* (New York: Vintage Books, 1989) 213.

passed over a structure which "consisted of seven spans of seventy-seven feet each, with two other spans of about fifty feet each."[26]

When the Warehouse Point Bridge was blown down, the stone abutments and piers were left intact. Amasa, along with D. L. Harris, rebuilt it in forty-five days by "using the whole output of saw mills up and down the valley and the services of many local boats."[27] Stone astounded the whole railroad world as well as himself. It was so well built that it lasted for nineteen years until replaced with an iron bridge in 1866. A local Springfield newspaper stated of the Stone/Harris bridge, "This bridge, which had stood until now, was of the same general plan as the first – the wooden truss bridge – but heavier and fastened so firmly to the piers by deeply sunken iron rods that there could be no danger of it blowing away like the first."[28] The bridge was all the more difficult and consumed an exorbitant amount of material because it was covered. The iron bridge that replaced it, almost at the exact time of the building of the Ashtabula Bridge, was "firmly bound together by iron plates at short intervals, also in the truss form, securing the two lower and the two upper stringers, and strong iron bars pass at short intervals from the upper stringer on one side to the lower one on the other. All is firmly riveted together, and in such a manner as to secure the greatest strength."[29] Sufficient riveting and plating for binding both the chords and braces together was not true of the Ashtabula Bridge.

According to Gasparini and Simmons, "The development of shop and field riveting equipment finally made riveted, gusset plate connections practical by the turn of the century."[30] It was also after the Ashtabula Bridge was built that bridge companies came into being, specializing in bridge building, displacing the entrepreneurial independent bridge designer, making bridge building a specialty, and eliminating Amasa Stone types, who seemingly had thousands of other responsibilities. It irked Stone as we will later see, that he had to show up in Ashtabula to advise A. L.

---

[26] Joblin, 461.

[27] "Warehouse Point R.R. Covered Bridge," *Connecticut River Valley Covered Bridge Society Bulletin* (Summer 1988) 6.

[28] "The New Railroad Bridge at Warehouse Point, Connecticut," unidentified newspaper, courtesy of Zoe Cheek, Massachusetts Historical Society, Springfield, Massachusetts.

[29] Ibid.

[30] Dario Gasparini and David Simmons, "American Truss Bridge Connections in the 19th Century, 1850-1900," *Journal of Performance of Constructed Facilities*, Vol. 11, Number 3 (August 1997) 130.

Rogers whom he had entrusted as the site superintendent. The Pennsylvania Railroad, almost always in the vanguard in anything pertaining to railroads by the 1870s, employed a bridge engineer "whose special duty it is to furnish plans for all new structures and attend to the inspection of those already built."[31] Bridge companies bid on projects, and thus, ensured "a proper quality of strength, and first-class workmanship, such as can be obtained in no other way."[32] The Industrial Age introduced "specialization," exponentially increasing from the building of the Ashtabula Bridge to this day. Bridge building as a part time endeavor would become a thing of the past.

Stone superintended the Warehouse Point Bridge with Daniel Lester Harris, who became president of the Springfield Railroad and a highly respected leader in the Springfield community. Though not a highly reflective person, the following letter from Stone to Harris's widow, written in July of 1879 demonstrates that he was not totally void of sentiment, theological perspective, and the ability for meaningful, if not profound prose,

Cleveland July 13, 1879

Dear Mrs. Harris:

I was pained to hear Friday morning of the death of your very worthy husband, although I had been informed of his rapid failure in health. When I learned last winter that he was much better I hoped that he would be spared to his family many years. I had felt that freedom from all cares would restore him to health; but the ways of Him who controls all events must not be questioned. I formed the acquaintance of Mr. Harris some 40 years ago, and from the beginning to the end I loved and respected him. I admired his unswerving honesty and integrity. I do not recollect that our friendship was ever broken by a ripple. It was long and continuous; therefore, from my own relations with him, I feel that I can realize what his loss is to you and your family. It always seemed to me that it was his purpose to perform every duty to his family; he also devoted much time to the interest of others, and especially the public. In matters of public interest, when he had come to the conclusion that a reformation upon any subject was needed, he would enter into it without regard to its popularity, or its cost to himself and labor and personal effort. He would persevere until his point was gained on the subject exhausted. I grieve that I shall never meet him again

---

[31] Ibid.
[32] Ibid., 131.

on this earthly sphere, and beg to tender to you and all of your family my sympathy in this your greatest bereavement and loss and hope and trust you will look to our Heavenly Father who alone can give you consultation and guidance. Had I been in good health, I should have been with you at the funeral. Again, desiring to express my deepest sympathies, I remain

Sincerely your friend,

Amasa Stone[33]

Stone left his industrial footprints wherever he went, and that was especially true of Springfield. Before Stone moved to Cleveland Richard Hawkins was an office boy for Harris and Stone. Hawkins was a quick study. In 1867, Harris sold out to Hawkins. Eventually with William H. Burrall, Hawkins employed two hundred men manufacturing bridge parts. The fabrication plant inundated New England, especially Vermont, with bridges. One journalist wrote, "It is impossible to travel over New England without coming across some of Mr. Hawkins' work, and wherever it is you are sure to find first-class work, not only reflecting credit upon him, but upon the city from whence it comes."[34] In 1885, Hawkins built a five-span structure across Lake Champlain, and in 1886, founded the Vermont Construction Company of which he became President. When the wooden bridge over the White River in Hartford collapsed, Hawkins replaced it with an all iron truss bridge. Stone's influence on bridge building in New England was formidable, to say the least.[35]

Stone's Ascending Prominence in the Northwest

As Stone continued to build his bridges and later inspect them, he discovered a flaw in the Howe truss design. While still in his twenties he experienced the following epiphany:

I came to the conclusion that something must be done or there must be a

---

[33] Henry M. Burt, ed. *Memorial Tributes to Daniel L. Harris with Biography and Extracts from His Journal and Letters* (Springfield, Massachusetts: Printed for the family for Private Presentation, 1880) 86-87.

[34] "R. F. Hawkins Iron Works: Bridge Building and Boiler Making for the World at Large," *Progressive Springfield: Successful Springfield Industries.* No. III, n.d., 89-90.

[35] Robert McCullough. *Crossings: A History of Vermont Bridges* (Barre, Vermont: Vermont Historical Society, 2005) 124-127.

failure, and it must not be a failure. The night following was a sleepless one at least until three o' clock in the morning. I thought and rolled and tumbled, until time and again I was exhausted in my inventive thoughts, and in despair, when at last an idea came to my mind that relieved me. I perfected it in my mind's eye and then came to the conclusion that it would not only restore the reputation of a Howe bridge, but would prove to be a better combination of wood and iron for bridges that then existed, and could not and would not in principle be improved upon. Sleep immediately came.[36]

The solution consisted of longitudinal keys and clamps in the lower chords and iron socket bracing instead of wooden for the braces and bolts.[37] The sockets were necessitated by the contraction of wood when pressed in by a bolt with a head on one side and the nut of the bolt on the other, thus constantly becoming loose over time. Stone would put a sleeve around the bolt so the head and the nut would be prevented from contracting the wood. Stone's racing imaginative mind, often in full gear after turning out the lights for sleep, could not be so easily turned off, and the ensuing life-long insomnia came at a great cost to his health.

By 1845, when Stone would have been only 27, his fame was spreading beyond New England. The railroad had opened up the Northwest and Cleveland, Ohio was growing as a prime shipping port for the coal in Pennsylvania, the grain in Illinois, the iron ore in Wisconsin and Minnesota as well as the many wood and agricultural products in Ohio.[38] The mouth of the Cuyahoga River provided a convenient port and harbor. (It would become all the more important with the discovery of oil in 1859 at Titusville, Pennsylvania some 100 miles east.) While superintendent of the New Haven, Hartford and Springfield Railroad, Stone came in contact with Frederick Harbach, the company's engineer, a man of similar instincts, initiative, and expertise. Harbach had built the first all iron version of the Howe truss, a span of only 30 feet on the Western Railroad, Pittsfield, Massachusetts. Stillman Witt, with whom Stone would later form a long working relationship, was president of the Western Railroad.

---

[36] Burton Smith Dow III. "Amasa Stone, Jr.: His Triumph and Tragedy." (Unpublished M. A. thesis, Western Reserve University, 1956) 4.

[37] Ibid.

[38] This area basically consisted of Ohio, Indiana, and Illinois, and would become known as the "Old Northwest" in contrast to the Northwestern states of Oregon and Washington today.

Harbach was contacted by the Cleveland, Columbus, and Cincinnati Railway to survey a route between Cleveland and Columbus. Stone, Harbach, and Stillman Witt formed a company and won the bid to build the railroad. The road was begun in November of 1849, and finished on February 18, 1851. The construction of this stretch between Columbus and Cleveland required the removal of, "3,511,903 yards of soil, the placing of 34,413 perches of masonry, and the laying of some 300,000 ties."[39] Because Stone took stock instead of full pay, according to biographer Burton Dow, this project became the beginning of the family fortune. After completion Stone was given due credit by the railroad's president Alfred Kelly, "To the vigilance, intelligence, and working ability of the Superintendent, A. Stone Jr., the company are largely indebted for the successful results of the last season's business. In him the company is fortunate in having a faithful and experienced man to occupy the most important place in their service."[40]

Until this time, Stone and his family had been living in Springfield, Massachusetts, but with almost all of his work taking place west of the Alleghenies, he moved his family to Cleveland in 1850. After the Columbus to Cleveland project, Stone's firm was contracted to build the Bellefontaine and Indiana Railroad, which was finished in July 1853.

The Lake Shore

Then came the project that for the rest of his life would be his primary occupational concern, and would forever link Amasa Stone to the Ashtabula Bridge disaster. In July of 1850, the Cleveland, Painesville and Ashtabula Railroad awarded Stone's firm the bid to build a 95-mile track from Cleveland to the Pennsylvania state line. (The CP & A was often referred to as the Lake Shore because it ran along the shore of Lake Erie.) This section would officially become the Lake Shore and Michigan Southern Railroad in 1869, extending to Erie, Pennsylvania in the east and linking to the Illinois Central in the west. Stone would hold so many positions with the Lake Shore that they are difficult to trace and identify: director, president, vice-president, general superintendent, manager, and above all, stockholder.

---

[39] Dow, 11.
[40] Ibid., 14.

No task daunted Stone. Building a railroad between Cleveland and the Pennsylvania state line required removal of 1,058,154 cubic yards of earth, bridging several gorges and rivers and the laying of some 7,548 tons of "T-Rail" at a cost of $11,357,913.[41] It was this project that primarily opened up and connected Cleveland with the east, and in particular New York City, the financial capital of the United States. For almost all of the projects which Stone superintended, or even marginally participated, he took stock rather than pay.

During the 1850s, Stone weathered and oversaw two revolutionary transitions. First, from December 7, 1853 to February 1, 1854, the so called "Erie Gauge Wars" took place. The Pennsylvania Railroad had retained a six-foot gauge. Obviously, this meant that freight and passengers would have to be transferred to a different train at the New York state line and the Ohio state line. The courts would eventually rule on a four foot, eight-and-a-half-inch gauge, which is the standard gauge for all American railroads to this day. Second, Stone replaced wood burning with coal-burning locomotives. Thus, Stone was in the coal business, the iron business, the banking business, and all other businesses necessitated by running and owning railroads. Amasa Stone may have suffered from insomnia, but he never suffered from inertia.

The Cleveland Entrepreneur

For the next quarter of a century, Amasa Stone would become the most important and influential business leader in Cleveland. What he accomplished in those 25 years staggers the imagination. His work ethic, having been honed on a hard-scrabble farm in New England, was unsurpassed. Though secretive, almost to the point of mystery, he was not beyond a little showmanship. He personally made his way to a train stuck in a snowstorm and commandeered the locomotive, driving it the rest of the way into Cleveland.

A cursory overview of Stone's life portrays a bewildering array of initiatives and involvements. He was a founder, builder, president, or director of 14 different railroads, periodically occupying most of these positions at the same time. He was founder or president of five different banks. He was founder or partial owner of seven different companies: the Mercer Iron and Coal Company, the Union Iron and Steel Company, the Cleve-

---

[41] Ibid., 17.

land Wire Mill Company, the Union Rolling Mill Company, the Cleveland Stone Dressing Company and the Union Steel Screw Company. The last was the first company in the United States to manufacture steel screws for wood. (These screws we now see packaged in every conceivable size at Lowe's and Home Depot.) His younger brother Andros Stone became President of the Cleveland Rolling Mill Company in 1863, from which the wrought iron was purchased for the Ashtabula Bridge, and called into question a conflict of interest. Amasa denied holding stock in the Cleveland Rolling Mill Company when he testified before the Ohio Legislative Committee in February of 1877.

Immediately upon arriving in Cleveland in 1850, Stone with other investors started the Cleveland Stone Dressing Company. The stones were shipped up the Cuyahoga River from Berea, Ohio, and used for buildings as diverse and distantly located as the First Presbyterian Church of East Cleveland, and the Ontario legislative building in Toronto, Ontario Canada. He served as a director on the board of Buckeye Insurance Company, and at the request of Cornelius Vanderbilt, was appointed a director of Western Union Telegraph. From 1850 to 1880, there wasn't much done in Cleveland, Ohio on which Amasa Stone did not leave his fingerprints.[42] For Stone's obituary, a *Plain Dealer* reporter asked someone supposedly in the know about Stone's business enterprises. He received the following answer:

> He had a controlling interest in the Chicago Union Iron and Steel Works, the Kansas City Rolling Mills, and Brown Bonnell and Companies Iron Works of Youngstown. He also had considerable interest in the Forest City Varnish, Oil, and Naptha company of this city and it the Hayden Brass Works at Elyria. These were the only manufacturing enterprises in which he had interest; I believe. He was a large stockholder in the First National, the Commercial National Banks of Cleveland and had stock in the Toledo Bank.... He was also a large stockholder in the Western Union Telegraph Company and in various railway companies – The Lake Shore, New York Central, St. Paul, Northwestern, Rock Island, New York and New Haven and Union Pacific. His other property consists of real estate, and government bonds of which he owned large amounts.[43]

---

[42] Much of the above information came from "Amasa Stone, American Industrialist:" datahub platform for finance.

[43] "Simple Services at the Funeral of Amasa Stone Today," *The Cleveland Plain Dealer* (May 14, 1883).

Stone's interests were truly diversified and eclectic. He was founder and director of the People's Mutual Fire Insurance Company in Springfield, Massachusetts.[44] Upon arriving in Cleveland, he purchased land for growing pear trees. According to the correspondence between Hiram Libby of Rochester, New York, and Stone, it doesn't seem that the horticultural endeavor was very successful.[45]

The average citizen in Cleveland would not have known of Stone's many industrial enterprises. He was primarily recognized as a banker. Cleveland historian James Harrison Kennedy identified Stone as a "Director in the Merchants Bank, The Bank of Commerce, and the Cleveland Banking Company. He was President of the Second National Bank, and for some years President of the Toledo branch of the State Bank of Ohio... "[46] In 1873, Stone was elected president of the Cleveland Bank of Commerce, the same year he was appointed by Cornelius Vanderbilt as managing superintendent of the Lake Shore and Michigan Southern Railroad. As I demonstrate in Chapter 5, Stone's controlling interest in the banking enterprises of Cleveland made John D. Rockefeller highly dependent upon him.

A. B. Stone

Shortly after Amasa Stone moved to Cleveland, his six-years-younger brother Andros Boyden joined him. These two brothers would work together until Amasa died in 1883. Andros would demonstrate a business acumen equal to his brother. He eventually became President of the Cleveland Rolling Mill Company, the Chicago Rolling Mill Company (The Union Iron and Steel Company,) and the Kansas City Rolling Mill Company. Thus, until Andrew Carnegie fired up his Bessemer furnaces in the 1880s, A. B. was responsible for providing more rails and I-beams for railroad construction than anyone in the U. S. An early history of Cuyahoga County stated of him, "It has truly been said that throughout his career, Mr. Stone has shown two marked characteristics which usually lead to suc-

---

[44] "The People's Mutual Fire Insurance Company of Boston," *The Springfield Republican* (January 13, 1849).

[45] Letter, Hiram Libby to Amasa Stone, May 21, 1863. Letter, Amasa Stone to Hiram Libby, May 28, 1863. Courtesy Melissa Mead, University of Rochester Archives.

[46] James Harrison Kennedy. "Bankers and Banks of Cleveland, Ohio," *Magazine of Western History* (July 1885) 283.

cess: a clear and thorough understanding of whatever he has undertaken, and unvarying respect for the rights and opinions of others."[47] This last characteristic was not necessarily true of his older brother, Amasa.

The Commendation and Condemnation by R. C. Parsons

In 1885, R. C. Parsons, who had been an Ohio representative to the United States Congress, wrote a letter to his close friend John Hay, recalling a conversation he had had with a Major Ellison who remembered working with Amasa, during his early days of construction.

My Dear Col. Hay: (10-20-1885)

It gives me pleasure to write a few words in regard to Mr. Stone while in Brazil. I met Major Ellison, Chief Engineer, and Superintendent of the Don Pedro Railroad. He was an early associate of Mr. Stone and knew him intimately as they learned their profession of bridge building, side by side in New England.

From the recollection of Maj. Ellison in the case of Mr. Stone, the boy was father of the man. In youth, he had the same persistent tenacity, self-reliance, courage, ambition, and practical common sense that distinguished him in later years. Never fond of social life and proud of possessing the dress code, quite a vanity in those days. He never sacrificed any duty or obligation for personal gratification, but first completed his work. And then only was ready for play.

I am told that Mr. Stone gave great promise of unusual mechanical genius and those analytical powers of mind which enabled him to grasp with ready intelligence the most complicated machinery. He was frequently called upon by his superiors in office to get his opinion upon special inventions who would have made a very clear and able patent lawyer, as natural to men on mechanics, logical method of thought, and broad powers of generalization enabled him to understand and explain readily new theories and problems....

But it is proper for me to say that after a close personal acquaintance with Mr. Stone of over 30 yrs., I can write with truth, he was a man whose

---

[47] Cristfield Johnson. *History of Cuyahoga County, Ohio. In three parts: Part First, - History of the County, Part Second, - History of Cleveland, Part Third, - History of the Townships. With Portrait and Biographical Sketches of Its Prominent Men and Pioneers* (Cleveland: D. W. Ensign & Co., 1879) 386.

business sagacity, wonderful self-reliance, large intelligence, conservative ways of thought, and power to bend the wills of other men to his own, amounted almost to genius. In his own special field of labor, he was as much a born leader as Napoleon.

....The logic of his mind was direct, straightforward, unbending, almost un-yielding. What he felt he knew so well, it was useless to question. So he went straight to his end with the calm confidence of superior wisdom and the absolute conviction that he was right.[48]

What Parsons did not realize, was that he had praised and condemned Stone in the same letter. Often our virtues are our vices: tenacity becomes stubbornness, self-confidence becomes arrogance, focus becomes tunnel vision, and success becomes pride. According to Parsons, Stone was un-yielding, beyond question and absolutely convinced that he was right. This self-perception and in some ways a self-deception, would be his undoing.

In 1880, Stone drew up plans for an industrial school building, which he would build. The following specifications just for heating the building, demonstrate his devotion to miniscule detail, his faithfulness to the children for whom it was built and leaving nothing to chance. This kind of exactitude makes the failure of the Ashtabula Bridge all the more puzzling.

Radiator – First Floor
In bathroom one   16 pipes
Family sitting room one   60 pipes
Boy's wash room one   16 pipes
Girl's wash room one   16 pipes
Parlor one   16 pipes
Family dining room one   60 pipes
Children's dining rm two E.L 60 pipes
Main hall one   80 pipes
Back hall one   48 pipes

Second Floor

Towels one   42 pipes
Du one   48 pipes

---

[48] Letter, R. C. Parsons to John Hay, October 20, 1885, Samuel Mather Family Papers, Container 22, Western Reserve Historical Society. The word "tenacity" in the second line in the second paragraph, I filled in for a word that could not be deciphered.

Family bath room one   20 pipes
Boys sick room one   20 pipes
Girls sick room one   20 pipes
Closet one   42 pipes
Du one   48 pipes
Lavatory one   16 pipes
Girls dormitory one   48 pipes
Main Hall one   60 pipes

Third Floor

Bath room one   20 pipes
Boy's dormitory one   80 pipes

Stone had listed 940 pipes for 22 radiators, and specified that all of them be "neatly bronzed," supplied with "nickel-plated, rosewood handled valves." Of course, "All material to be used in this steam-heating apparatus to be of the best quality and the labor to be done by first-class mechanics and in the most thorough manner."[49]

A correspondent to *The Cleveland Leader* visited Stillman Witt's and Amasa Stone's office, to persuade one of them to visit the local school in Cleveland and observe the students taking examinations. Only Stone was present and protested that he had no children in the school. But after being reminded that he paid heavy taxes, enabling children to attend the school, he consented to give the visit one-half hour. He became so interested in the enterprise, that the one-half an hour passed unnoticed, and he insisted staying all afternoon until the exercises were finished.[50]

Amasa Stone was a complex person, a man of many personas, not easily categorized.

---

[49] "Specifications for the Pressure Steam Heating Apparatus for the Industrial School Building, 1880" Amasa Stone, Jr. papers, 1874-1881, Container 2, Western Reserve Historical Society.

[50] "Amasa Stone Visiting Schools." *New Ulm Review* (August 18, 1883). New_ulm_review_1883-08-15-5.pdf.

Chapter 3

Piety and Greed in America's Centennial Year

The Centennial Year

On December 31, 1875, Americans celebrated as never before or perhaps since; almost everyone left their house at midnight to witness the fireworks, to grab musical instruments and make whatever noise possible. A huge bonfire was lit at city center in Amasa Stone's hometown of Cleveland. Dwight Moody and Ira Sankey held a victorious watchnight service in St. Paul, Minnesota, and ushered in the New Year with thanksgiving and optimism.[1] It would be the greatest year of Moody's popularity in America and numerical success in his city-wide crusades. With the Ashtabula disaster and Philip Paul Bliss's death, the year would end on a far different note.

The Ashtabula Bridge disaster was a tragic denouement for America's centennial year and at the same time parabolic. Philadelphia spared no expense in celebrating American technological advancement and economic power on its one-hundredth birthday by exhibiting the latest inventions including the telephone and offering artifacts from all over the world. The main attraction was the nineteen-foot-high Corliss steam engine which stood in the Machinery Hall. To be sure, steam powered ships, trains, and all kinds of industrial machines had been the most revolutionary change in the nineteenth century. Maury Klein writes, "Power held the key to the steady march of industrialization — more power from more and big-

---

[1] William Randel. *Centennial: American Life in 1876* (New York: Chilton Book Company, 1969) 1.

ger machines to perform ever more tasks with ever greater efficiency. The steam engine was the revolutionary source of that power, and the Corliss engine displayed in striking fashion how far its development had come in a remarkably short time."[2]

The fair, which ran for six months from May 10 to November 10, hosted nine million visitors. The reviews were quite mixed. Russell Lyon wrote three-quarters of a century later, "Critics look back upon the Centennial Exhibition as an architectural and artistic calamity that produced not a single new idea, but rather the epitome of accumulated bad taste of the era that we called The Gilded Age, the Tragic Era, the Dreadful Decade or the Pragmatic Acquiescence - depending on which epithet you thought searing."[3] *Harper's Weekly*, on July 15, quoted Japanese Fukui Makoto, "The first day crowds come like sheep, run here, run there, run everywhere. One man start, one more thousand follow. Nobody can see anything; nobody can do anything. All rush, push, tear, shout, make plenty noise, say damn great many times, get very tired, and go home."[4]

Humankind was no longer subject to the vicissitudes of the winds and tides and whether a river flowed east or west. Steam had enabled railroads to become the biggest business in America. The lead bishop in the Methodist Episcopal Church, Matthew Simpson, as did almost all other Protestants, interpreted this technological advance as God's blessing on America and providential instrument in spreading the Gospel. Simpson preached, "Look at the iron bands, which have traveled the Atlantic and the Mississippi and will soon bind the Mississippi and the Pacific. Look at those telegraph wires on which men whisper and their words ought to be words of light and love. What is all of this? It is Jesus conquering the world."[5] As Simpson stood before Pittsburg's Christ Methodist Episcopal Church in 1874, he rhetorically asked:

> Have you thought of it? No railroads outside of Christianity? Why, if the Mohammadans build a railroad across the Black Sea, it is with Christian's money. If the Suez Canal is built, it is done by Christianity. Christianity develops civilization, improves the nations, joins them together, for the

---

[2] Maury Klein. *The Power Makers: Steam, Electricity, and the Men Who Invented Modern America* (New York: Bloomsbury Press, 2008) 4.

[3] Website: libwww.library.phil.gov/cencol/exh-testimony.tem.

[4] *Harper's Weekly* (July 15, 1876).

[5] George R. Crooks, ed. *Sermons by Bishop Simpson* (New York: Harper and Brothers, 1885) 190.

Church though it does not make railroads, nor erect telegraph wires nor tunnel mountains, yet it sends the spirit abroad that accomplishes all these and joins the ends of the earth together.[6]

We who over a century later wince or even stand aghast at the air-tight compartmentalization between professed Christianity and shyster business practices readily note that a culturally conformed Church was not a solution to the problem but represented the problem. Herbert Gutman wrote, "Legal and political theory, a moral 'social science' and institutional Protestants emphasized that in industrial America interference with the entrepreneur's freedom violated 'divine' or 'scientific' laws and historians have given much attention to the many ways Gilded Age social thought boasted the virtues of 'Acquisitive Man.'"[7] Engineering could construct a bridge over a river, but the Church could not construct a bridge between a biblical theology of ethics and the everyday moral confusion of America's sprawling industrial empire, the likes of which had never been witnessed in the history of the world. A ruthless business environment was no place for a sensitive Christian conscience. Henry Flagler, another ardent churchman who never worked on the Sabbath because he was attending church with Amasa Stone at First Presbyterian Church, was the most influential business partner whom John D. Rockefeller ever had. He kept a sign over his desk, "Do Unto Others As They Do Unto You, but Do It To Them First."

The Church adopted the very pharisaical teaching that Jesus had warned against; poverty was the result of sin. And what pastor was going to stand before his congregation and rail against capitalism when the likes of an Amasa Stone or a John D. Rockefeller sat in his congregation, persons who were responsible for more than their fair share of their shepherds' salary? A perusal of the sermons preached at Old Stone (First) Presbyterian Church reveal an exaltation of capitalism never addressing its contradictions. Stone may have listened to A. C. Aiken's sermon, "The Moral View of Railroads," preached at Old Stone in 1851 in which he stated,

> The hand of God is in all this. Some look with gloomy eye upon the "iron horse," as destined to subvert the laws of God and man, introducing moral and political anarchy, but we are not to be troubled by such spectres. To

---

[6] Darius L. Salter. *"God Cannot Do without America;" Matthew Simpson and the Apotheosis of Protestant Nationalism* (Wilmore, Kentucky: First Fruits-The Academic Press of Asbury Seminary, 2017) 553.

[7] Herbert Gutman. *Work, Culture and Society* (New York: Vintage Books, 1977) 81.

view the railroad as a mere auxiliary to increase wealth is very superficial. That is a consideration for the economist, but there are higher moral and social aspects of the railroad's advent. It will prove a barrier against frequent wars, by bringing nations together and creating more sympathy for and knowledge of each other, thus promoting a spirit of brotherhood. [8]

The Religion of the Robber Barons

The "Gilded Age," a moniker bequeathed to us by Mark Twain and Charles Dudley Warner, was the most industrially explosive in American history. It was also the most ostentatious and at times seemed pathologically amoral. The creative and highly-talented individuals who established America as the premier industrial and technological nation in the world were notorious for watered stock, short-selling stock, creating monopolies, destroying competition by conspiratorial practices, drawing kickbacks, insider trading, and fabricating stock certificates worth no more than the weight of the paper on which they were written. The rise of the American industrial titans during the last quarter of the nineteenth century has left us with a fascinating legacy which we will never tire of examining, condemning and even exalting.

No era, other than that of the giants of the early Republic who wrote our founding documents and freed us from British oppression, has left more colossal and larger than life figures, than the likes of Vanderbilt, Rockefeller, Carnegie, Gould, and Morgan. It could be claimed that these industrial and financial giants are the best or worst examples of unfettered capitalism ever witnessed in the history of the United States. Their idiosyncrasies, business shenanigans, and straight-out cut throat ruthlessness have embedded them forever into the American consciousness.

No "robber baron" outdid John D. Rockefeller, the richest man America ever produced, for serving both God and mammon. His legalistic code of ethics would have been on par with the pre-Christian Paul, "a Pharisee of the Pharisees," or as in John D.'s case, "a Baptist of the Baptists." His religious habits were akin to his penny-pinching scrupulosity. He kept a financial ledger for much of his life, jotting down every penny he spent such as 5 cents for a bridge toll and 5 cents for a stamp. He never traveled on the Sabbath, and was in church every time the door was opened. Even

---

[8] Arthur G. Ludlow. *The Old Stone Church: The Story of a Hundred Years, 1820-1920* (Cleveland: Privately printed, 1920) 138.

in the very beginning of his financial endeavors, he did not just tithe to the Euclid Avenue Baptist Church when he lived in Cleveland; he matched the pledges of everyone in it.

Though Andrew Carnegie was not an overt born again evangelical, he more than any other Gilded Age industrialist illustrated Max Weber's well-argued thesis in his classic, *The Protestant Work Ethic and the Spirit of Capitalism.* Though transporting his Scottish Presbyterianism to America, he strayed far from its orthodoxy; his religion became a mishmash of Swedenbourganism, mysticism, and spiritual séanceing, and above all exemplifying a "Christian work ethic." Carnegie's *Gospel of Wealth* found its roots in Herbert Spencer's evolutionary "survival of the fittest," rather than the teachings of Christ. As biographer David Nasaw suggests, it seemed Carnegie was listening to Spencer when he did not give in to the demands of employees.[9] Spencer wrote, "It seemed hard that a labourer incapacitated by sickness from competing with his stronger fellows, should have to bear the resulting privations. It seems hard that widows and orphans should be left to struggle for life or death. Nevertheless, when regarded not separately, but in connection with the interest of universal humanity, these harsh fatalities are seen to be full of the highest beneficence."[10]

Carnegie's philanthropy, the likes of which America had never seen, exemplified John Wesley's exhortation to "gain all you can, save all you can, and give all you can."[11] But the Methodist Bishop Hugh Price Hughes labeled Carnegie an "anti-Christian phenomenon, a social monstrosity and a grave political peril," representative of a system that created "millionaires at one end of the scale while making paupers at the other end."[12] Unfortunately, Hughes was correct in that while Carnegie was giving away his millions by building libraries all over America, (many are still in existence today) his workers were feeding coke into Bessemer furnaces in12-hour work days and a new employee lost as much as 50 lbs. in his first three months on the job. The 1892 Homestead strike (on the outskirts of Pittsburgh) is one of the best remembered in the United States along with the 1886 Hay Market Massacre in Chicago, and the 1892 Pullman Car Revolt led by Eugene Debs at the time of the Homestead Strike.

---

[9] David Nasaw. *Andrew Carnegie* (New York: The Penguin Press, 2006) 351.

[10] Ibid., 251.

[11] John Wesley, "The Use of Money," in *The Works of John Wesley*, Vol. VI (Kansas City: Beacon Hill Press of Kansas City, 1978) 124-136.

[12] Nasaw, 352.

While Carnegie was ensconced in his baronial castle Skibob in Scotland, he had left the dirty work to his company president, Henry Clay Frick. Frick hired 300 Pinkerton detectives, who in a showdown with thousands of strikers killed six employees from the Homestead plant. The six labor martyrs died in vain, as Frick could afford to shut down the plant, since there would not be much need for steel during the 1893 recession. Frick and Carnegie would become bitter enemies, providing one of the most intriguing, yet saddest postscripts to the Gilded Age. As the two men lay dying in New York, Carnegie wrote a note to Frick, requesting reconciliation. Frick responded, "See you in Hell because that is where we are both going."[13] The two men died within four months of one another in 1919.

No one outdid Daniel Drew for compartmentalizing his Sunday religion and his Monday business shenanigans. He may have been the first person in America to make a million dollars by manipulating stock, which mainly consisted of the Erie Railroad. He copiously repented on Sunday with abundant tears and then on Monday went out on the New York streets and did his best to spread rumors that the stock of the Erie Railroad was rising. One story has it, possibly apocryphal, that he stepped into a New York tavern and fumbling for his handkerchief, intentionally dropped a piece of paper with "Buy Erie" written on it. These hot stock tips made Drew rich, while those who fell for his chicanery ended up with worthless shares of the Erie.

Supposedly, Drew attended a Methodist camp meeting and was sitting next to someone who did not know him. During testimony time, an attender confessed the worst sins imaginable. Drew leaned over to his pew companion, and asked, "Who is it that is confessing all of those crimes?" The person answered, "I don't know unless it is Daniel Drew." Drew endowed the founding of American Methodism's first seminary in Madison, New Jersey. Drew Theological Seminary began in 1866 with his initial gift of $500,000. He reneged on a second $500,000 because he was wiped out in the 1873 Panic and died broke.

One of Pierpont Morgan's biographers, George Wheeler, characterized the supreme financer as one who evidenced delight "in the trappings of Christianity, leading the hymn sings on Sunday in a very public manner, attending annual diocesan conventions, and entering whole-heartedly into

---

[13] Les Standiford. *Meet You in Hell: Andrew Carnegie, Henry Clay Frick, and the Bitter Partnership that Transformed America* (New York: Three Rivers Press, 2005).

the discussion of faith, morals, and the Episcopal Church."[14] Wheeler argues that Morgan outdid Drew in watering stock. Morgan had stock certificates printed for his U. S. Steel Company for $1,402,000,000, when the company was worth only half of that. In other words, the market value of the component companies making up this "hydraulic wonder" was estimated by the U. S. Bureau of Corporations at no more than $726,846,000.[15] Morgan had to struggle with this great bag of water, which was like "some sagging animal on a Mardi Gras float. He had plenty of help in getting the great hydrocephalic object to stand on its own water-logged feet."[16]

A group of somewhat irreverent Episcopal clergymen dubbed Morgan, "Pierpontifiex Maximus."[17] Many of Morgan's deals were worked out on his yachts, the last one being 600 feet long, the length of two football fields. For all of the seemingly frictionless interface between religion and wealth, Jay Gould's biographer, Maury Klein, notes that "Wall Street was never much to look at, a narrow strip that began east of Broadway opposite the site of Trinity Church and wound its crooked way down to the East River. Cynics of a later age found it altogether fitting that this major artery of speculative finance should start at the doorstep of a church and plunge downhill to a watery grave."[18] Gould professed no religion; he may have been a scoundrel, but he wasn't a hypocrite. He flaunted neither the piety he did not have, nor the immense wealth which he did have.

The Response of the Church

The Gilded Age was the golden age of revivalism, personified by the 300 lb. Dwight L. Moody. Moody was the first prominent American evangelist to formulate a business model for conducting city-wide campaigns. Efficiency was important, and there were revivalists who advertised that they could save a soul for $2.00, as opposed to those who could do no better than $3.50 lavished on a conversion. As some have argued, Moody would have been successful at anything he tried, just as his contemporary business tycoons. He without a doubt believed God had called him into

---

[14] George Wheeler. *Pierpont Morgan and Friends: The Anatomy of A Myth* (Englewood Cliffs, New Jersey: Prentice-Hall, Inc., 1973) 59.

[15] Ibid., 210.

[16] Ibid.

[17] Ibid., 209.

[18] Maury Klein. *The Life and Legend of Jay Gould* (Baltimore: The Johns Hopkins University Press, 1986) 68.

the "soul-saving business," but he was not more convinced than John D. Rockefeller who believed God called him into the money-making business. Moody and Henry Ward Beecher were contemporaries. The former preached with a humble, modest, anecdotal style and the other with a dramatic, oratorical, bombastic proclamation. Moody appealed to the heart and Beecher to the mind with pithy statements and dramatic gestures for entertainment purposes. Both were theologically shallow and exemplified the wide latitude that American Christianity granted its primary platform performers. Neither adequately addressed systemic evil, but were in keeping with America's individualistic ideology.

How would rich industrialists alleviate their consciences when they were vaguely aware that there was an increasing gap between their accumulated riches and a new class of working poor, hopelessly crowded into tenement housing described by Jacob Riis in his *How the Other Half Lives: Studies Among the Tenements of New York*. As I have noted elsewhere, "If someone suggested that the poor needed attention, the bourgeoisie threw a Christmas party at Madison Square Garden, wore tuxedoes and gowns, while they sat in box seats and watched the 20,000 urchins below scramble for food and presents."[19] Ministering to the poor did not mean that one had to mix with the poor. As Sven Beckert wrote, "If elite churches wanted to remain fashionable, filled with gay parties and ladies with feathers and mousseline de Lame dresses, they had to move with their congregations and this principally meant relocating further uptown."[20]

It is easy to assume that the above Gilded Age snobbery, greed, and cut-throat ruthless competition would have put a damper on the Church and retarded its growth. Not so, argued eminent Church historian Arthur Schlesinger, Sr. in his seminal article, "A Critical Period in American Religion: 1875-1900." "Despite the many difficulties, theological and practical which beset the path of religion, the last two decades of the century witnessed a substantial gain in church membership. Protestant communicants increased from ten million to nearly eighteen million; the Roman Catholic population from well over six million to more than ten; the number of

---

[19] Salter, *"God Cannot Do,"* 662.

[20] Swen Beckert. *The Money Metropolis: New York City and the Consolidation of the American Bourgeoise* (New York: Cambridge University Press, 2001) 60.

Jews from less than a quarter of a million to approximately one million."[21] Schlesinger further observed that the growth "was a striking testimonial to the vitality of organized religion" and "was proportionately greater than the general advance of population, though, as might to be expected, less than the rate of increase of the urban wage-earning class."[22]

The Dysfunctionalism of Euclid Avenue

Stone's home described in Chapter 1, was among the first of many mansions built upon what many thought to be the most elegant street in America. A travel guide cited Euclid Avenue as the "Showplace of the World," and Mark Twain recommended the elm-lined street as an essential tourist destination for travelers from Europe.[23] These were strange claims for my generation, who often heard Cleveland referred to as the "Mistake on the Lake." (This is an undeserved moniker, in that Cleveland for the fifth time has been voted as having the most beautiful park system in America for 2021.)

In post-bellum America, Cleveland was anything but a mistake. It was a leading port in America for the shipping of iron ore, coal, and oil on Lake Erie to manufacturing cities in America. And much of the iron ore coming mainly from Wisconsin, Minnesota, and Ohio was transformed into track rails, sewing machines, telegraph wiring, screws, nails, and just about anything else requiring metal fabrication and formation. All of this shipping and manufacturing made Cleveland the banking center of the Midwest before Chicago could establish a firm footing in the 1870s. Some of the homes were truly elegant, such as the one Stone built for himself and another one for his son-in-law, John Hay. And the house built by Samuel Mather, another Stone son-in-law, which is now part of the Cleveland State University campus, was majestic. Industrialists Anson Smith and Stillman Witt built beautiful Greek-Revival homes.

But as the reader can imagine, several got caught up in who could build the biggest house, which resulted in monstrosities of quickly gained wealth. Samuel Andrews built a 100-room mansion which required 100

---

[21] Arthur Schlesinger, Sr. "A Critical Period in American Religion, 1875-1900," *Religion in American History: Interpretative Essays,* John M. Mulder and John F. Wilson, eds. (Englewood Cliffs, NJ: Prentice-Hall, Inc. 1978) 312.

[22] Ibid.

[23] Dan Ruminski and Alan Dutka. *Cleveland in the Gilded Age: A Stroll Down Millionaires' Row* (Charleston, South Carolina: History Press, 2012) 13-14.

employees: cooks, servants, and maintenance workers, just to keep the place from falling down. Charles Brush, who invented and developed the arc light, built a 40,000 square-foot house decorated with Tiffany designed stained glass windows and lights and a massive organ with pipes that extended from the first floor all the way to the third-floor ball room. It may have been one of the first sustainable energy homes in America, as Brush built a 52-foot diameter windmill in his backyard to supply power for his lights and the numerous electrical contraptions with which he experimented. Jeptha Wade, President of Western Union, did not think ahead when he built his mansion in proximity to St. Paul's Church. The ringing of the bells so irritated him that he paid the church $1,500 to cease pealing out the Good News of the Gospel over his exclusive neighborhood, if not the total environs of Cleveland. The bells never rang again during Wade's lifetime. Never was Christianity muffled with so little money.[24]

In spite of or maybe because of all of the parties, balls, and entertainment of world-wide dignitaries, Euclid Avenue was not a happy place. Chaotic dysfunctionalism shrouded over the masquerade of pretentious materialism and opulent ostentation. Lurking behind the manicured lawns and majestic porticos, doors, and towering gables were sexual scandals, betrayals, domestic violence, and even a swindler named Cassie Chadwick, (alias Cassie Hoover) who claimed to be the illegitimate daughter of Andrew Carnegie, professing that he had promised her $5,000,000 to right the wrong. After trying to cash a fraudulent $5,000,000 check, this Euclid Avenue resident was sentenced to fourteen years in prison, but died after only three years.[25]

No narrative was sadder than that of Euclid Avenue resident James Potter who poisoned himself, his wife, and his two children on January 18, 1928. He had purchased a coal mine in West Virginia that yielded little coal, and he failed to win a critical railroad contract. The once wealthy banker attempted to deal in real estate but came up empty. Unfortunately, within the paradigm of Euclid Avenue values, monetary failure was worse than moral failure and there was plenty of that; the sexual scandals were far too numerous to sort through here.[26]

---

[24] Ibid., 45.

[25] Alan F. Dutka. *Misfortune on Cleveland's Millionaires' Row* (Charleston, South Carolina: The History Press, 2015) 85-95.

[26] Ibid., 55-58.

The Ashtabula Bridge disaster and Stone's subsequent suicide were not anomalies, but representative events of industrialism run amuck with no encompassing theology or philosophy, much less adherence to the Hebrew prophets and the Christ who had plenty to say about wealth built on the backs of oppression. But the inhabitants of Euclid Avenue were not stupid, hoarding money as did Silas Marner and Ebenezer Scrooge with no other values. They lived within the angst of compartmentalization and rationalization, defense mechanisms that provided insufficient security from the upheavals and turmoil that roiled through post-bellum American economics. A few persons were aware of at least some of the contradictions.

In 1851 Stillman Witt, probably the closest and most revered business partner that Amasa Stone ever had, built one of the first mansions on Euclid Avenue. Stately but subdued, it was among the most beautiful in Cleveland, with graceful Ionic columns, avoiding the ornate and gaudy. He was a devoted Baptist seemingly without the defects of some of his fellow entrepreneurs. He co-founded an orphan asylum and the Cleveland Female Seminary. He was a lay director of the American Baptist Foreign Mission Society and built a "missionary church" for the First Baptist Church of Cleveland. In order to recover his health, he took a trip to Europe in 1875. During a storm he was thrown from his deck chair, injuring his head. He died the next night at 67 years old.[27]

Poverty or Wealth Written into the "Laws of Nature"

Historian Henry May argued that laissez-faire capitalism was creedal for nineteenth century American Protestantism. For the establishment to which Old Testament scholar Walter Brueggemann would later refer as the "royal conscience," one's poverty or wealth was due to the inequities that were written into the laws of nature. For those who had "made it," there did not exist systemic circumstances forbidding one from realizing his or her own potential by obeying God's prescriptions for industriousness and frugality. This apriori faith in a just stratification of society prevailed in both Church and Academy. The pews at Old Stone Church where Amasa

---

[27] "Death of Stillman Witt," and "Stillman Witt Dead," *Cleveland Plain Dealer* (May 4, 1875) 1.

Stone's family attended, rented from $400 to $1,000 annually.[28] Standing before the communion rail was not the ultimate leveling ground but rather a sanctification of one's place in society. How could the monetary value placed on a pew announcing one's social status, be more antithetical to the Gospel?

Stone's son-in-law Samuel Mather, who died the richest man in Cleveland, attended Trinity Episcopal Church. According to Cleveland historians Ruminski and Dutka, "Anybody attempting to sit in the Mather pew was always asked nicely to move, for Mather expected his pew to be available to him and his family at all times. Samuel earned this little benefit as a board member. If the church was in the red at year's end, Mather and fellow board members would divide the debt themselves and each write a personal check to return the church to debt-free status for the upcoming year."[29] May wrote that, "American scholars worked out a school of political economy which might well be labeled clerical laissez-faire. For at least a generation and in many institutions far longer, this body of doctrine dominated American teaching."[30]

May reflects that for most Americans, "the men who were builders of the railroads and the steel mills were national heroes."[31] If they had arrived at this preeminent stage of life, they deserved it. And those who had fallen between the cracks, deserved their basement or tenement dwelling as well. Nowhere did John Calvin and Adam Smith converge together more frictionlessly and smoothly than in America. The *Watchman and Reflector* declared that "Labor is a commodity, and, like all other commodities, its condition is governed by the laws of supply and demand."[32] The writer further claimed that labor was simply a "matter of barter and sale."[33] The same periodical stated in 1871, "In this country every industrious, prudent, skillful, healthful laborer can acquire a handsome competence by the time he is fifty years old, if that is what he desires."[34]

---

[28] Jeanette Tuve. *Old Stone Church: In the Heart of the City Since 1820* (Virginia Beach, VA: Donning Company, 1993) 29.

[29] Ruminski and Dutka, 20.

[30] Henry F. May. *Protestant Churches and Industrial America* (New York: Harper & Brothers Publishers, 1949) 14.

[31] Ibid., 51.

[32] Ibid., 55.

[33] Ibid.

[34] Ibid.

This book suggests and possibly even demonstrates that the surest route to this "handsome competence" was in some facet of railroad building, directing, and investing. May observes that for the religious press, "It was quite usual to print advertisements for railroad stocks in close proximity to articles extolling the progress of the railroad, explaining the reasons for its soundness, and even pointing to its moral beneficence."[35]

Strains of individualism ran through all of America's ruling, ecclesiastical and political institutions; from republicanism to transcendentalism, from Jeffersonian idealism to Jacksonian populism, no nation before America had so exalted the self-made man. Every man carried his own guilt for his failure and could take pride in his success. Wherever one landed, it was inevitable. It was all a strange mythology, a disconnect which few questioned, much less dissected, a dialectic of self-initiative and the unseen hand of determinism.

The titans of the pulpit were to do or say nothing in post-bellum America to upset the prevailing social order. From his majestic pulpit in Boston's Copley Square, Phillips Brooks proclaimed, "There could be no doubt, I think, of whatever puzzling questions it may bring with it, that it is the fact of privilege and inequities among men for which they do not seem to be responsible, which makes a large part of the interest and richness of human existence. . .!"[36] There was no incongruity between the inalienable rights of life, liberty and pursuit of happiness, and the fact that the prevailing economic conditions, at least for some, did not conjure portraits of happiness. As the Five Points slums reeking of garbage and sewage wafted only a few miles from him, Henry Ward Beecher from his Brooklyn pulpit shouted, "The things required for prosperous labor, prosperous manufacturer, and prosperous commerce are three. First, liberty; second, liberty; and third, liberty."[37] Beecher further stated, "Even in the most compact and closely-populated portions of the East, he that will be frugal and save continuously, living every day within the bounds of his means, can scarcely help accumulating."[38] The bottom line for Beecher, Brooks, and almost all Protestant clergy was "that no man in this land suffers from poverty unless it be his sin."[39]

---

[35] Ibid.
[36] May, 65.
[37] Ibid., 69.
[38] Ibid.
[39] Ibid.

The above leads May to conclude that "in 1876, Protestantism presented a massive almost unbroken front in its defense of the social status quo."[40] It is not unimaginable that we would suggest the Church was at least complicit, however unwittingly, in the disaster that capstoned America's centennial year and troubled the conscience of those most involved in the tragedy, if not indicting a much wider circle of accomplices.

Amasa Stone's Religion

How did Amasa Stone sift out from all of the above?

If we believe Flora Stone Mather, Amasa Stone's youngest daughter known for her piety, philanthropy, and a multitude of good works, the Stone family was in church Sunday morning, Sunday night, and for Friday evening prayer meeting. Flora recalled her own childhood and teenage years as a ceaseless round of church socials, entertainment, and ministry to the poor. What kind of parents would produce children who raised funds for church missions, sewed garments for the poor immigrant girls, and read to them inspirational materials? According to Flora's biographer, the young lady was raised in "an atmosphere of religious teachings within her home, school, and church."[41] Amasa Stone was the leading financial founder and supporter of Cleveland Academy where "religion was integral to the curriculum, the Bible was used as a textbook with direct instruction and examination of its contents."[42]

When one of Stone's former pastors Anson Smyth was asked, "Was Mr. Stone a Christian?" he replied,

> Again, I must answer that I do not *know*. I hope that he was, though he was not a communicant in the church. For twenty-five years he was an active member of the Board of Trustees of our First Presbyterian Society, and he ever, by word and deed, expressed a deep interest in its prosperity. In the building and repairs of the church he not only contributed freely to the required funds, but busy as he was with his personal interests, he spent much time in supervising and directing the work. He was a regular attendant upon the Sabbath services of the church. While in the city, and not detained by illness, he was almost always seen in his pew in the evening as well as the morning worship, and in this regard he was an example, if not a reproof, to many church members. His life was one of strict morality.

---

[40] Ibid., 91.
[41] Haddad, 30.
[42] Ibid.

True, he relished keenly the excitement of business, and no doubt enjoyed the process of accumulation, but he used his wealth freely and generously. Within the last year a man who owed him $15,000 failed in business under circumstances which showed that his misfortune was not his fault, and Mr. Stone cancelled his claim and encouraged the man to make a fresh start in life.[43]

He was not a "communicant." This meant that he did not take communion, because communion requires the examination of conscience, and we can only conclude that Stone believed himself unworthy of the Church's most distinct and defining sacrament. Stone was aware of the contradictions in his own existence. As far as we know, no one ever accused him of being a hypocrite. He was not a saint. One story has it that he froze Irishmen out of his machine shops, because of their demand for $14 per week. On a trip to Europe, he recruited Poles to emigrate to Cleveland promising them $7 per week, which many of them did. He may have been faintly aware that Jesus said, "One cannot serve God and mammon at the same time." He may have been even more aware of this tension in his conscience than were some of his fellow industrialists. It is to his credit that in the still of the night, the ethical issues that others seemed to trample or dismiss, were never completely quieted. Sadly, he never found peace within the turmoil of the numerous enterprises that constantly beckoned for his attention. As is often the case, in attempting to control them, they controlled him. His only escape was death.

There are many reasons to attend Church, among them social capital, and there was plenty of that at Old Stone Church. The attenders read like a "Who's Who" of Cleveland: Henry Flagler, Amasa Stone, Edwin Higbee, and Leonard Case, Jr. Unfortunately, there is probably truth in Jeanette Tuve's assessment:

> Few churches were as cosmopolitan as Old Stone and so proud of it. It included among its members some of the oldest, wealthiest and most cultured families of Cleveland, along with many newcomers and "strangers in our midst." Some people properly thought of it as a bastion of wealth and antiquity in the heart of Cleveland, and remembered pews filled with Cleveland's elite, women in fancy hats and fur coats and straight-backed men who exuded confidence, ability, and sometimes arrogance.[44]

---

[43] Anson Smyth, "To the Memory of Amasa Stone," *New York Evangelist*, (July 12, 1883) 54.

[44] Tuve, 76.

But no doubt, Stone attended church for personal reasons other than status. He loved hymns, the streamlined American service and a thoughtful sermon. He was engaged in worship with both heart and mind. Above all, the Church was a moral guide. He was fully aware of the Ten Commandments and took them seriously. He was Protestant through and through, as well as American through and through. He condemned the aristocracy of Europe and the formality of the Anglican Church which were one and the same. In 1882 he wrote John Hay, while the latter was in England:

> I can fully appreciate & believe all your stats as to British landlords. From this time forth, the British farmers will require all, but a small percent of the product of the land, in the maintenance of a comfortable existence and landed aristocracy in England will soon be a thing of the past. As to ritualism in their church, I have no patience with it - is it flummery and does not invite the culture of music or the intellect. The matter of prayers, sermons, and music should be studied and got up for the occasion to be in keeping with the progress of this age.[45]

This spiritual sincerity of the Stone family can hardly be questioned. Henry Adams, the nineteenth century's most astute social critic and perhaps John Hay's closest friend, referred to Clara as Hay's "stout and pious wife."[46] As Adams stood in the "gallery of machines," at the 1900 Paris exhibition, he referred to the "dynamo" as "humanity's Faustian bargain with technology."[47] In his autobiography, *The Education of Henry Adams,* Adams believed there to be a spiritual force much more infinite and meaningful than these gawdy and sometimes destructive inventions. After reading her copy, Clara Hay had the temerity to suggest that Adams could transcend his obsession with a dynamo and seek Jesus. "But it seemed to me that you have studied too much to find that 'single force.' You are still seeking. Why instead of all of those other books, you have gone to — to find it — did you not go back to the Bible....I have waited so long to tell you this but have not had the courage."[48]

Clara was reflecting the values of the Stone home, and in particular her father. He desired his daughters to have the best of both worlds, well

---

[45] Letter, Amasa Stone to John Hay, September 28, 1872. Courtesy of John Hay Library Archives, Brown University.

[46] David S. Brown. *The Last American Aristocrat: The Brilliant Life and Improbable Education of Henry Adams* (New York: Scrivner, 2020) 178.

[47] Ibid., 335.

[48] Ibid., 370.

really, the best of three worlds, piety, domesticity, and intellectualism. In piety Clara was not lacking. Samuel Langhorne Clemens admired John Hay, but was flummoxed and perhaps intimidated by his wife. One Sunday morning Clara happened in on the profane conversation of her husband and Clemens. Twain would later write "Mrs. Hay, gravely clad, gloved, bonneted, and just from church, and fragrant with the odor of Presbyterian sanctity" caused the two men to spring to their feet. Clara coolly greeted Twain with a stern, "Good morning Mr. Clemens." Hay, apologetically fumbling for words explained to his friend that his wife was, "very strict about Sunday."[49]

In these convictions, Clara and Flora mirrored their parents. This is not to say that there were not moral contradictions in the Stone household, and in particular, Amasa's aggressive materialism which at times may have seemed ruthless. But the family's integrity and commitment to God transcended the scandals and dysfunctionalism endemic to many of Euclid Avenue's residents.

As the reader is well aware, motives are multifarious, hardly understood by ourselves much less those who observe us. Obsequies are just that - words spoken in deference or flattery and reserved for an obsequy, a funeral, but there is usually a kernel of truth spoken on behalf of one who is deceased. Rev. Arthur Mitchell stated of Stone:

> In his relations as a worshiper, there were few more attentive, or more disposed to condemn the spirit of carping and criticism on religious matters. One remark that he made has touched me to the heart. It was that what he wished was not so much to hear the religious life described, as how he might attain unto it. He turned with aversion from this so-called philosophy which makes us doubt as to the existence of God and the inspiration of His Holy Scriptures. He would not brook the mention nor listen to the repetition of the sophistries which attempt to turn away men from Jesus Christ. Said He: "Doubt the efficacy of prayer? How can anyone refuse to believe it? We know God hears and answers prayer." Let his integrity, his benevolence, and his childlike trust in his Redeemer be an example to young men, that they, when they go to their tomb, may be followed by such fragrant memories, that their departure may be so regretted, and that

---

[49] Taliaferro, 56.

so many may rise up and call them blessed.[50]

Obviously, Stone had his detractors, but he may have had an equal number of admirers.

---

[50] John Hay (ed.) *Amasa Stone: Born April 27, 1818, Died May 11, 1883* (Cleveland: De Vinne Press, n.d.) 3. This volume, published after Stone's death, includes a biographical sketch by John Hay, a newspaper obituary, and several memorials to Stone by his contemporaries.

Chapter 4

Organizing Chaos

The Growth and Influence of the Railroad

In 2021, there were 804,398 startup companies or businesses in the United States. Though many or most of the initiators may not have had a degree in business administration, they would no doubt have read books in the field of business. The present American economic market consists of thousands of books on entrepreneurship, finances, organization, administration, motivation and leadership. One of the largest sections in any given Barnes and Noble bookstore consists of books by Peter Drucker, John Maxwell, Thomas Friedman, and biographies of Bill Gates, Steve Jobs, Jack Welch, Warren Buffet and dozens of others who are the current leaders in the tech, communication and investment professions.

The first degree in "business administration" was offered by the Wharton School of Business in Philadelphia, Pennsylvania in 1881. The idea that business could be studied as a specific curriculum with universal principles, organizational charts, management skills and leadership knowhow, much less motivational and success material, was foreign to Amasa Stone as well as other railroad titans in the post-Civil War Era. Many entrepreneurs failed. They were driven out of business by men with more intuitive instincts, innate business savvy, observational insight, competitive drive, and perhaps most of all, sheer greed. In 1865, there were 35,000 miles of railroad in America and in 1910, 240,000 miles.[1] No industry in the history of the world would be faced with so many unforeseen exigencies.

---

[1] John F. Stover. *American Railroads* (Chicago: The University of Chicago Press, 1997) 134.

No industry before or after the railroad explosion of the latter half of the nineteenth century would ever be so successful and so poorly managed.

No aspect of American life was untouched by the railroad. Railroads built new cities, founded financial empires, became the driving engine for Wall Street, and provided markets for farmers. Smaller, almost countless ancillary manufacturing enterprises devoured undreamed quantities of iron, oil, coal, as well as other underground resources. Railroads multiplied banks providing capital, made millionaires above and beyond any other industry, regulated exact time for scheduling into four U. S. zones, distributed manufactured goods on a mass scale, and became the foremost industry that united America. Negatively, the railroads confiscated land, dislodged farms, polluted virgin soil, accidentally killed people, intentionally killed Native Americans, abused and impoverished the Chinese and Irish, corrupted individuals, corporations, and government, ruined speculators, and created scandals on such a mass scale that tabloids had no lack of sensational news, provided by the likes of Daniel Drew, Jim Fisk, Jay Gould, Cornelius Vanderbilt, and others who provided constant entertainment, if not embarrassment. It is not that Gould was more dishonest than E.L. Harriman, Henry Flagler, Leland Stanford, Collis Huntington, as well as other larger than life personalities; his genius for dishonesty and subterfuge were simply better than anyone else's. As an individual, Gould would own more miles of railroad than any individual ever had, or ever would in the history of America, almost 30,000 miles.[2]

And it was not just railroads. Maury Klein, Gould's most recent and comprehensive biographer, states "By 1884 there emerged the image of a sinister figure who controlled not only a transportation empire but a communication monopoly embracing the national telegraph system, all overseas cable, several leading newspapers and the associated press."[3] Klein astutely appraises Gould, as well as the other railroad magnates, which would have included Amasa Stone:

> Gould's drive was elemental, a force of nature churning relentlessly onward, unstoppable, except by death. It was a quality common to the titans of business, a magnetism that made them at once compelling and repugnant figures. Journalists gestured vaguely at it, but only the novelists - Frank Norris, Robert Herrick, and above all Theodore Dreiser - captured

---

[2] Maury Klein. *The Life and Legend of Jay Gould* (Baltimore: Johns Hopkins University Press, 1986) 464.

[3] Ibid., 308.

the heroic dimensions, the epic proportions of their deeds, the complexity of their personalities, their inexorable drive.[4]

An Incomprehensible Behemoth

The above is simply to say Amasa Stone was caught up in a game for which there were almost no regulatory rules, no prohibitive laws, no Interstate Commerce Commission, and no Securities and Exchange Commission. The railroads experienced an unprecedented explosion of growth, with no one fully comprehending the colossus that was more chaotic than organized. A developmental evolution had leaped beyond management skills already in place, unable to brace itself against unforeseen contingencies. Budget planning for things that had never been built before by people who had never built them before was a definitive exercise in myopia and sheer guesswork. From railroad owner Henry Hillard's perspective, "The trouble was not only that the cost was much greater than was expected but that it was almost impossible to know at any one time what it would be. Hence it was unpractical to make provision in advance for current requirements and no measure of permanent relief could be resorted to until the whole of the road was actually completed."[5] Railroad historian Stewart Holbrook summarizes:

> (The railroads) were built at a time when men proceeded by trial and error; there were no rules, no sort of science. Neither the grading of a roadbed nor the laying down of iron rails had any precedent. Nor had the operation of trains. The stress laid on rolling stock and rails due to curvature was still somewhere in the future, in an unworked algebraic equation. So, to a large extent were the laws of traction that govern the starting of a load on wheels and its continued motion. As to grades nobody knew how much of a hill a ten-ton Rogers locomotive pulling twenty little freight cars could surmount; nor what would happen when such a train started down a hill, what with the crude brakes then in use.[6]

The railroads happened, and the ownership tried to figure out what had happened. Accounting procedures were so primitive that railroad owners

---

[4] Ibid., 432.

[5] Richard White. *Railroaded: The Transcontinentals and the Making of Modern America* (New York: W. W. Norton and Company, 2012) 222.

[6] Stewart H. Holbrook. *The Story of American Railroads* (New York: American Legacy Press, 1947) 56.

in the West did not know whether they were making or losing money. Financial and administrative practices were necessitated by unforeseen problems. Today, 2022, the smallest piece of mail can be located with a tracking number. During Stone's day, at times, an owner or administrator could not even track his million dollar train. An Erie operator explained, "Formerly, the utmost confusion prevailed in this department, so much so that in the greatest press of business, cars in a perfect order have stood for months, upon switches without being put to the least service and without it being known where they were. All these reforms are being steadily carried out as fast as the ground gained can be held."[7]

The early railroads were the purest form of Adam Smith's laissez faire capitalism ever witnessed in America. According to Smith, everything would simply self-regulate: supply would be calculated by demand and vice versa; wages would be rightly fitted to labor, and labor would be content with their wages. Unregulated industrialism would produce the "wealth of nations" with everyone in their proper place. All of this fallacious thinking would go up in smoke and leave spilled blood at the 1877 Baltimore and Ohio Railroad Strike, Haymarket Riot, the Homestead Strike, the Pullman Strike, and many other confrontations between labor and management.

The Management Disconnect

In a book on railroad economy written by James Whiton in 1856, *Railroads and their Management*. Whiton wrote, "The management so far as the duties are executed should be vested in a single head. On a line of large extent, the manager must see through the eyes and act with the hands of others to a great extent. Here the ruling power should be despotic and absolute."[8] The inherent contradiction found in this piece of advice is that such autocratic top-down management would not trust the eyes and ears of others. The disconnect was further widened with the inability to communicate in any meaningful way with on-site employment, and to require regular reporting, or take time to read the reports. The complaints about lack of communication were legion.

---

[7] H. W. Brands. *American Colossus: The Triumph of Capitalism 1865-1900* (New York: Doubleday, 2010) 22.

[8] Thomas C. Cochran. *Railroad Leaders 1845-1890: The Business Mind in Action* (Cambridge, Mass: Harvard University Press, 1953) 83.

The complexity of the railroad called for and created a management revolution in the nineteenth century, argues A. D. Chandler in his *The Visible Hand: The Managerial Revolution in American Business*, winner of both the Pulitzer and Bancroft prizes. As late as the 1840s, there were no middle managers in the US, and nearly all top managers were owners; they were either partners or major stockholders in the enterprise they managed. By 1876 that had radically changed for most railroads, but not for the Lake Shore. Chief engineer Charles Collins was immediately answerable to Charles Paine, the General Superintendent of the Lake Shore and Michigan Southern Railroad. Who were the people between Charles Paine and the President, Cornelius Vanderbilt? It would have been Amasa Stone if he had not resigned as manager in June of 1875.

John Newell replaced Amasa Stone, and from all indications, he was a competent railroad man.[9] No evidence exists that Newell paid any attention to the Ashtabula Bridge Disaster. He did not visit the site. He was summoned by neither the Coroner's Jury nor the Joint Assembly of the Ohio Legislature who investigated the accident. New York Central historian Alvin Harlow, in narrating the Ashtabula accident, makes no reference to Newell. Harlow states of Newell, "He would ride over the road at the tail of a train for hours, watching the track, envisioning it as it would be, when he had remolded it to his heart's desire."[10] We know for sure that he did not "remold" the Ashtabula Bridge. Ironically, Newell was known for the "perfection of alignment and grade of the Lake Shore and Michigan Southern Railroad…"[11]

What was the job description of Charles Collins? One was not provided as far as we know by either the Lake Shore and Michigan Southern or the New York Central. We can only infer from job descriptions provided by other railroads. According to Ray Morris in his book *Railroad Administration*, written in 1910:

> On the Norfolk & Western the organization is similar; on the Louisville and the Nashville it is prescribed that the chief engineer shall sit as assistant to the general manager in all matters connected with the maintenance of way, bridges, and buildings, and shall have the direction of any construction work that may be placed in his charge; he shall prepare plans and estimates for the construction or repair of all bridges, and other structures

---

[9] See Harlow, 337-338.
[10] Ibid.
[11] Gary, 137.

necessary to be constructed or repaired, and it shall be his duty to make periodical examinations of all bridges and track and report upon their condition to the general manager.[12]

The Annual Report for the LS and MS Railroad in 1875 gives us some clues as to Collins' responsibilities. Collins oversaw 1,852 miles of railroad track that laid straight would have reached across two-thirds of the United States. There were 548 miles of main line, and 635 miles of branches and tributaries to the main line. The remainder of the mileage was leased from industries that did not want to be responsible for railroad tracks. Two hundred thirty miles were dual tracks. The report for the Chief Engineer's department listed operational responsibilities for new side tracks, Ashtabula Harbor docks for coal, real estate purchased, steel rail, re-rolled iron rail (21,033,081 tons,) repaired iron rail, cross-ties renewed, fences built, track ballasted with stone and gravel, and purchase of wood and coal. Plus, he was responsible for the upkeep of 52 stations.[13] The Bylaws of the Lake Shore and Michigan Southern Railway Company stated in article XIX:

> The General Manager, Superintendent, Engineer, Freight Agent, Pay Master, Ticket Agent, and all other subordinate officers, agents and employees, shall perform such duties as usually pertain to their respective offices or positions, and such as may from time to time be prescribed for them by the Board of Directors, the President, or the Executive and Finance Committee.[14]

Several general statements can be made concerning Collins' job description. He had far too much on his plate to discover there was a problematic fracture in a bridge. In fact, bridge inspections as a separate entity were not even listed under Collins' responsibilities. We do not know the total number of bridges within the over 1,800 miles of track. What we do know is that the following amounts were spent on bridge masonry: $5,504,217 in 1872, $2,329,959 in 1873, $1,043,723 in 1874, $275,481

---

[12] Ray Morris. *Railroad Administration* (New York: D. Appleton and Company, 1920) 106-107. https://archive.org/details/railroadadminis01morrgoog/page/n147/mode/2up?ref=ol&review=theater.

[13] *Annual Reports: Lake Shore and Michigan Southern Railroad Company, 1875:* https://quod.lib.umich.edu/r/railroad/0549714.1875.001?rgn=main;view=fulltext. 1876: https://quod.lib.umich.edu/r/railroad/0549714.1876.001?rgn=main;view=fulltext.

[14] The Penn Central Transportation Company Records, 1835-1960, Oversize Vol. 24, Courtesy of Bentley Library Archives, Ann Arbor, Michigan.

in 1875. The diminishing amounts were not because there were fewer bridges to be maintained, but because there was *less money to be spent on the bridges*. Also, because of the 1873 Panic, during 1874 the employees of the Lake Shore went from 12,318 to 10,747 – there were simply fewer people to do the work.[15]

Chandler, the foremost authority on nineteenth century railroad administration and management, described the engineer of track, bridges, and buildings "as one who acted as an inspector of work and supervisor of plans and methods and will also have charge of such other engineering matters as the general superintendent may direct."[16] In order to fulfill these responsibilities, "There should be an assistant engineer in charge of each 100 miles of main-line track, with nothing else to look after, and directly responsible to the assistant superintendent of the subdivision as well as to the engineer or division engineer of tracks, bridges, and buildings."[17] These intermediate engineers did not exist for the Lake Shore and Michigan Southern. We should not be surprised that Collins testified he could give no concrete details as to regular inspections of the Ashtabula Bridge.

The investigation by both the Coroner's Jury and the Legislative Committee (which we will later examine in detail) was almost entirely focused on the past. What had Amasa Stone done wrong in the building of the bridge? Who was most to blame? As we will argue in Chapter 14, instead of one person or one cause, there were all kinds of contributing factors, failures, and complicit persons who were responsible for the disaster. Chandler states concerning the historians of the Gilded Age, "They have argued as to whether these founding fathers were robbers, or industrial statesmen, that is, bad fellows or good fellows."[18]

There were no good or bad persons who caused the bridge to collapse. The descriptors "good," or "bad," do not tell us anything about the people who contributed to the horrific death of 98 individuals. The citizens of Ashtabula had no emergency plan in place, no medical facilities, no fire department specifically trained as to how to respond to a major disaster,

---

[15] Paul F. Laning, "The History of the Lake Shore and Michigan Southern Railway in Ohio," (Unpublished M. A. thesis, Ohio State University, 1938) 77.

[16] Alfred D. Chandler, Jr. *The Railroads: The Nation's First Big Business* (New York: Harcourt, Brace & World, 1965) 120.

[17] Ibid., 122.

[18] Alfred D. Chandler, Jr. *The Visible Hand: The Managerial Revolution in American Business* (Cambridge: The Belknap Press of Harvard University Press, 1977) 5.

no telegraph message warning that the bridge was nearly impassable on December 29, 1876. No annual reports for the Lake Shore and Michigan Southern Railroad warned of depreciating changes in the Ashtabula Bridge. No one reported the bridge was a shifting structure. Paint clearly showed that it was brushed on metal not previously painted because the diagonal compression braces had moved. In order to turn the braces, as we will later examine, the lugs for the purpose of holding them in place had been sheared off.

None of the annual reports to the shareholders noted that insufficient time and money were being spent on the bridge, and worse yet, there was no daily log or regular reporting as to the changing realities of the bridge's condition. Cornelius Vanderbilt, owner of the Lake Shore and Michigan Southern, was concerned with dividends, but not ongoing detriments, or what was needed to be in place to prevent a disaster. Adam Smith had not dealt with these unforeseen contingencies, and unfortunately his "invisible hand" was no hand at all for an industry so new and gargantuan that it called for many hands, and a new administrative paradigm, a philosophy of operation calling for specialization in the exponentially growing facets of railroad operations.

Both Stone and Vanderbilt were caught in a managerial time warp, double entry accounting, and content with the bottom-line indicating profit and loss based on the difference between income and expenditures. William Vanderbilt, who took over his father's empire after his death, became notorious for saying, "The public be damned, my only concern is for the stockholders."[19] The most consistent action of the Board of Directors during the time of Amasa Stone's service was to increase the total shares for each director. How they were determined is a mystery. In 1870, each director held 260,000 shares; in 1875, 318,000 shares; in 1876, 361,000 shares; in 1877, 375,000 shares; 1878, 399,000 shares, in 1879, 427,000 shares; and in 1880, 434,000 shares. There was no correlation between the earnings and worth of the railroad and increase in shares per director.[20]

Railroads, the largest business enterprise in the world, "created new needs and opportunities. On the large roads it became increasingly difficult for the engineer (Charles Collins) and his assistants to oversee the

---

[19] Stover, 102.

[20] The Penn Central Transportation Company Records, oversize vol. 24, pp. 114, 142, 181, 261, 263, 279, 311, 349, 381, 413. Courtesy of the Bentley Library, Ann Arbor, Michigan.

work of many small contractors. Labor and equipment often became hard to find at the time they were most critically needed."[21] These people needed to be already in place. With technological advancements came liabilities demanding ongoing questions: "What could possibly go wrong with this device, structure, or technique?" "Well just about anything." If the Ashtabula Bridge was "experimental," as confessed by Collins, such risks needed to be constantly monitored with research and discovery, an ongoing process that was foreign to the management ownership, board of directors, and stockholders.

Attempting New Paradigms

All of the above people were more concerned about economic growth than improving what was already in place. The only alarms which sounded were set off by economic factors rather than security systems which would constantly attempt to discover flaws in the growing array of stations, railroad sheds, construction methodologies, technical innovations and ever-expanding mileage. In 1846 Louis McLane, president of the Baltimore and Ohio, stated "the great augmentation of power and machinery demanded by the increasing business" called for "a new system of management."[22] Its chief engineer, Benjamin Latrobe, described the objectives of the new plan for the Baltimore and Ohio, which he put in place:

> [They] consisted in confining the general supervision and superintendence of all the departments nearer to their duties, and, by a judicious subdivision of labor, to insure a proper adaptation and daily application of the supervisory power to the objects under its immediate charge; in the multiplication of checks, and to effecting a strict responsibility in the collection and disbursement of money; in confining the company's mechanical operations in their shops to the purposes of repairs, rather than of construction; in promoting the economical purchase and application of materials and other articles needed in every class of the service; and in affecting a strict and more perfect responsibility in the accounting department generally.[23]

No similar document could be found in the reports and records of the Lake Shore and Michigan Southern. This is not to cite the Baltimore and Ohio as a perfect example, because no system can prepare for all unseen

---

[21] Chandler, *Visible Hand*, 93.

[22] Ibid., 99.

[23] Ibid.

contingencies; the system has to be continually tweaked and improved as an evolutionary process. This evolution calls for trust and delegation, as opposed to micro-managing control by a single person. Problems of railroading were far too complex for Stone's primitive instincts, and multiple reporting was simply too much to funnel into a singular line of command that went in only a hierarchical direction. Chandler indicts the New York Central, owner of the Lake Shore, as a railroad which gave "little attention to the problems of organization."[24]

One of the solutions would have been to separate maintenance and repairs from operational expenses, as did Albert Fink in 1875 as Senior Vice President of the Louisville and Nashville Railroad. There were three separate headings designating the L&N's accounts: Maintenance of Roadway and General Superintendence, Station Expenses Per Train Mile, and Movement Expenses Per Train Mile. There are 73 different expenditure items, and exact formulas for predicting the costs for the items such as those under the category "Movement Expenses Per Train Mile:" the projected costs of repair items included adjustment of track, cost of renewal of rails, labor replacing rails, locomotive repair, washing and cleaning, passenger car repairs, sleeping car repairs, freight car repairs, damage to freight and lost luggage, damage to stock, wrecking account, damage to persons, and of course, law expenses.[25] The Annual Report for the LS & MS did not offer a similar overview, and it was difficult to see the forest for the trees. What immediately stands out in the Baltimore and Ohio report is that details, rather than being obfuscated or hidden in the many facets of keeping a railroad in proper running order, are the many items to which questions can be addressed.

Rapacious Greed and Competition

The cutthroat competition resulted in backstabbing, price fixing, insider trading, and outright deceit, creating behemoth monopolies by displacing scores of honest small-town manufacturers, leaving in its wake resentment, bitterness, jealousy, and revenge. Ida Tarbell would never forgive John D. Rockefeller for putting her small-town "cooper" (barrel maker) father out of business because Rockefeller could both make and ship a

---

[24] Ibid., 107.
[25] Ibid., 118-119.

barrel more cheaply. (He came up with the simple idea that if he allowed the barrel staves to dry before shipping, the freight weight would be less.) She would take out her revenge on Rockefeller with a two-volume expose on the oil titan that would leave the richest man in the history of America forever tarnished, and make her famous.

One of Gould's competitors, Henry N. Smith, became so angry at him that he shook his finger at Gould, vociferating, "I will live to see the day, sir, when you have to earn a living by going around these streets with a hand-organ and a monkey." The unperturbed Gould quietly retorted, "Maybe you will Henry, maybe you will, and when I want a monkey, Henry, I'll send for you."[26]

Solvent and healthy organizations pay more attention to themselves and their own defects, rather than constantly eyeing the competition. But most railroads were concentrating on gaining new territory rather than improving service to the territory which they already possessed. As Chandler says, "Never before had a very small number of very large enterprises competed for the very same business."[27] Authors Maury Klein and Harvey Kantor summarize:

> All these uncertainties threatened the well-being of everyone and drove most businessmen to a common conclusion: cutthroat competition was anathema in industry. It was fine in theory but ruinous in practice. It bred waste, inefficiency, and instability. As a form of industrial warfare it too often brought Pyrrhic victories. It was far too rugged a form of individualism. To survive, much less to prosper, businessmen had to eliminate or at least minimize competition in their industry, though of course they might continue to extol its virtues for other industries.[28]

Facing rate cuts, competition for the same territories, and counterproductive expansion by stock manipulators such as Jay Gould, railroad operators began to form associations, namely the South Railroad Rates Association, the Transcontinental Association, and the Western Traffic Association, but they did little to bring order out of the chaos. When railroad historian Charles Francis Adams (grandson of John Quincy Adams) attended a joint executive committee of a "federation" organized by Albert Fink, it struck him "as a somewhat funeral gathering. Those comprising

---

[26] Klein, *Gould*, 132.

[27] Chandler, *Visible Hand*, 134.

[28] Maury Klein and Harvey A. Kantor. *Prisoners of Progress: American Industrial Cities 1850-1920* (New York: Macmillan Publishing Co., Inc., 1976) 44.

it were manifestly at their wit's end... Mr. Fink's great and costly organization was all in ruins... they reminded me of men in a boat in the swift water above the rapids of Niagara."[29]

As business contracted through the '73 Panic, competition became even more pronounced. It truly was survival of the fittest. Stone, as manager of the Lake Shore, felt the pinch. On February 13, 1875, he wrote New York Congressman George W. Patterson, "You ask 'will not the competition east of Chicago in freight and passengers prove ruinous to the stock of all roads' – I reply that it is possible. The sharp competition and consequent low rates are quite discouraging."[30]

Absentee Ownership of Cornelius Vanderbilt

On December 11, 1867 the stockholders voted Cornelius Vanderbilt in as president of the New York Central. He immediately fired all of the directors except Henry Baxter. He then appointed 11 others including his son-in-law Horace Clark, his son William, and Amasa Stone Jr., a "practical railroad man." Biographer Wheaton J. Lane summarized the six charges Vanderbilt gave the Directors: "1. Buy your line. 2. Stop the stealing that went on under the other man. 3. Improve it in every practical way within a reasonable expenditure. 4. Consolidate with other lines for the sake of efficiency. 5. Water its stock. 6. Make it pay a large dividend."[31] Lane assessed that the summary steps were correct. But the actual chronology of the steps Vanderbilt took was different than those he ordered. Before he did anything, he "watered the stock."[32] The term comes from the myth that Daniel Drew, after driving his cattle from wherever denied them water until just before market, then let them drink their fill.

But the constant watering of stock was not apocryphal on Wall Street. On December 17, 1867 the New York Central capitalized at $28,537,000, which was already more than it was actually worth. On Monday, December 21, Vanderbilt announced a public offering; on the following Saturday the stock soared from 134 to 165, finally closing at 154. By sometime in January, 1868, the New York Central had sold stock for almost $52

---

[29] Chandler, *Visible Hand*, 142.

[30] Letter, Amasa Stone Jr. to George W. Patterson, February 13, 1875, courtesy of Melissa Mead, University of Rochester Archives, Rochester, New York.

[31] Wheaton J. Lane. *Commodore Vanderbilt: An Epic of the Steam Age* (New York: Alfred A Knopf, 1942) 227.

[32] Ibid.

million. There was so much "water" that Charles Francis Adams Jr. estimated "according to the books of the company, over $50,000 of absolute water had been poured out for each mile of road between New York and Buffalo."[33]

In June of 1869 Vanderbilt rode from New York to Chicago, "inspecting the line for himself," He found nothing of which he could complain. Does inspection mean he got off the train and examined all the bridges, in particular the one which crossed the Ashtabula River? Or, did he simply look out the window? *The Cleveland Herald* editorialized on June 22, "We understand the Commodore was well satisfied with the trip."[34] He was satisfied enough that when he stopped in Cleveland on the return trip on June 12, he met with the stockholders of the Lake Shore which included Amasa Stone. We have no record of what was said.

Chandler blames Vanderbilt for the New York Central's dysfunctionalism: "After Gould forced Vanderbilt to take over the Lake Shore, the Commodore did little to integrate the operations of that road with those of the New York Central."[35] Chandler summarizes:

> One result of this loose organization was that the New York Central was unable to obtain the economies of scale provided by the staff units in the general office. There was no standardization or testing laboratories for the system as a whole comparable to those set up on the Pennsylvania in 1875 and on the Burlington in 1876. Nor could the Vanderbilt system benefit from the advantages derived from centralized purchasing, a centralized legal staff, or a centralized management of insurance and pension funds for workers.

> More serious was the lack of a central office to evaluate the performance of the operating units and to plan and allocate resources for the system as a whole. The statistical data reviewed by the board and its committees were financial rather than operating. The finance and executive committee looked at the balance sheets and operating figures or cost accounting data that flowed into the office of the different presidents and on which evaluation of managerial performance had to be based.[36]

---

[33] Ibid., 233.

[34] T. J. Stiles. *The First Tycoon: The Epic Life of Cornelius Vanderbilt* (New York: Alfred A. Knopf, 2009) 487.

[35] Chandler, *Visible Hand*, 157.

[36] Ibid., 182.

Cornelius Vanderbilt, who owned the Lake Shore at the time of the bridge disaster, incarnated almost all the problems of railroad management. As Thomas Cochran points out he was an amateur, had no hands-on experience other than riding on a train, and knew almost nothing about the technicalities of the actual machinery. To be sure the tycoon had unsurpassed business savvy and had a broad exposure to all kinds of transportation from row boats to sailing boats, from canal boats to steamships, but his broad experience could not make up for absentee ownership. Thus, when the Commodore bought the Lake Shore in 1869, he conferred managerial responsibility on his son-in-law, Horace Clark, who knew less than his father-in-law. "Clark was a well-educated New Englander, a graduate of Williams College in 1833 who preferred planning a financial empire from his office in NYC rather than attending to the details of a railroad from the Lake Shore office in Cleveland."[37] Upon being given his new job, Clark in turn appointed John Henry Devereux as Managing Director. Devereux had much experience as a "railroad engineer," but with the economic crisis building between the gold bust of 1869 and the panic of 1873, he did not know how to take initiative in warding off or solving the mounting financial crisis. Thus, when Clark died in 1873, Vanderbilt replaced Devereux with Stone, the man who knew more about the Lake Shore than any other human being. Stone was critical of Clark's leadership and adopted a conservative financial policy, in his mind the only way to save the railroad. On April 28, 1874, he wrote again to George Patterson, "It is due to the Commodore to state that he has practiced a very conservative policy, but nothing could be more reckless than the general policy of the parties who controlled the management a year ago."[38]

Vanderbilt epitomized a railroad president whom Cochran describes as the entrepreneurial owner. This person was faced with constant decision making in the absence of any kind of executive committee, and certainly not with the convenience of a conference call or a Zoom meeting. "Bargains had to be snapped up when they were available, competition had to be forestalled by prompt building or rate cutting. Construction had to proceed in the face of unforeseen obstacles."[39] And when an entrepreneur owned more than one railroad, what was good for one railroad was not good for

---

[37] Cochran, 27.

[38] Letter, A Stone Jr. to George Patterson, April 28, 1874. Courtesy of Melissa Mead, University of Rochester Archives.

[39] Ibid., 65.

another. Thus, the manager may have been caught in a no-win situation, being unable to come to a mutually satisfying decision between, or among several railroads. When the board of the Michigan Central clashed with the New York Central, the director of the former, Henry Ledyard, wrote Cornelius Vanderbilt on September 1, 1873, "At the meeting of the boards of both companies in June last, two lines were submitted. The one marked A, I consider the best, but the one marked B was ordered to be adopted."[40] Everybody lost on this decision. Vanderbilt's absentee ownership was the rule rather than the exception. Railroad historian Marshall Kirkman wrote:

> In the early history of railroads the president of a company rarely partici-
> pated in its physical management. Some matters were left to the superin-
> tendent. The president was usually a man of wealth, chosen with a view
> to the favorable effect his name would have upon investors rather than
> because of his knowledge of railroad details. His responsibility, therefore,
> was merely nominal; the operating officer was expected to consult with
> him and listen to his views when he had any to express. He performed
> this dignified office of chairman of the board of directors, and was also
> a member ex officio of its various committees. He signed its bonds, only
> in some cases, its certificates of stock. He was in many cases merely a
> figurehead... he had advisory powers and direction over all the affairs of
> the company, but he exercised these duties only in a general way. The trea-
> surer looked after its interest in one direction, while the superintendent has
> entire charge of local matters.[41]

The management picture was further clouded in that according to *The Cleveland World*, Cornelius Vanderbilt and son William purchased the Lake Shore only on the condition that Amasa Stone replace Cornelius's son-in-law as acting manager. Even though Horace Clark was the son-in-law of Cornelius, they seemingly were at constant odds with one another. In 1873, the acute beginning of the financial panic, Stone accepted his new responsibilities. The writer for *The Cleveland World* assessed that Stone's extreme conservatism resulted in the Lake Shore always finish-ing in the black for each calendar year. Unfortunately, this also meant "If things went ill improvements were stopped until they went better. All this

---

[40] Ibid., 68.

[41] Ray, Morris. *Railroad Administration* (New York: D. Appleton and Company, 1920) 147. https://archive.org/details/railroadadminis01morrgoog/page/nmode/2 up?ref=ol&review=theater.

is directly opposite to modern railroad management."[42] Lake Shore and Michigan Southern historian Ferdinand Gary concurs: "All projects for the enlargement and betterment of the system were, for the time, held in abeyance, and strict conservatism dominated its policy until the storm blew over. The conversion to cautious methods was timely, and saved the company from impending disaster."[43]

The Cleveland World contributor also claimed Amasa Stone was the "father" of the Lake Shore and Michigan Southern "and made it the most solid property in its way, the sun shines on." Further, "If Amasa Stone had not grasped the reins with a masterly hand when, on Clark's death in 1873, ol' Commodore Vanderbilt placed them in his hands, the Lake Shore and Michigan Southern would have either been wrecked or eaten up by its rivals."[44] At the end of his first year of management Stone offered his resignation, which would be effective a year hence. Two weeks later, he wrote the following letter to Vanderbilt:

Cleveland, July 13, 1874
C. Vanderbilt, Esq:

I am in receipt of yours (letter) of the 9th, and am satisfied with your expressions. I had expected that you would send someone here by this time with whom I could spend some two weeks before I should take some recreation with my family, but as it is, I will so arrange matters that I can leave about the 25th for Sharon Springs, which will be within reach of all of our business by wire. It may be that you will be at Saratoga about the final of August, at your home when we can talk the matters over as you suggest.

Whoever occupies the position I do should have the fullest confidence of yourself and your son William which at times I have felt I did not possess, but which I felt I was fully entitled to -

My nature is such that when it is intimated that I am not faithful to others interests entrusted to me, it causes me pain as/so far greater than the loss of any money could when especially I know it was without the slightest foundation. In this circumstance, I will explain steel rail contract. I was

---

[42] "How the Lake Shore Railroad Became Great" *Conductor and Brakeman*, March 1898, 158-163, retrieved 9/11/2020.

[43] Ferdinand Gary. *Lake Shore & Michigan Railway System and Representative Employees* (Chicago: Biographical Publishing Company, 1900) 65.

[44] Ibid., 159.

never consulted as to that contract – although I knew that it was the lowest price 1st class rails could be laid down here at that time – for the contract was for 20,000 tons to be delivered during the year 1873 and no agreement was made that our old rails should be taken in part payment – when I took charge of the road I exercised all of my ingenuity to get through with the contract on the most favorable terms to the Company which I should have done at a gain of a full $100,000 to our Company had not the new contract for 5,000 tons been made with other parties and I have no doubt I could do the same thing now if that contract could be disposed of although as you intimated we should have enough to test the comparative quality – I never allowed myself to own a dollar in the rolling mill until about the time I went to Europe and until I supposed I should have no further direct management of railroads – and my interest since then has been comparatively small.

The product of those works has been at or about $7,000,000 in a single year, and they have given us a very large tonnage and which would continue if we are on friendly terms with them. I am convinced they have such advantage that they can furnish steel rods at the best quality at less rates than any other works, and will do so if a proper course is taken with them.

As to this road it is a very large and complicated concern – it has upon it many extraordinary, good men but they need on the ground with them someone that they have respect for with whom they can daily consult and who fully appreciates them, which to them is more than money – In view of this I have had much anxiety, as to whom you should appoint to fill my place. If a mistake should be made they will scatter away and great inconvenience would be caused to the management of the road –

I hear that you may deem it best to make a dividend for the last six months; while this would be personally agreeable to me, I deem it very inexpedient that it should be done – if the debt is funded as of the first of this month, can the net earnings of the next six months be divided next January, this will give you the proxies for the next election, and will keep the affairs of the co. in a solid basis - all these thoughts are written as confidential and may or may not have weight with you, but are nonetheless my convictions and I think in your interest.

I hope you will favor me with a reply to this.

Very truly yours,

A. Stone[45]

Several conclusions can be drawn from the above. Stone felt slighted by William who gave him an informal vote of no confidence; he did not appreciate having his integrity questioned; a contract was made with other railroad manufacturers on which he was not consulted; the company with which the New York Central had been previously doing business (no doubt the Cleveland Rolling Mill, of which his brother A. B. was President), was still best for the NYC; believed that a contract with another company was a betrayal of confidence and a loss of money; did not perceive that the company should pay a dividend for the last half of 1874; believed that after his departure the NYC would need people on the ground to regain the confidence of the Cleveland Rolling Mill management.

Stone was President of the Cleveland, Painesville, and Ashtabula Railroad, August 1858 - March 1869,[46] which built the bridge. In 1868, the CP & A was renamed the Lake Shore, and upon merging with the Michigan Southern and Northern Indiana Railroad in 1869, became the Lake Shore and Michigan Southern Railway. He served on the Board of Directors 1853-November 29, 1882; General Superintendent December 1852-July 1853; General Manager under the appointment of Vanderbilt July 1, 1873-June 30, 1875. Charles Paine was General Superintendent April 1, 1872-August 16, 1881 at a salary of $7,000 / year and Charles Collins was Chief Engineer April 1, 1872-January 14, 1877 at $6,000 / year.[47] *Other than being a Director, Amasa Stone had no official position with the railroad, nor managerial responsibilities when the bridge failed, December 29, 1876.*

At the time of the bridge failure, Charles Paine was 46 years old. He possessed extensive railroad experience. When only 15, he began working as a rodman (a person who operates a leveling rod for surveying) in a surveying party for the Vermont Central Railroad for four years. By the age of 19, he had become division engineer in charge of construction. He then took the same position for the Vermont and Canada Railroad, moving on to engineer and manager of reconstruction and extension of the Cham-

---

[45] Letter, Amasa Stone to Cornelius Vanderbilt, July 13, 1874. Courtesy of John Hay Library Archives, Brown University.

[46] Gary, 54.

[47] Dates are also from Ferdinand Gary, and the salaries from The Penn Central Transportation Company Records, 1835-1960, Oversize Vol. 24, p. 19. Courtesy of Bentley Library Archives, Ann Arbor, Michigan.

plain and St. Lawrence. He then served several railroads in Wisconsin. For 23 years, he consecutively fulfilled positions of superintendent of the Michigan Southern and chief engineer of the Northern Indiana. In 1872 he became general superintendent of the Lake Shore & Michigan Southern. He eventually became president of the American Society of Engineers. *The Railroad Gazette* offered the following tribute:

> Mr. Paine's character was of a type that is now infrequent. We read of such men in the Bible and in some other books, men who were invulnerable to every form of temptation that can come to a man of position and power. We have such men now; we hope more of them among corporation officers in the future, for sturdy moral character in a severely tempted man is an accomplishment usually got under stress of public opinion. He was a learned man, with fine literary tastes and capacity for expression; a sound engineer of good judgement; and a safe railroad officer who enforced discipline and inspired loyalty.[48]

Stone, Paine, and Collins all lived in Cleveland and worked out of the Lake Shore office. How often they were physically present is unknown. We do not have any reason to believe the three of them had anything other than an amicable working relationship. Railroad historian Marshall Kirkman suggests that it was particularly difficult for the superintendent's and the civil engineer's departments to work together. "If an accident occurs at a point when the division superintendent has been complaining of the track to the maintenance department, the department, finding itself in a corner will do its best to fix the blame on the operating department."

For the year ending May 1, 1876, the Lake Shore and Michigan Southern Railroad listed 30 directors including Amasa Stone, who had no other title at that time. Cornelius Vanderbilt was listed as President and William Vanderbilt, his son, as Vice President. Charles Collins was listed as Chief Engineer. The railroad had spent $121,297.73 for new side roads. $31,880.03 for the Ashtabula coal docks, $24,293.04 for real estate, new buildings, $90,918.18, and bridge masonry $7,295.87, for a total of $275,481.85 in 1874. But the year before, 1873, almost five times as much had been spent on the railroad, and almost five times as much in 1872 as 1873.[49]

---

[48] "In Memoriam, Charles Paine," *Journal of the Western Society of Engineers Vol. XI* (July - August 1906) 472-473.

[49] *Annual Reports.*

The May 1, 1876 report stated that shares within the company were worth $50,000,000. This kind of round number figure causes us to suspect there was no accurate computation of exactly what the company was worth It was probably no more inaccurate than the reports of Worldcom in 2001. But to be sure, a railroad had more physical property to evaluate than did a company marketing cellular phone miles.) And the railroad had earned money, at least according to the report: $5,140,414.45 after subtracting expenses. The gross income from freight had been $11,918,349.78 and the gross income from passengers had been $3,170,234.[50]

The report did give a few statements of explanation. "Mr. Amasa Stone your Managing Director resigned July 1, 1875 on account of his health and private business. Mr. John Newell, formerly president of the Cleveland and Toledo Railroad Company (now part of the Lake Shore) and later president of the Illinois Central Company was upon Mr. Stone's resignation appointed GM of the line. Great credit is due these gentlemen and their subordinates for efficiency and economy in operating the Road and managing its affairs."[51] It also explained the paucity of earnings, at least an attempted rationalization: "The bitter conflict with each other inaugurated by other lines in 1874, extended through this quarter of the year 1875 into the sharp competition resulting from the conflict the company was unavoidably drawn; and to this may be added however, by the general depression of business, especially the great protraction of the iron and coal business, causes beyond the control of its management."[52]

The All-Consuming Goal: Profit

The goal of management was three-fold: 1. Profit, 2. Profit, and the reader can guess the third. Cochran writes, "An unquestioned part of the role of executive was the duty to make profits for the stockholders. This involved both proper use of the investors' capital and just allocation of earnings. It was also assumed by many, if not all investors that social and economic efficiency resulted from executives directing their policies toward the goal of maximum profit."[53] Charles Perkins, president of the C B & Q, wrote to his general manager, C. Holdings, July 13, 1888 that "The only

---

[50] *Reports.*
[51] Ibid.
[52] Ibid.
[53] Cochran, 109.

object of a railroad is to get *net* earnings."[54] The pressure on a president or a managing director to show a profit in each annual report would obviously conflict with a project that would not show immediate return, but instead maximize profit at a later date or over a decade rather than a year.

James Joy, president and director of several railroads, accepted as a matter of fact that "all human beings are selfish, and that all mankind appears to act from selfish motives, either in hopes of gain, or to gratify the will."[55] Personal achievement and profit were often one and the same. Henry Villard, after purchasing control of the Northern Pacific, boasted "I can truly say that it is the most extraordinary success ever accomplished here in a financial way.... I have really carried Wall Street by storm."[56] Others tried to keep their own ego subordinated to what was financially best for the whole, sort of a Kantian categorical imperative of finance. J. C. Clark, vice president of the Illinois Central, wrote William Osborne, of the St. Louis and New Orleans, on January 22, 1880 "I have now delegated to me all the power that I desire to exercise. I do not want to be able to say that I have a vast number of miles to manage but I would rather be able to say that I am managing a very moderate extent of railroad at the same time trying to make it pay."[57]

Amasa Stone was fully aware that trains, boasting of speed and frequency of schedule, might be monetarily counterproductive. On July 24, 1873 he wrote to Charles Paine "When you make up your new time schedule, I think you had better cut down upon pass train schedule 1,000-1,500 miles per day. I shall however be quite ready to restore the trains that are abandoned as soon as I can be convinced that the company's interest will be promoted thereby."[58] By "interest" he meant profit.

Cornelius Vanderbilt cared nothing about the Lake Shore other than its profits. When it came to regulations demanded by law, which hardly existed, Vanderbilt repudiated with, "Law! What do I care about law? Hain't I got the power?"[59] George Westinghouse patented his air brake in 1869 at the age of 22, and when he presented the idea to Vanderbilt the old man responded, "Do you pretend to tell me that you can stop a train with wind?

---

[54] Ibid., 126.

[55] Ibid., 212.

[56] Ibid., 213.

[57] Ibid.

[58] Ibid., 467.

[59] Stover, 102. Stover says, "Undoubtedly apocryphal."

I will give you to understand young man, that I am too busy to have my time taken up talking to a damned fool."[60] In that men were still standing on the tops of freight cars to open and close individual brakes by turning a wheel at the risk of life and limb, Westinghouse had come up with one of the great inventions of the nineteenth century.

The Railroad Ignorance of Railroad Boards

The greater problem was with the board of directors rather than the owner-president. The directors were chosen only because of their stock investment and seldom ever had hands-on experiences with the intricacies of railroading. In short, they were incompetent. John Murray Forbes, president of the Chicago, Burlington, and Quincy, said of his directors "we have plenty of law, finance, etc. but no railroad skill whatever."[61] From the directors' perspective, more important than skill was political leverage, perhaps lobbying in Washington. Charles Pickering, vice president of the Chicago, Burlington, and Quincy, wrote Forbes on June 10, 1879, nominating James F. Wilson to the board with the rationale "He may want to be Senator and might not care to come in. But if he would come in, he could be of service to us out here and in Washington."[62] Richard White describes the incestuous nature of the railroad boards:

> While the makeup of a single railroad's board of directors changed from year to year, powerful families moved from board to board, sat on multiple boards at once, or had multiple family members sit on the same board at the same time. Certain families maintained influence on railroad boards of directors for decades. They used the boards less to direct the railroads than to manage their investments, culling inside information, and picking the ripest and lowest speculative fruits and insider contracts that the railroads had to offer.[63]

The Financial Resurrection of the Lake Shore and Michigan Southern Railroad

When Vanderbilt bought the Lake Shore at a fire sale in 1869, the financial situation only became worse until in 1873 it was in receivership. In

---

[60] Ibid., 141.
[61] Cochran, 76.
[62] Ibid.
[63] White, 51.

Vanderbilt's estimation only one man on earth could make the Lake Shore solvent. That man was Amasa Stone. When Vanderbilt laid bare the financial condition of the Lake Shore on December 1, 1874, according to reporters, Vanderbilt did not use a scalpel, but rather a "blunt instrument."[64] Vanderbilt recalled, "When I was elected your president on July 1, 1873, I found the financial condition to be as follows, capital stock 50 million dollars, all issued; funded debt $29,730,000, floating debt $6,277,485, including a dividend due August 1, $2,008,315 and bills and pay-rolls for June 1, $478,686."[65] There was not a dollar in the treasury.

In 1874 a fierce rate war was exploding between the railroads, and freight dropped down to ½ cent per ton per mile. Some 40 railroad executives showed up in July at the resort town of Saratoga, New York in order to cut their losses. Among them were Cornelius Vanderbilt and Amasa Stone. Even though the executives signed the "Saratoga Compact" it did little good. They met again at the Commodore's office at New York's Grand Central Station, the most magnificent station in the world. It was an especially futile gathering. By 1876 the rates were so low that the profits would not pay for the "axle grease." The rates did not increase until 1877, but the Central continued to pay dividends at 8%. According to Wheaton, "The road revealed an extraordinary earning power despite the rates war in which its revenue per ton mile declined from 1.46 to .88 cents. Explanation is to be found in the great efficiency with which the road handled the increased volume of traffic."[66] Most of that "great efficiency" was due to Amasa Stone.

The truth is, Amasa Stone almost single-handedly saved the Lake Shore and Michigan Southern from both bankruptcy and going completely out of business. Gould, beginning in 1869, attempted to corner the gold market which sent Wall Street into a frenzy of unloading stock.

Gould thought that he had Grant in his pocket through the President's brother-in-law, Abel Corbin. Gould persuaded Corbin into influencing Grant to keep federal gold from the market. With a limited gold supply, Gould believed that he could "corner" the gold market. He drove the price of gold up, but Grant changed his mind when agricultural prices dropped.

---

[64] Alvin F. Harlow. *The Road of the Century: The Story of the New York Central* (New York: Creative Age Press, Inc., 1947) 291.

[65] Ibid.

[66] Wheaton, 299.

The President dumped four million dollars into the market, and the price of gold plummeted.[67]

On Wall Street, perception is greater than reality, and the perception was that the problem was not ultimately Jay Gould, but Jay Cooke. The American public perceived Cooke as the financial Rock of Gibralter, because he had financed the North during the Civil War by selling bonds. But Cooke was not satisfied with being a bond salesman and bank president. He decided to build a railroad, the Northern Pacific, from Duluth, Minnesota, to Tacoma, Washington, 1871-1873. The project faced too many Indians, mountains, rivers, sloughs, mosquitoes, locusts, droughts, and everything that Cooke never faced and never encountered in his refined setting of Philadelphia. Cooke never visited the territory through which his railroad was to traverse, a war zone protected by hostile, well-equipped and mobile Indians. By the summer of 1873, Cooke was bankrupt, and on September 18, the bronze doors of his First National Bank closed, "with a heavy thud."[68] This seismic quake on Wall Street would not be outdone until October of 1929.

Vanderbilt had purchased the Lake Shore in 1869, when he himself, according to railroad historian Michael Hiltzik, was "strapped. The collapse had driven down shares of his (Vanderbilt's) three major holdings, the New York Central, the Lake Shore and Michigan Southern which ran between Chicago and Buffalo, and the telegraph company Western Union by more than $50 million combined." The Union Trust Company called in its "1.75-million-dollar loan to the Lake Shore." Hiltzik succinctly assesses the 1873 panic as a "classic bubble, born of frenzied growth in the railroad industry that had long-term financial obligations, but funded them with short-term debt and the issuance of grossly overvalued securities."[69]

All his life Amasa Stone had been proactive with the unbounded faith of American optimism. As manager, he would cut his and the shareholders' losses. He stopped all new construction. He sent agents to entice European immigrants coming into New York City to move west by boarding

---

[67] Jean Edward Smith. *Grant* (New York: Simon and Schuster Paperbacks, 2001) 489.

[68] John M. Lupetkin. *Jay Cooke's Gamble: The Northern Pacific Railroad, the Sioux, and the Panic of 1873* (Norman, Oklahoma: University of Oklahoma Press, 2006) 282.

[69] Michael Hiltzik. *Iron Empires: Robber Barons, Railroads, and the Making of Modern America* (Boston: Houghton Mifflin Harcourt, 2020) 86.

the New York Central, which would get them as far as Chicago. He gained freight business by dropping rates. He also constantly tweaked the system, making it more attractive for the customers. On July 16, 1874 Stone wrote detailed instructions to George Gates:

> I have carefully watched matters relating to the policy of our Company in the matter of drawing room or sleeping cars from this point, which is that we should have a drawing room or sleeping car on each of the three express trains through the year except perhaps during the coldest weather… If seats and berths cannot be secured in fresh cars here there is a great noise, and as the A. & G. W. [Atlantic and Great Western] Road always have a car from here, large numbers will stay away from us if they have to take a car that comes from some point west…I think also that our uptown office should have charts for the sale of berths and seats.[70]

By 1876 the directors could report that the floating debt had been paid off and Mr. Vanderbilt's stock all returned to him. The LS & MS had at last become a typical Vanderbilt railroad. Alvin F. Harlow, one of two comprehensive NY Central historians, gives no credit to Stone, but heaps praise on Vanderbilt.

> The Street (Wall) now changed its tune and high praise was heard of Mr. Vanderbilt for risking his own property to save both railroad and bank and for the skill with which he was accomplishing the rehabilitation of the former. The Union Trust Company was back in business again in three months. As for the railroad, its earnings were carefully applied on debts, extravagance was eliminated, and yet out of income, good new steel was bought to replace much of the old era track.[71]

Fortunately for the railroad the report was made on December 1, 1876. Less than a month later all hell would break loose.

Harlow made at least one questionable claim: "Cut out all extravagance." In the summer of 1874 Vanderbilt got caught up in a race as to who could deliver the Sunday morning paper the fastest from Saratoga, New York to Chicago. By building and using a powerful locomotive with 61-inch drives (an extravagance) the normal time of 36 hours was cut down to 27. Harlow wrote, "The countryside stood open-mouthed as it shot past. A news writer claimed that, 'the suction of air caused by it was like a tornado' and that people on the track when it operated leaped fences and

---

[70] Cochran, 446.
[71] Harlow, 293.

threw themselves into ditches to escape its deadly embrace."[72] On the last lap of heading into Chicago at such a fast pace, the Engineer fainted upon arrival at the LaSalle Street station.[73]

Stone or Vanderbilt?

Harlow placed all of the blame for the Ashtabula disaster on Stone. He described the exchange between Tomlinson and Stone:

> When the Ashtabula Bridge was authorized, he drew the plan for the 150-foot span as a Howe Truss. An engineer on his staff who had kept abreast of the times remonstrated with him, saying that the Howe Truss was practicable for a shorter timber structure, but it should never be attempted in iron; in that length, its own weight made it dangerous. Stone, whom success had made arrogant, purpled with rage. Tell him, the president of the company and a veteran bridge builder, that he didn't know his business, eh? In furious terms he ordered the man out of his office and struck him from the payroll.[74]

As I will later show, Harlow's narrative was a gross over-simplification, an assessment based on hearsay rather than knowing what actually transpired between Joseph Tomlinson ("an engineer on his staff") and Stone.

In his history of the New York Central, Edward Hungerford does not mention the Ashtabula disaster, neither does he inform his reader that Stone was brought on board to restore its financial fortunes. He gives full credit to Vanderbilt, "It was none other than Commodore Vanderbilt that saved the Lake Shore from ignominious bankruptcy at that critical time. After the death of Horace F. Clark, his son-in-law, who was president of the Road until 1873, things went very badly with the property. (How could Hungerford have been so wrong?) The Commodore came to the rescue, by his own great fortune he restored credit to the Lake Shore and brought it back to its dominating railroad position."[75]

Mark Twain was not so generous toward Vanderbilt. "First you seem to be the idol of only a crawling smarm of small souls, who love to glorify your most flagrant unworthinesses in print; or praise your vast possessions worshippingly; or sing of your unimportant private habits and sayings and

---

[72] Ibid., 294.

[73] Ibid., 296.

[74] Ibid., 297.

[75] Edward Hungerford. *Men and Iron: The History of the New York Central* (New York: Thomas Y. Crowell, 1938) 273.

doings, as if your millions gave them dignity." And of course, he was not beyond sarcasm. "You rob yourself of restful sleep and peace of mind, because you need money so badly. I always feel for a man who is so poverty ridden as you." His final verdict: "You observe that I don't say anything about your soul, Vanderbilt. It is because I have evidence that you haven't any."[76]

When Vanderbilt died six days after the Ashtabula disaster, all of New York City paid homage to its best-known citizen and the richest man in the world. "A great cortege followed his body by boat down the harbor and along miles of snowy road to the little Moravian hillside cemetery on Staten Island where the Vanderbilts are still buried."[77] Not everyone in New York state paid the Commodore such homage. The citizens of Syracuse were so weary of his antics and his trains careening through their town that four days after the Commodore's death the city superintendent of Syracuse ordered that snow be shoveled on the tracks and tamped down. The pile was sufficient to derail the New York Central locomotive and delay traffic for five hours; it took five engines to clear the track.

Lane, though sympathetic to Vanderbilt, states that in building a railroad empire "Selfishness and more than a touch of vanity guided his career and in some respects his policies were contrary to public interest."[78] On his death watch, as Vanderbilt's life ebbed and flowed, stock of the Lake Shore went down and up. When the hearing on the details and dispersal of Vanderbilt's will was begun in a lower Manhattan court, November 12, 1877, Amasa Stone was the first person to testify Vanderbilt was "of sound mind" when he made the will. Whether Vanderbilt knew of the Ashtabula disaster six days before he died is a matter of controversy. Some say it hastened his death, but Vanderbilt was the kind of person who never let such matters slow him down.

---

[76] Stiles, 496-497.
[77] Harlow, 300.
[78] Lane, 300.

Chapter 5

Stone and Rockefeller

Early Personality Traits

Around 3 pm, December 18, 1867, a 28-year-old man could be seen running after a train at the Cleveland station. He wasn't late for the train, because John D. Rockefeller was late for hardly anything. Being on time was a life motif; it was the train that was late. After boarding the New York Express he took a seat, placing his valise beside him. When the train did not leave on time, Rockefeller stepped back out on the platform to investigate the matter. He fell into conversation with a friend while the train left with his attaché. Fortunately, the train did not wait for absent-minded passengers. Three hours later the train derailed on a bridge in Angola, New York, killing 50 of the passengers, and injuring dozens of others. Upon learning of the disaster that evening, Rockefeller telegraphed his wife, "Thank God I am unharmed. The 6:40 train I missed had a bad accident."[1] The train just happened to be owned by the Cleveland, Painesville and Ashtabula Railroad of which Amasa Stone was President.

Punctuality, orderliness, frugality, and keeping track of every penny he spent, a practice extending through his early business years, were all attempts to provide coherency and compensate for an incoherent childhood. John Davison Rockefeller was born to Eliza Davison and William Avery Rockefeller July 8, 1839, in Tioga County, New York. It was left mostly to his mother, Eliza, to raise John and his four siblings, because the dad was hardly ever at home. The family never knew when he was

---

[1] Charity Vogel. *The Angola Horror: The 1867 Train Wreck That Shocked the Nation and Transformed American Railroads* (Ithaca, NY: Cornell University Press, 2013) 139.

going to show up, how long he would stay, and when he would leave. By the time John was 16 his family had lived in six different places, finally settling in Parma, Ohio, a community just outside of Cleveland. It was then, at the age of 16, John left home to attend E. C. Fulson Commercial College, where according to biographer Grace Goulder he studied principles of banking, methods of business transactions, and general commercial subjects. "Bookkeeping, both single and double entry systems was emphasized. Rockefeller excelled in the drill – he would be in essence a bookkeeper all of his days."[2] One of his antagonists referred to Rockefeller as that "bloodless, Baptist, bookkeeper."[3]

John D.'s life fits the simplistic, but nonetheless true deterministic theory of the "bad father." William Rockefeller was a flim-flam man, a snake oil salesman of whatever elixir he could prescribe to ignorant bumpkins, always peddling an answer for all physical maladies, in particular cancer. John D's father was so mysterious to his biological family that he acquired another wife and family, living in two distinct worlds, and died under an assumed name, Dr. William Livingston. Obviously, this was an embarrassment to his eldest son John, but in no way would the son honestly deal with his father's identity. William Rockefeller, in his son's mind, would simply be, if not an affable and garrulous rake, at least a good businessman who taught his son valuable financial lessons.

If there is any truth to Freud's hypothesis that religion is a neurotic search for the perfect father, this would have been at least partially true for John D. Rockefeller. (Freud borrowed the idea from Ludwig Feuerbach.) One of Freud's disciples, Erik Erikson, believed there to be a correlation between the individual's ego crisis and the cultural crisis, the microcosm with the macrocosm. John with his turbulent childhood entered into the turbulent industrial revolution, the "Gilded Age," the era of graft, greed, and gargantuan egos. John would bring order to his own life and to his surroundings. Early Rockefeller biographer John Flynn wrote that his subject was a "large, intelligent lion who knew what he was doing in a vastly altered jungle and a whole race of mice, disorderly, disorganized, running about in circles without any notion of the changes that had taken

---

[2] Grace Goulder. *John D. Rockefeller: The Cleveland Years* (Cleveland: The Western Reserve Historical Society, 1972) 14.

[3] Hildegarde Dolson. *The Great Oildorado -The Gaudy and Turbulent Years of the First Oil Rush: Pennsylvania 1859-1880* (New York: Random House, 1959) 256.

place in the forest."[4] In Erikson's understanding, the identity formation of ending adolescence and finding one's vocation involves "constitutional, idiosyncratic libidinal needs, favored capacities, significant identification, effective defenses, successful sublimations, and consistent roles."[5] Though examining every receipt to the last cent and accounting for every item may be idiosyncratic micromanagement, a sublimation for the unwillingness to face the disturbing elements of one's past, it can also be a valuable business practice. One of the most prevalent theories of psychohistory is that all achievements are driven by some defect or disorder brought forward from the past.

Rockefeller and God

John D. Rockefeller believed that not only did God's providence extend to protecting his life, but to making him rich. Of all the "robber barons," Rockefeller was the most religious and the most compartmentalized. While belief in God's providence engenders gratitude, confidence, fortitude, and many other admirable traits, it can also be egocentric and exclusive, enriching the self at the expense of others. The Rockefeller home was regulated by religious scrupulosity: no square dancing, no alcoholic drink, and no Sunday work. According to Rockefeller's latest biographer, Ron Chernow, "Sabbath observance was so rigid that when he wrote to a colleague when he should have been in church, he tended not to put the real date on his letter."[6] Chernow argues, "John D. Rockefeller was the Protestant work ethic in its purest form, leading a life so consistent with Weber's classic essay that it reads like his spiritual biography."[7] In simple terms, Rockefeller believed God desired him to be rich.

The same missionary vision that Rockefeller had for his own welfare routinely produced myopia, if not total disregard for the welfare of others. Thus, the man who would become the richest individual in the world seemingly would not let the rights of others equally motivated stand in his way. He bought out rivals by underselling them, bribed congressmen to

---

[4] John T. Flynn. *God's Gold: The Story of Rockefeller and His Times* (Westport, Conn: Greenwood Press, 1932) 5.

[5] Erik H. Erikson. *Childhood and Society* (New York: W. W. Norton and Company, 1963) 312.

[6] Ron Chernow. *Titan: The Life of John D. Rockefeller, Sr.* (New York: Random House, 1998) 177.

[7] Ibid., 55.

block passage of a competing pipeline and formed a cartel to corner freight rates. Rockefeller salved his conscience with "God is on my side," even to the extent of believing those who did not come into the shelter of his oil monopoly, which mauled through everything in its path, were assigned to perdition. Rockefeller really did believe he was offering economic salvation to whosoever will. Chernow sums up what seemed to be, at least to his rivals, his brazen and sanctimonious hypocrisy: "Where Rockefeller differed most from his fellow moguls was that he wanted to be both rich and virtuous, and claim divine sanction for his actions. Perhaps no other businessman in American history has felt so firmly on the side of the angels."[8] He would rescue "sinful refiners from their errant ways."[9]

With his mother's strict piety and discipline John fortified himself with elaborate defense mechanisms: rationalization, compartmentalization, and obsessive orderliness, all serving to legitimize his questionable ethics and quiet his conscience, while failing to mitigate the vile contempt poured out on him. Above all, the man was focused; nothing could distract him from making money and attending the Erie Street Baptist Church, where he faithfully worshiped in every service and faithfully tithed all the money he made. By 18 years of age, John had become a trustee of the struggling "mission church," which would have gone under with its $2,000 debt had the young trustee not personally raised the money to get the church out of hock. All of this was in contrast to Amasa Stone's prestigious Old Stone Presbyterian Church, the most noted ecclesiastical structure in town

Chernow argues that Christianity and capitalism formed the twin pillars of Rockefeller's life.[10] Rockefeller's undiluted faith in capitalism provided both an economic philosophy and blinders for ignoring or destroying anything that got in his way. His Messianic complex combined with the connivance and manipulation modeled by his father, even to the point of believing his rapacious behavior to be redemptive, and the salvation for an inchoate and chaotic randomness of a new discovery in 1859. "Oil was God's gold," and his religiosity gave him divine right to it. Flynn wrote, "He was a model of the Christian businessman. Profound, almost funeral seriousness, fervent attention to prayers and the hymns, and the lesson - outside that a deep and solemn composure. In the Bible lessons he read to his class, he found many points he was able to apply to himself and

---

[8] Ibid., 152.
[9] Ibid., 153.
[10] Ibid., 51.

incorporate in his own policy."[11] The diligence, frugality, and hard work prescribed in the book of Proverbs (a pre-Christian book which could be utilized for manifest destiny and acquiring wealth in the kingdom of America) provided theological direction and confidence. "Seest thou the man, diligent in his business, he shall stand before kings."[12] According to scholar Rebecca Edwards, Proverbs provided exegetical support for all Americans who touted success as a result of "determined will, industry, perseverance, economy, and good habits."[13]

And if biblical balm was not sufficient salve for Rockefeller's conscience conflicted by trouncing less talented competition (if it ever was conflicted), Herbert Spencer's evolutionary and agnostic "survival of the fittest" provided further rationale for forging and foraging ahead. Without ever reading Spencer, (Rockefeller did not read anybody.) he adopted a philosophy of almost all of the titans of the Gilded Age. He stated, "The growth of a large business is merely survival of the fittest…. This is not an evil tendency of business. It is merely the working out of a law of nature, and a law of God."[14]

Gilded Age historian Sean Dennis Cashman states, "The public persona of the self-made man was based on a cult of outward modesty and respectability. As a rule, robber barons were puritanical, parsimonious, and pious."[15] Cashman further claims that when Rockefeller boasted "God gave me my wealth" he was in the spirit of Cotton Mather.[16] But the integrity of Mather's genetics did not descend to Rockefeller. Cashman argues Rockefeller's activities were "criminal," and that "his methods lead to a public outcry that consumed more time in courts and legislatures over a period of 50 years than any previous or subsequent controversy in American business." (We have yet to evaluate Donald Trump.) But "Standard Oil was impregnable against both state and federal governments whose regulatory powers were minimal. By bestowing favors and giving bribes, it created friends and allies in every important legislature, including Congress, and it employed the most skillful lawyers to defend its positions.

---

[11] Flynn, 132.

[12] Ibid.

[13] Rebecca Edwards. *New Spirits: Americans in the Gilded Age 1865-1905* (New York: Oxford University Press, 2006) 78.

[14] Ibid., 153.

[15] Sean Dennis Cashman. *America in the Gilded Age: From the Death of Lincoln to the Rise of Theodore Roosevelt* (New York: University Press, 1984) 41.

[16] Ibid.

Its income was greater than that of most states and its profits were large enough to finance further expansion."[17]

Stone and Rockefeller Intersect

By the time the Rockefeller family moved to Cleveland in 1852, Stone had already established himself as the up-and-coming business man in Cleveland. Stone would have had no reason to cast more than a glance at the "commission house." (A middle man for all kinds of dry goods, agricultural products, and the developing iron and coal trade.) In a warehouse at 32nd Street with Morris B. Clark, who was about ten years older, the two went into business in 1859, obviously a year famous for something far more momentous, the first oil successfully forced from Mother Earth in Titusville, Pennsylvania. Finally, Edwin Drake had extracted a flow of oil which could be barreled and marketed so that it would turn a profit. One preacher, as the story goes, cornered Drake on the street and condemned him for stealing oil from the earth, "Don't you know that you're interfering with the Almighty Creator of the Universe? He put that oil in the bowels of the earth to heat the fires of Hell. Would you thwart the Almighty and let sinners go unpunished?"[18]

The first time Stone and Rockefeller met, other than in passing, may have not been in Cleveland, but in New York, when he and his brother William met with Cornelius Vanderbilt, owner of the New York Central, April 19, 1868. A year later Vanderbilt would purchase the Lake Shore. Rockefeller believed himself to have impressed both Vanderbilt and Stone: "I made a proposition to draw 100,000 barrels to Mr. Stone, desired to meet with Van last eve at a Manhattan Club room and Will engaged to meet them at 9 o'clock. We talked *business* to Amasa and guess he thinks we are really prompt men."[19] Allan Nevins, the most extensive biographer of Rockefeller until Chernow, narrated this encounter and used the occasion to excoriate Stone. He described Stone as,

> A stout, powerful-built man, with iron jaw and stern eyes, Stone had shown for years the autocratic and domineering traits which led a novelist of the day to present a savage report of him under the name of Aaron

---

[17] Ibid., 48.

[18] Dolson, 27.

[19] Allan Nevins. *John D. Rockefeller: The Heroic Age of American Enterprise* Vol. I (New York: Charles Scribner Sons, 1940) 294.

Grindstone. Everybody feared his arbitrary ways, hard temper, and biting tongue. "Amasa Stone," exclaimed one observer, "the richest man in Cleveland and he will have the smallest funeral!" – and in fact, his life ended in a terrible tragedy. He over-rode his engineers, dictated an old-fashioned type bridge at Ashtabula, and lived to see it collapse under a crowded passenger train. Some years later he committed suicide. At this time, he viewed Rockefeller's fast, increasingly power sourly.[20]

Nevins, as the title of his book indicates, viewed Rockefeller through rose-tinted glasses and exaggerated Stone's malfeasance. The bridge was anything but old-fashioned, but to be sure Stone felt threatened by Rockefeller's ascendancy.

Henry Flagler and Standard Oil

If Rockefeller and Henry Flagler were not Siamese twins joined at the hip, they were alter-egos. Flagler was almost ten years older than Rockefeller, and served as a mentor. John D. gave him due credit. If Rockefeller owed his success to any one individual, it was Henry Flagler, who proved to be his best friend as well as the best business partner he ever had. In his *Reminiscences*, Rockefeller recalled,

> For years and years this early partner and I worked shoulder to shoulder; our desks were in the same room. We both lived on Euclid Avenue, a few rods apart. We met and walked to the office together, walked home to luncheon, back again after luncheon, and home again at night. On these walks, when we were away from the office interruptions, we did our thinking, talking, and planning together. Mr. Flagler drew practically all our contracts. He has always had the faculty of being able to clearly express the intent and purpose of a contract so well and accurately that there could be no misunderstanding, and his contracts were fair to both sides.[21]

Flagler came up with the idea that would propel the success of Rockefeller, and conversely make his name anathema to his competitors, if not the entire business world. Flagler conceived a corporation under the name of Standard Oil, for the purpose of monopolizing the refinery business in

---

[20] Ibid., 388.
[21] John D. Rockefeller. *Random Reminiscences of Men and Events* (New York: Sleepy Hollow Press and Rockefeller Archive Center, 1984) 23-25.

Cleveland. Rockefeller recalled, "I wish I had the brains to think of it. It was Henry Flagler."[22]

The original charter for the Standard Oil Company, written on cheap looking legal paper, faded yellow, and made of poor material, enabling the Standard Oil Company the right to engage in business, led Rockefeller to boast, "The Standard Oil Company will someday refine all the oil and make all the barrels."[23] He wasn't far from right. As the bottom dropped out of the oil market heading towards the 1873 panic, Standard Oil gobbled up 22 of the 26 Cleveland refineries in a little over a month, February 17-March 28, 1872. Standard Oil's voracious cannibalism became known as the "Cleveland Massacre."

Amasa Stone was not aloof to the machinations of the newly formed Standard Oil Company. Standard Oil needed access to large piles of money, and entrance into a bank's vault could be provided by only the bank's president. Thus, on January 1, 1872, Stone became a director for Standard Oil. This did not mean Stone was going to throw open his safes for Rockefeller to buy worthless refineries. Fallout between the two men quickly ensued. When Rockefeller approached Stone as President of the 2nd National Bank with his expansion scheme, Stone balked. But rather than responding with a flat out "No" he entrusted the issue to two of his bank's directors, Stillman Witt and Colonel Oliver Hazard Payne. Even though Stone reminded the two men who had appointed them (Stone himself), they ruled in favor of Rockefeller. What really piqued Stone was that in a gesture of generosity, Witt brought out a cash drawer full of greenbacks, implying that he had so much faith in Standard Oil that, if possible, he would give him all the money that the bank possessed.

One upped, Stone was so galled that he refused to act on a Standard Oil option which soon expired. Realizing the foolishness of his pride, which would eventually cost him millions of dollars, Stone groveled before Rockefeller, begging for reconsideration. Flagler advocated on Stone's behalf, but Rockefeller would not budge. Stone compounded his error when he unloaded all the Standard Oil stock he owned. The resentment between the two men continued to intensify, because within the world of Cleveland finances, consisting of railroad extension, iron manufacturing, oil refining,

---

[22] David Leon Chandler. *Henry Flagler: The Astonishing Life and Times of the Visionary Robber Baron Who Founded Florida* (New York: McMillian Publishing Company, 1986) 62.

[23] Chernow, 132.

and all kinds of overlapping endeavors, there was no way for the two men to escape one another.

The Rebate Scheme

The railroad rebate scheme preceded the Cleveland Massacre. Though it was Flagler's idea, the episode further launched or debauched Rockefeller's reputation. Again Stone, because he was President of the only east-west railroad serving Cleveland, was the victim or the accomplice, whichever interpretation one prefers. Flagler came up with the idea of asking for a 36% kickback, or rebate, to be kept secret from the other railroads. Stone needed the business, and again found himself cornered by Rockefeller. It was actually Vice President John Henry Devereux who sealed the deal, because Stone was in Europe recovering from his carriage accident. Flagler approached Devereux, Vice President of the Cleveland, Painesville, and Ashtabula Railway proposing that if the railroad would ship oil for the rate of $1.65 per barrel, rather than the public's rate of $2.40, Standard Oil would guarantee the CP & A sixty carloads of refined oil daily. According to Chernow, "From that moment, (the spring of 1868) the railroads acquired a vested interest in the creation of a gigantic oil monopoly, that would lower their cost, boost their profits, and generally simplify their lives."[24]

There was probably no one who knew more about moving anything on a railroad than John Henry Devereux.

> In 1862, as a colonel, he was in charge of all Union rail lines in Virginia, in disarray because of damage inflicted by Confederates and conflicts between various Army and government departments using the lines. Devereux improved efficiency, organized inspection and repair units, obtained equipment, enforced use rules, and smoothed differences between departments. Under his supervision, the trains moved large amounts of troops, artillery, and the sick. Devereux resigned as a general in the spring of 1864.[25]

This informal handshake seemed to work until the formal cartel known as the South Improvement Company was created in 1872. A freight agent at the Lake Shore office failed to tell his subordinate there was a secret

---

[24] Chernow, 113.

[25] "Devereux, John H.," *Encyclopedia of Cleveland History* (https://case.edu/ech/)

arrangement, and he perceived that the slashed rates were for public consumption.[26] Thus, the word got out and competitors finally understood why they could not compete in the oil refinery business.

The muckraker Ida Tarbell condemned the clandestine agreement as the "original sin," from which all others were conceived. She narrated a typical confrontation that a Mr. Hanna and a Mr. Baslington had with John Devereux over rebate rates.

> They were told that the Standard had special rates; that it was useless to try to compete with them. General Devereux explained to the gentlemen that the privileges granted the Standard were the legitimate and necessary advantage of the larger shipper over the smaller, and that if Hanna, Baslington and Company could give the road as large a quantity of oil as the Standard did, with the same regularity, they could have the same rate. General Devereux says they "recognized the propriety" of his excuse. They certainly recognized the authority. They say that they were satisfied they could no longer get rates to and from Cleveland, which would enable them to live, and "reluctantly" sold out. It must have been reluctantly, for they had paid $75,000 for their works, and had made thirty per cent a year on an average in their investment, and the Standard appraiser allowed them $45,000. "Truly and really less than one-half of what they were absolutely worth, with a fair and honest competition in the lines of transportation," said Mr. Hanna, eight years later, in an affidavit.[27]

Tarbell wrote, "Under the combined threat and persuasion of The Standard, armed with the South Improvement Company scheme, almost the entire independent oil interest of Cleveland collapsed in three-month's time. Of the twenty-six refineries, at least twenty-one sold out. From a capacity of probably not over 1,500 barrels of crude a day, the Standard Oil company rose in three-months-time to one of 10,000 barrels."[28] The agreement was not the sole reason for Rockefeller's success in that, as Chernow points out, "As in other industries, the railroads developed a stake in the growth of big business whose economy of scale permitted them to operate

---

[26] Steve Weinberg. *Taking on the Trust: The Epic Battle of Ida Tarbell and John D. Rockefeller* (New York: W. W. Norton & Company, 2008) 222.

[27] Ida Tarbell. *The History of the Standard Oil Company* Vol. I (New York: The MacMillan Company, 1925) 67. Robert Hanna was brother of the more famous Mark Hanna.

[28] Ibid., 67-68.

more efficiently - an ominous fact for small, struggling refiners who were gradually weeded out in the savage competitive strife."[29]

The financial survival of Amasa Stone, as well as other bankers and railroad operators, was becoming increasingly precarious. The big "squeeze" was on, and John D. Rockefeller was the chief operator of the vise. "Without doubt, the Lake Shore deal marked a turning point for Rockefeller, the oil industry, and the entire American economy."[30]

The South Improvement Company

The front or holding company for the rebates and rate cuts on freight was the South Improvement Company, the brainchild of Tom Scott, president of the Pennsylvania Railroad. An admirer described Scott as the most sagacious of politicians, especially with anything that had to do with money. In terms of underhanded tactics, he may have been outdone by only Jay Gould. In order to abolish other railroads, Scott ran his own company with "Machiavellian strategy." He posted columns and letters in newspapers, and bribed editors to back his legislation. The Union Army during the war paid Scott's railroad 80 percent more for baggage than it charged a passenger.[31] One who did not join the South Improvement Company, "a device for the extinction of commerce that was the cruelest and most deadly yet conceived," would be driven out of business.[32] Stone was one of the charter members. Rockefeller opined that Scott was "the most dominant authoritative power that ever existed, before or since, in the railroad business of our country."[33] He also called him "a daring dare-devil, bold and courageous man," and that he was capable of a "sledge hammer stroke" on his friends in order to do business with his enemies.[34] One refinery owner, John H. Alexander, explained the devouring of his own refinery:

> There was a pressure brought to bear upon my mind, and upon almost all citizens of Cleveland engaged in the oil business, to the effect that unless we went into the South Improvement Company we were virtually killed as refiners; that if we did not sell out we should be crushed out. . . .It was said

---

[29] Ibid., 114.

[30] Ibid.

[31] Jack Beatty. *Age of Betrayal: The Triumph of Money in America 1865-1900* (New York: Alfred Knopf, 2007) 239-241.

[32] Ibid., 270.

[33] Chernow, 135.

[34] Beatty, 271.

that they had a contract with railroads by which they could run us into the ground if they pleased.[35]

The members actually swore and signed a conspiratorial covenant, "I _____ do solemnly upon my honor and faith as a gentleman that I will keep all transactions which I have with the Corporation known as the South Improvement Company."[36] Dolson trenchantly assessed: "It had been thought up by all three major lines to the oil fields - Vanderbilt's New York Central, Tom Scott's Pennsylvania, and Jay Gould's Erie, with their three subsidiary lines - and it was as ruthless and ruinous to nonmembers as anything ever devised by rogues."[37]

The collusion would provide the leverage to bring down freight prices while receiving rebates on the freight shipped by other Cleveland refineries. Non-members simply could not compete. Of course, the knowledge of such bullying would eventually leak out. C. E. Bishop, the editor of the Oil City newspaper *The Derrick*, referred to the South Improvement Company as an "Anaconda Octopus." In the early days and weeks of 1872, he published a "black list" on the mast head of his paper, including eight names in bold letters, "P.H. Wilson, Charles Lockhart, W.P. Logan, R.S. Warring, W.G. Warden, John Rockefeller, Amasa Stone."[38] The governor of Pennsylvania outlawed "The Anaconda," and the South Improvement Company collapsed.

Some decades later, in 1894, Rockefeller was asked if he had ever been a member of the South Improvement Company to which he responded "no," which may have been technically correct, but was false if evaluated by intention. The self-righteous Rockefeller had been involved in subterfuge if not outright deception. The oil war of 1872 turned Cleveland society upside down; several who had made easy fortunes in oil refining and built splendid mansions on Euclid Avenue found themselves bankrupt and forced to sell. "Whether it was Rockefeller or the slumping oil market that forced them to sell their refineries at distress sale prices, they chose to see Rockefeller as the author of their woes."[39]

---

[35] Chernow, 143.
[36] Flynn, 154.
[37] Dolson, 260.
[38] Ibid., 166.
[39] Chernow, 147.

The Tension Between Stone and Rockefeller

In personality Stone was reserved and taciturn, exuding a self-confidence that needed to do little talking. Above all, he did not call attention to himself, was intolerant of differing ideas, and had no time for frivolities. His sole enjoyment in life was his family, a pious, submissive wife, two beautiful daughters, and until 1865 a handsome son. Life had been far better to him than he ever dreamed, and a quiet evening at home was sufficient reward for the hard labor that occupied his early years, and his many responsibilities in the rough and tumble financial world of a frontier town feeling growing pains. Aside from the Case incident, (mentioned in Chapter 11) no one would ever question the integrity of Amasa Stone, and if a businessman wanted something accomplished in Cleveland or beyond, Stone was the first person to seek out.

All of us adopt a persona, if not several, that we believe to be to our advantage in whatever situation we find ourselves. Gladys Haddad, Flora Stone's biographer, stated that "Amasa Stone personified the model of the nineteenth century's big giver. Stone, powerfully built, competitive, aggressive, and domineering was the archetypical nineteenth century business tycoon whose life was underscored by struggle, achievement, and exercise of power. Like Vanderbilt and Rockefeller, ambition was his companion."[40]

There was a distinguishing difference between Rockefeller and Stone. Stone had come up the hard way: backbreaking work on a farm, and later as a bridge builder. Stone never gave an order to do anything that he had not himself done. No doubt resentment and even jealousy grew within Stone as the young pencil pushing bookkeeper, slightly built with lily white hands, kept closing in on Stone's territory. After all, it was Stone who would buy stock from Rockefeller, and not the other way around. But Rockefeller needed money for expanding his oil refining enterprises, and Cleveland's leading banker was his best source.

After Witt and Payne opened their vault and sided with Rockefeller, Flagler expressed surprise that the fledgling entrepreneur had stood up to Stone. Rockefeller quietly responded, "But he is mistaken." Of course, history would prove Rockefeller correct, that is, if success is measured by financial pragmatism. According to Nevins, when Stone believed himself to be betrayed by his business partners, he briskly, "Thumped angrily

---

[40] Haddad, 66.

downstairs into the street."[41] No one ever took more delight in one upping a competitor than did Rockefeller, and he was compelled to tell the story to Henry Chisolm, another Stone business partner, founder and first president of the Cleveland Rolling Mill, who "guffawed" saying, "There were many who truckled to Amasa Stone, but I could never see any reason for it."[42]

No man was more representative of the Gilded Age than John D. Rockefeller. He was religious, self-righteous, rich, acquisitive, and completely compartmentalized, even more so than Daniel Drew. At least Drew repented of his sins on Sunday, even though his scandalous, deceiving behavior continued on Monday. Drew did not deceive himself; Rockefeller did. The most "successful" man of the Gilded Age, and eventually the wealthiest, saw no need for honest introspection and repentance. He was a holy man, at least in his own eyes, if not of his contemporaries, and more critically, the Christ of the New Testament, whom he claimed to serve, but to whom he gave little attention. Rockefeller was legitimized by the Jehovah of the Old Testament, who breathed out fire on the disobedient, and rewarded the diligent and obedient. Even as the Old Testament Israelites, Rockefeller was one of the elect; God had chosen him to make money.

Above all, Rockefeller was persuasive. His aura and determination brought the most powerful under his influence. Rockefeller, at least in his 20s and early 30s, needed Stone more than Stone needed Rockefeller. The younger man needed the older man for contacts, for legitimacy, for leverage, and for money. At every point of the journey Rockefeller hitched Stone to his wagon. He put Stone on Standard Oil's original board; he and Flagler assailed Stone's railroad, the Lake Shore, for the first railroad rebates, borrowed money from Stone's bank, and most unfortunately, talked Stone into being a member of the South Improvement Company, a secret cartel for a half dozen railroad owners, for the purpose of cornering the rebate market, and freezing out all competitors.

Even the collapse of the South Improvement Company was not the end of the subterfuge for which Rockefeller recruited Stone. Richard Parsons, who had practiced law in Cleveland and was known to be a persuasive lobbyist in Washington, was elected to the United States House of Representatives in 1873. He was a person of importance in Cleveland, having served as "Collector of Internal Revenue." He furthered his influence by serving as editor of the *Cleveland Daily Herald*. When he ran for

---

[41] Nevins, 390.
[42] Ibid., Note 3.

Congress, a circular went out to all Standard Oil employees, "We deem the election of R.C. Parsons vital to your interest as well as ours." It was signed by Standard Oil directors Amasa Stone, Stillman Witt, and Truman Handy.[43] (We met Parsons in Chapter 2.)

The Ark Analogy

Rockefeller likened his Standard Oil Company to the ark which Noah built. "The organization of the Standard Oil Company was a safe and benevolent institution for all competitors when in good faith and loyal devotion joined their interest in it."[44] Those who did not enter the safety of the "ark" were people whom the Lord Almighty cannot save. "They don't want to be saved, they want to go on and serve the devil and keep on in their wicked ways."[45]

Thus, Rockefeller provided a theodicy for his greed, rooted in the predestinarian Calvinism of his Baptist heritage. One of the problems with the ark metaphor was there was room for only one captain, and this floating zoo needed a whole lot of deck hands. Stone Jr. was increasingly at the mercy of the man steering the ship, and his resentment grew. The leader who had been recognized as the most powerful businessman in Cleveland was now living in the shadows of a financial genius who evoked whispers and drew stares wherever he went. Rockefeller had become Cleveland's leading citizen, and the displacement caused Stone's gut to churn all the more. With the reduced freight rates, turning a profit on the Lake Shore had become all the more difficult. Stone began to despise Rockefeller.

But Rockefeller himself was not all that comfortable in the ark. Those in it are caught in the dilemma of not being able to stand the stench on the inside nor the storm on the outside. And even taking into consideration Ida Tarbell's personal vendetta against Rockefeller, she in all likelihood, gave an accurate description of the oil titan's presence at church. She suggested that he was paranoid, at least that would be the diagnosis today. "They say in Cleveland, Mr. Rockefeller always sits with his back to the wall when it is possible. So many things can happen behind one's back in any assembly." He seemed to be uneasy. "Throughout the church service which followed, this same terrible restlessness agitated him. He sat bent forward

---

[43] Ibid., 175.

[44] William O. Inglis. *John D. Rockefeller Interview: 1917-1920* (Sleepy Hollow, New York: Meckler Publishing/The Rockefeller Archive Center) 134.

[45] Ibid.

in his pew, for a moment, his eyes intent on the speaker, then with a start he looked to his right, searching the faces he could see, craning his neck to look backward. Then his eyes would turn again to the speaker, but not stay there."[46]

When the Russian Grand Duke visited Cleveland in 1872, Stone hosted the aristocratic celebrity with an elaborate reception and banquet at his home. He invited Flagler, but not Rockefeller.[47] The snub may have not even registered on the oil mogul. The puritanical teetotaler was not that much into partying.

Rockefeller did not build a mansion on Euclid Avenue, at least not as massive or stately as the others. He purchased a house in 1868, impressive enough but not as ostentatious as Stone's. Eleven years later he added an adjacent house, thus doubling his acreage, and distancing himself from his neighbors. He would further seclude himself when he moved into the seventy-nine-acre Forest Hill Estate in 1878. Rockefeller was now literally and metaphorically king of the hill. Stone took notice, but by this time he was besieged by so many problems he really did not care.

In 1882, the Rockefellers moved to a mansion on West 54th Street in New York City. "Along Fifth Avenue near the Rockefeller home, the palaces of the rich – notably the fantastic turreted confections of William K. Vanderbilt at Fifty-first Street and Cornelius Vanderbilt II at Fifty-eighth Street stretched uptown in gaudy profusion."[48] The Rockefellers rarely returned to Cleveland other than to vacation at Forest Hill. It is a well known fact that the Cleveland which had provided him with his wealth received little of his philanthropy. His money would be poured into Chicago for his grand University.

Rockefeller died in 1937 over a half century after Stone's death. Paranoid in life, paranoid in death, Rockefeller was buried in a bombproof tomb. Above it rises a sixty-five-foot granite monolith. Approximately fifty feet away lies Amasa Stone and family. Having started from very different places in life, they wound up in the same location, Section 10 of Lake View Cemetery, Cleveland, Ohio. There ought to be a parable buried somewhere between them, if not several. There could not be a more serene place for that kind of reflection.

---

[46] Weinberg, 230.
[47] Goulder, 109.
[48] Chernow, 220.

Chapter 6

Amasa Stone and John Hay

Early Years

John Milton Hay, 1838-1905, may have been short in stature, about 5 feet 2 inches, but not short in ambition. His grandfather, John Milton Hay, for whom our subject was named, migrated from Scotland to Fayette County, Kentucky where he set up a prosperous cotton mill utilizing slaves. Always ill at ease with this source of labor, the grandfather moved to Sangamon County, Illinois in 1850, the same year that Abraham Lincoln's family arrived in the same location. John Milton Hay's son, Charles, stayed behind in Lexington, Kentucky where he attended medical school at Transylvania, eventually ending up in Salem, Indiana. Here he not only practiced medicine but was also the editor of the Whig newspaper. Charles was an avid reader, "stuffy in his own literary tastes," and a firm believer that "education by any means, was the key to advancement and well-being."[1]

When John Milton was three years old his father moved the family to Warsaw, Illinois, a frontier town on the Mississippi River. The meager environment did not deter the family's library acquisitions, and John learned Greek, Latin, and German. John was only five when in 1854 vigilantes broke into the Carthage jail, just some 20 miles away, and murdered the Mormon leader Joseph Smith and his brother Hiram. Where Charles the father stood in this incident is not entirely clear, but he snidely remarked that for the trial the town found people "ignorant enough and indifferent enough to act as jurors."[2]

---

[1] John Taliaferro. *All the Great Prizes: The Life of John Hay, from Lincoln to Roosevelt* (New York: Simon & Schuster Paperbacks, 2013) 17.

[2] Ibid., 22.

At the age of 13, John went to live with his Uncle Milton in Pittsville, Illinois, to attend a private classical school and there made a lifelong friend, John George Nicolay. After a year in Pittsville, Milton Hay moved with his nephew to Springfield, Illinois. All of these moves and relationships were aligning themselves with the most important life transforming job in which Hay would ever be employed, undersecretary to Abraham Lincoln. Before that, he would matriculate at Brown University in Providence, Rhode Island. Hay excelled academically, and a friend Billy Norris jocularly recalled the myth that, "In those days, all text was memorized and it was the general opinion that Hay put his books under his pillow and had the contents thereof absorbed and digested by morning, for he was never seen 'digging' or doing any other act or thing that could be construed into hard study."[3]

Upon graduation (1859), Hay moved back to Springfield to practice law with his Uncle Milton. They set up shop directly across the street from the law firm of Lincoln and Herndon. Hay's early expertise would prepare him for another critical relationship in his future. "He continued to work in the courts of the Eighth Judicial Circuit, settling estates, and untangling the affairs of merchants, railroads, and the local gaslight company."[4]

After Lincoln's presidential nomination in 1860, he hired John Nicolay at $75 per month to help him with his correspondence. Nicolay, becoming overwhelmed with letter writing, asked Lincoln for help which led to the hiring of John Hay. In this capacity Hay served until Lincoln's death, a span of five years, and became so familiar with his boss that he paradoxically lost his own identity but also found himself.

Entrance into the Stone Family

Obviously, after Lincoln's death Hay lost his job. One might say he was fired by Mary Todd Lincoln with whom he was always in disfavor. Hardly anyone got along with the President's widow. He moved to New York, finding a job with Horace Greeley's *New York Tribune*, with Whitelaw Reid as managing editor.[5] There he met a young lady, who in 1872 was visiting her uncle, A. B. Stone. Clara, daughter of Amasa Stone, was then 22 years old. Clara was not a beautiful woman, but fleshy and about a half

---

[3] Ibid., 24.

[4] Ibid., 30.

[5] Horace Greely had founded the newspaper, but died shortly after his loss to Ulysses Grant in the 1872 election.

foot taller than her future husband. Hay was more attracted to her mind than her body, Clara having been afforded the best of a classical education, at least the best afforded in Cleveland. He gave an honest appraisal to his friend Nicolay: "She is a very estimable young person, large handsome and good."[6] He commented on her "firm and inflexible Christian character."[7] By letter Hay asked for Amasa Stone's consent, and the father responded with, according to the soon to be son-in-law, with the "kindest and nicest character."[8]

When Abraham Lincoln's son Robert caught wind of the engagement he wrote John Nicolay, "I have a letter from New York today which says Hay is about to marry Miss Stone of Cleveland, whose father will one day be obliged to leave to JH and one other fellow from six million dollars to eight million dollars—which will make John to write with a first class gilded pen."[9] The other fellow would be Samuel Mather, who married the younger sister Flora. Lincoln's prophecy came true, though the amount was considerably less than he predicted. Even before the marriage, as Amasa Stone navigated the treacheries of the 1873 panic, Hay noted of Stone as he piled up millions in losses, "He is a man of great courage and energy, and may be able still to retrieve himself.... He will not do any wrong in the meantime."[10] Hay's appraisal was accurate in that Amasa Stone was stoical and taciturn, rarely letting his emotions show unless he was particularly perturbed. Of course, what emotions Stone did not release on the outside ate away at him in the inside, as he was already developing severe digestive disorders.

The wedding between John and Clara took place at the Stone mansion on February 4, 1874, and the couple moved to New York to live in Hay's small apartment. Even living that far away from Cleveland he was anxious about being an adequate provider for the daughter of an empire builder, especially after receiving the following gentle, yet sobering exhortation from Clara's father. "Your life and habits have been such that it would have been quite easy for you to have fallen into idle habits. I presume experience has taught you that such habits would only lead you to ruin. I doubt whether anyone can enjoy true happiness who is not reasonably

---

[6] Taliaferro, 152.

[7] Ibid., 160.

[8] Ibid., 162.

[9] Ibid., 163.

[10] Ibid., 164.

industrious and feels that he is doing something for himself and his fellow beings."[11]

Amasa Stone, who had worked 14-hour days as a construction foreman and no less as a businessman, was writing to an individual who had never done a hard day's work in his life. The 35-year-old clerk, secretary, editor, and would-be poet with fair skin, already bulging waistline, and non-calloused hands would need to prove himself. There was even more pressure on Hay that he would replace the son whom Amasa no longer had. Amasa tested Hay's financial acuity by gifting him with two Lake Shore Railway bonds totaling $10,000.

Years in Cleveland

The pressure of his new family, a child born within the year, and the newspaper's low pay brought on what may have been psychosomatic illnesses, vertigo and loss of clear vision. The father-in-law perceived the couple's dire straits, and made an offer that Clara and John could not refuse. If they would move to Cleveland, he would build them a house and introduce John to the world of industry and finance. The house would be nothing less than a mansion next door to Amasa and Julia on Euclid Avenue with sandstone as its exterior and sumptuous interior wood carvings by the gifted craftsman from Germany, John Herkomer. Joseph Ireland was the architect.[12] Euclid Avenue historian Jan Cigliano gives evidence that John and Clara gave much attention to the details of the design and construction of their new house.[13]

Until the Ashtabula disaster Hay's duties for Stone were light filing, keeping ledgers, examining financial accounts, and making sure the bills were paid. He was little more than an executive secretary, but after the bridge failure and the Ohio Legislature hearing, Amasa needed to escape the steady torrent of public scorn, and with his family took a 15-month sojourn to Europe.

---

[11] Ibid., 164-165.

[12] Jan Cigliano. *Showplace of America: Cleveland's Euclid Avenue, 1850-1910* (Kent, Ohio: Kent State University Press, 1991) 125.

[13] Ibid., 125.

The 1877 Strike

Hay was left with the unofficial title (as we would designate him today) of CEO of Stone enterprises. All hell broke loose when in July of 1877, Baltimore & Ohio Railroad workers went on strike. Shortly thereafter, July 23, employees of the Lake Shore walked out while demanding a 20% wage increase. Hay believed the strike to be an international conspiracy of "demagoguing labor organizers."[14] On August 23, 1877, he wrote to his father-in-law that "these subversive activities have been motivated by the very devil… into the lower class of working men."[15]

Obviously, John Hay knew very little about "lower class working men." The dirt, soot, sweat, smoke, and back-breaking toil of mill and foundry labor, much less the constant threat of loss of life and limb especially on railroads was entirely beyond his comprehension. To Hay, the strike was nothing more than the selfish whims of "thieves and tramps waiting and hoping for a riot."[16]

The malignancy of the recession ate away at the railroad business above all other industries. According to economic historian Philip Foner, "It was not unusual for a railroad worker to be unemployed, and therefore, unpaid for as many as four days of the week while the company expected him to be prepared to work at all times."[17] Expectation by employers became unbearable and irrational: "One worker of the Lake Shore living in Collinwood, Ohio, was paid 16 cents in wages to take a train to Cleveland, but then had to report back to his superior at Collinwood, at a cost of 25 cents in fare."[18] Over the two-year period 1873-1875, the wages of some railroad workers had been cut as much as 45 percent, amounting to a little more than $1.00 a day. While salaries were being cut, the New York Central, owner of the Lake Shore, paid 8 percent in cash dividends in 1873 and 1874, and 10 percent in 1875.[19] Railroad workers went on strike and trains came to a grinding halt.

The Lake Shore felt the pinch. The President, William Vanderbilt, then worth ninety million dollars, told the employees of the Lake Shore that

---

[14] Taliaferro, 175.

[15] Ibid.

[16] Ibid.

[17] Philip Foner. *The Great Labor Uprising of 1877* (New York: Pathfinder, 1977) 21.

[18] Ibid., 21.

[19] Ibid., 25.

their 10 percent salary cut would not be restored since their railroad could not afford to manage it. William Vanderbilt had learned from the best, swallowing wholesale the stubborn ruthlessness of his father with a total inability to empathize with his employees. In fact, he may have been less empathetic since the Commodore had known long hours of back-breaking toil and sacrifice, a climb to success that his son William did not have to make. Incomprehensibly he stated:

> Our men feel that, although I may own the majority of the stock in the Central, my interests are as much affected in degree as theirs, and although I may have my millions and they the rewards of their daily toil, still we are about equal in the end. If they suffer I suffer, and if I suffer they cannot escape.[20]

Hay may have not had an overall perspective on the situation but he was well aware that the inability to move goods had put a stranglehold on Cleveland. The Cleveland Rolling Mill, in which Amasa and A. B. Stone were heavily invested, had to shut down and in despair he wrote his father-in-law on July 25 that "the railroad would probably have to surrender to the demands of the strikers even though he felt it was disgraceful."[21] But no worries; those with money can usually stare down those without it, and the Lake Shore employees went back to work on July 30.

After the strikers were quickly replaced, or forced back to work, Hay wrote Stone on August 23 "The prospects of labor and capital seem gloomy." But with a bit more optimism, he wrote again on September 3, "I am thankful you did not *see* and *hear* what took place during the strike. You were saved a very painful experience of human folly and weakness." He confidently added "All your investments look reasonably safe and snug. I make no new ones except with ample margins."[22] Giving Hay due credit, he had kept his poise during a turbulent summer without being blown sky-high out into Lake Erie.

Politics

For over a decade, Hay became the face of Amasa Stone's empire, exuding expertise, civility and gentrification beyond that which can simply be gained by building and living in a mansion. On July 31, 1880 he

---

[20] Ibid., 126.
[21] Ibid., 261.
[22] Taliaferro, 175.

spoke to two thousand people on behalf of James Garfield in Cleveland's most opulent setting, the Euclid Avenue Opera House. There was no doubt where he stood, as well as his father-in-law, in his devotion to the Republican Party and his condemnation of the Democratic Party.

> On the one side is a record of glory and good repute which sheds something of lustre on the declining days of every man who fought that desperate battle against slavery and treason. On the other it is a shameful story of half-hearted loyalty or open rebellion, of ignorant and malicious opposition to light and knowledge, of blind and futile defiance to the stars in their courses.[23]

Henry Adams believed his friend John Hay could be a "loyal Republican who never for a moment conceived that there would be any other ideals possible....with Hay, party loyalty became a phase of being, a little like the loyalty of a highly cultivated church man to his church."[24] Stone exalted the Republican Party and condemned the Democrats. Concerning Garfield, he wrote on June 18, 1880, "It will be him as a dark horse with no record. A dark horse with no record will make the most popular nominee. Any record of a Democrat is a bad record, as against Garfield."[25] Upon meeting Garfield, Stone "was much impressed by his stalwart manliness. It seems to me that he will meet the expectations of the people, but I doubt whether (he) will (meet) the expectations of the politicians, any more than Presd. Hays has done."[26]

Stone was correct in that he did not meet the expectations of the politicians, especially Roscoe Conkling, who was the most powerful party boss in the Republican Party. Garfield appointed William H. Robertson, Conkling's bitter enemy, as Collector of the Port of New York. [27] Conkling was tough to stand up to. "Self-consciously vain, he stood six foot, three inches tall, with an athletic build, erect posture, and well-muscled arms

---

[23] Ibid., 193.

[24] Henry Adams. *The Education of Henry Adams* (New York: Barnes & Noble Incorporated, 2009) 256.

[25] Letter, Amasa Stone to John Hay, June 18, 1880. Courtesy of John Hay Library Archives, Brown University.

[26] Letter, Amasa Stone to John Hay, January 19, 1881. Courtesy of John Hay Library Archives, Brown University.

[27] Kenneth D. Ackerman. *Dark Horse: The Surprise Election and Political Murder of President James A. Garfield* (New York: Carroll & Graf Publishers, 2003) 348.

and shoulders from daily exercise at the punching bag."[28] Conkling and James G. Blaine were bitter enemies and to Conkling's chagrin, Chester Arthur appointed Blaine as Secretary of State. Blaine said of Conkling, "The contempt of that large-minded gentleman is so wilting, his haughty disdain, his grandiloquent swell, his majestic, super-eminent, overpowering, turkey-gobbler strut has been so crushing to myself and to all the men of the House, that I know it was an act of the greatest temerity for me to venture upon a controversy with him."[29] Overestimating his strength to walk through a snowstorm, Conkling was found unconscious on a street in New York City. After falling ill with fever and delirium, he died a few weeks later at age 58.

If Stone was impressed with Garfield's manliness, he thought, as many others, that Chester Arthur was "oppressed and timid," but all who belittled Arthur mistook their man. He forever did away with "patronage," (at least officially) by instituting a Civil Service exam. He also restored the U.S. Navy, which had dwindled to almost nothing after the Civil War. "Accidental President" historian, Jared Cohen, claims that Arthur is responsible for the "birth of the modern-day Navy."[30] The man who at one time had been a wheel in New York machine politics, taking more than his fair share as collector of the New York Customs House, found redemption in the White House by demonstrating resolve and integrity that had not been expected of him. Upon Arthur's death, the *New York Times* stated, "His memory will have the benefit of a record as President which raised him steadily in the esteem and respect of the Nation and left him a distinction which at the age of 50 had had no ground for expecting and which no one could have predicted for him." [31] No partial term, accidental President, accomplished as much. He, too, escaped the hands of Conkling, a nonpartisan feat for which Stone did not think he had the courage.

Stone always attempted to steer his son-in-law away from politics. He interpreted political life as diminishing domestic life and even more importantly to Stone, mitigating business life. Politics preempted quality time with family. As Hay was being offered an opportunity to run for

---

[28] Ibid., 2.

[29] Scott S. Greenberger. *The Unexpected President: The Life and Times of Chester A. Arthur* (New York: DaCapo Press, 2017) 68.

[30] Jared Cohen. *Accidental Presidents: Eight Men Who Changed America* (New York: Simon & Schuster: 2019) 181.

[31] Greenberger., 238.

Congress or a place in a presidential cabinet, Stone wrote on December 1, 1880,

> It should be certainly very pleasant to you to know that you command this confidence of both the powers that are, as well as the powers that are to be.

> I presume it is difficult to tell what effect it might have upon you, should you be offered a place, a "peg higher." This however could not be expected when there are throughout the country so many "shining lights" that are ready to sacrifice not only themselves but an immense retinue of "swimming" for the place. [32]

Stone later wrote on December 28, 1880, "I have no doubt but General Garfield & wife would be an agreeable associate, then comes the other side of the question; it takes you out of business channels…and will so far tend to divert your mind that you might regret the result in the end…"[33] A month later, "In a material point of view, I have no doubt you have acted wisely in not continuing a diplomatic life. If the end to gain is to have a 'good time,' that life would promise well." Further Stone said, "But it seems to me that you could mark out and pursue a life about your home that would result in the end much to the advantage of the welfare of yourself and family."[34]

When Hay ran for Congress, the local Republican party approached Stone for financial help, to which he stubbornly responded, "Not a dollar shall you have of me."[35] William Roscoe Thayer speculate, "The shrewd millionaire suspected the whole love affair was a ruse for tapping his barrel rather than for honoring his son-in-law."[36] Hay dropped his bid for Congress. He wrote to James Garfield that he "refused definitely and forever to run for Congress…. The constant contact with meanness, ignorance, and the swinish selfishness which ignorance breeds, needs a stronger heart and a more obedient nervous system than I can boast."[37]

---

[32] Letter, Amasa Stone to John Hay, December 1, 1880. Courtesy of John Hay Library Archives, Brown University.

[33] Letter, Amasa Stone to John Hay, December 28, 1880. Courtesy of John Hay Library Archives, Brown University.

[34] Letter, Amasa Stone to John Hay, January 31, 1881. Courtesy of John Hay Library Archives, Brown University.

[35] William Roscoe Thayer. *The Life and Letters of John Hay* Vol. I (Boston: Houghton Mifflin, 1915) 437.

[36] Ibid., 438.

[37] Ibid., 443-444.

Hay was essentially correct, at least according to his closest friend, Henry Adams. Adams believed politics had poisoned him, causing an early death at 65. According to Adams' biographer, David S. Brown, "Henry supposed that Hay, a poet, novelist, and historian, stood no chance of surviving in the contemporary climate; his demise seemed to offer proof that the 'best' of his generation, though sensitive, literary, and intellectual, were being summarily ground down in the process of empire building, and fortune making."[38]

The Bread-Winners

In spite of his respect for the hard work and business acumen of his father-in-law, Hay never lost his aristocratic heart, and desire for more elitist companions. He longed for a cosmopolitan ambience contrasted with the rough and tumble Cleveland, displaying its nouveaux riches with its garish facade. He craved companions with more refined tastes, or at least companions who thought they possessed more refined tastes. In his novel, *The Bread-Winners*, Cleveland became "Buffland," and Euclid Avenue became "Algonquin Avenue." Hay biographer Tyler Dennett described *The Bread-Winners* as "among other things, a patronizing description of pretentious wealth."[39] Hay constructed a typical social gathering in Buffland, people who did not know what to do with themselves, or could not converse about what really matters.

> It was the usual drawing-room of provincial cities. The sofas and chairs were mostly occupied by married women, who drew a scanty entertainment from gossip with each other, from watching the proceedings of the spinsters, and chiefly, perhaps, from a consciousness of good clothes. The married men stood grouped in corners and talked of their every-day affairs. The young people clustered together in little knots, governed more or less by natural selection - only the veterans of several seasons pairing off into the discreet retirement of stairs and hall angles.[40]

The novel was well-plotted with lustful love stories, surprise twists, an accusation of murder, a foiled killing, sexual tension between different so-

---

[38] Brown, 358.

[39] Tyler Dennett. *John Hay: From Poetry to Politics* (New York: Dodd, Mead, 1933) 105.

[40] John Hay. *The Bread-Winners: A Social Study* (Ridgewood, New Jersey: The Gregg Press, Inc., 1967) 103.

cial strata (the quintessential Romeo and Juliet theme) and a social-climbing gold digger, finally finding herself by ultimately marrying the poor and uneducated carpenter who sincerely loved her. The lead character, Arthur Farnham, marries the much younger Alice Belding, with whom he had been in love all along. But the novel lacked subtlety and stereotyped the social strata, especially the strikers, lead by ruffians that are no less than the scum of the earth. When Farnham was accosted as he rode his horse over a deserted road, the ruffians identified themselves as "Labor Reformers" and the spokesman boasted, "We represent the toiling millions against the bloated capitalists and grinding monopolies." Hay left nothing to the imagination:

> He was a dirty looking man, young and sinewy with long and oily hair and threadbare clothes, shiny and unctuous. His eyes were red and furtive, and he had a trick of passing his hand over his mouth while he spoke. His mates stood around him, listening rather studidly to the conversation. They seemed of the lower class of laboring men. Their appearance was so grotesque in connection with the lofty title their chief had given them, that Farham could not help smiling, in spite of his anger.[41]

When the strike finally did break out, the laborers attacked naïve and innocent Algonquin Avenue. Hay described the leader as a "long haired, sallow looking pill, who was making as ugly a speech to a crowd of ruffians as I ever heard."[42] His demented character was surpassed only by the wrong headedness of the self-deception of which he proclaimed.

> "Yes, my fellow-toilers" - he looked like he had never worked a muscle in his life except his jaw tackle, "the time has come. The hour is at hand. The people rule. Tyranny is down. Enter in and take possession of the spoilers' gains. Algonquin Avenue is heaped with riches wrung from the sweat of the poor. Clean out the abodes of blood guiltiness." [43]

The only solution for the laborers was to hang every "aristocrat from every lamppost."[44]

Worse yet was the ringleader, Andrew Offitt, who attempts to murder Farnham, and blames the murder on the poor, somewhat demented carpenter who did the maintenance at Farnham's house on Algonquin Avenue.

---

[41] Ibid., 166.
[42] Ibid., 218.
[43] Ibid.
[44] Ibid.

Again, Hay caricatured the antagonist: "He called himself Andrew Jackson Offitt - a name which in the West is an unconscious brand. It generally shows that a person bearing it is the son of illiterate parents, with no family pride or affection, but filled with a bitter and savage partisanship, which found its expression in a servile worship of the most injurious personality in American history." Offitt was the offspring of "a small farmer of Indiana who had been a sodden, swearing, fighting drunkard," who became converted by an "attack of delirium tremens at a camp-meeting and resolved to join the church, he and his household."[45]

Hay had not been evasive about either his political or religious opinions. In his defense, he did clarify that for the most part the strikers were of "singular good-nature of almost all classes. The mass of working men made no threats; the greater number of employees made no recriminations."[46] But there was a negative proposition underlying this assessment. Good working men were those who did not complain, and would accept any working conditions, while being content with their lot in life. During the years between the 1873 panic and the strike, almost everything had become worse for labor: pay, hours, and demanded production. All of these trends pushed the working men into ghetto shanties, further removed from Algonquin (Euclid) Avenue.

In the end the strikers would return to work and fulfill their "pre-determined" place in society. Those who lived on Algonquin Avenue "deserved" to be there, having thrived within American aristocracy by thrift, perseverance, business acuity, and all the premises of meritocracy. But Hay's first and only novel was amateurish, painted in stark colors of white and black, with little gray in between. It lacked both the dialectic tension and profound redemption imagined and penned by his genius contemporaries, Dostoevsky and Hugo. He was not even a match for the cleverness of his closest friend, the wordsmith Henry Adams. Hay was spread too thin, resulting in the difficulties of simultaneously being a business man and a novelist. In fact, no two vocations could hardly be more incongruent; financial ledgers rarely demand creative reflection.

As *The Bread-Winners* was serialized in *The Century*, it received a few positive reviews such as, "*The Bread-Winners* is the work of a very clever man; it is told with many lively strokes of humor; it sparkles with

---

[45] Ibid., 90.
[46] Ibid., 216.

epigrams; it is brilliant with wit and it has depth."[47] But the positive reviews were quickly buried under an avalanche of criticism. The plot and characterizations were unrealistic, insulting to working men and women. The Transcendentalist *Dial* stated that the novel is "worse than a failure; it is deliberately insulting to working-class men and women and to all who sympathize with them."[48] The bottom line was the snobbish tone of the novel: "The author of *The Bread-Winners* will never turn out permanently valuable work, so long as he represents a legitimate force in the interest of a false political economy and an antiquated spirit of caste."[49]

Hay could not completely shake the criticisms and complaints of the struggling workers during the 1877 strike. Thus, his novel, *The Bread-Winners* served as both catharsis and self-justification. In response to the charge of aristocracy, Hay defended himself in *The Century.*

> I hardly know what is meant by an aristocratic point of view. I am myself
> a working man, with a lineage of decent working men; I have been accus-
> tomed to earning my own living all of my life with rare and brief holidays.
> I have always been in intimate personal relations with artisans and with
> men engaged in trade. I do not see how it is possible for an American to be
> an aristocrat; if such thing exists, I have never met one.[50]

By retaliating Hay only dug himself into a deeper hole. "The inner circle of petty tyrants who govern the trade unions expressly forbid the working man to make his own bargain with his employer; his boys may become thieves and vagabonds, his girls may take to the streets; but they shall not learn his trade without the consent of the union."[51] To make the leap from a desire for a shorter working day and better wages, to thievery and prostitution, demonstrated both Hay's refuge on Euclid Avenue, and his non-empathy for the working class. They were not breadwinners but bread usurpers, deserving only a place in a caste system which was stratified and permanent. The title *The Bread-Winners* resulted from Hay's creation of a poorly organized union called The Brotherhood of Bread-Winners.

> The brothers, who had taken seats where they could find them, on a dirty
> bed, a wooden trunk, and two or three chairs of doubtful integrity, grunted
> a questionable welcome to the new-comer. As he looked about him, he

---

[47] Dennett, 110.

[48] Ibid., 111.

[49] Ibid., 112.

[50] Ibid., 113.

[51] Ibid.

was not particularly proud of the company in which he found himself. The faces he recognized were those of the laziest and most incapable work-men in the town - men whose weekly wages were habitually docked for drunkenness, late hours, and botchy work. As the room gradually filled, it seemed like a roll-call of shirks.[52]

Obviously, the anonymous author did not reveal that he had married into wealth and had been gifted with a mansion on Euclid Avenue; thus, the following from his pen was true, "I'm engaged in business in which my standing would be seriously compromised if it had been known that I had written a novel."[53] Anonymity was the only way to protect the Stone family, and most importantly, disguise the fact that Hay had been given wealth and not earned it. Dennett condemned Hay's anonymity:

> In *The Bread-Winners,* Hay had again mounted the pulpit and put himself forth as a moralist a role which it is not legitimate to play anonymously. As a preacher he showed himself a better master of invective than of the facts on which he rested his case. He sincerely thought that he was a believer in democracy, but he meant political democracy only. Industrial democracy was to him incomprehensible.[54]

There was wide speculation as to who had written the novel. Flora was reading the *Bread-Winners* in 1881, but did not know who wrote it.[55] Hay's authorship was not revealed until 1907 by Clara, after her husband's death.

In the end, Amasa Stone had more influence on his son-in-law than the son-in-law had on him. Hay, later to become Secretary of State under William McKinley, had been converted to pure capitalism, if not the dark underbelly of American nationalism defined by manifest destiny and im-perialism. The most often found quote from Hay regarded the 1898 Span-ish-American conflict as a "splendid little war." In an insightful article by Granville Hicks, "The Conversion of John Hay," the author wrote:

> The posthumous collection of addresses (speeches which Hay had given throughout his professional life) shows that, though he could still speak on literary themes, he was at his most eloquent when praising the Republican Party as the creator of American prosperity. Not a doubt assailed him as he

---

[52] Hay, 82. Note that the largest railroad union was the Brotherhood of Loco-motive Engineers with a membership of 14,000 in 192 branches. Foner, 32.

[53] Dennett, 113.

[54] Ibid., 115.

[55] Letter, Flora to Sam, October 4, 1881. Flora Stone Mather correspondence, Container 9, Western Reserve Historical Society.

lauded McKinley, interpreted our occupation of the Philippines as a sacred mission and defended the seizure of Panama as a just and beneficent act.[56]

The Money-Makers and the Crucifixion of Amasa Stone

Hay should have expected critical reviews, which were more acerbic and vociferous than warranted. Henry Keenan's muckraking novel, *The Money-Makers,* stunned both the Hay and Stone families. Keenan served as an editor along with Hay on Whitelaw Reid's *The Tribune.* Did Keenan have a vendetta against Hay? Did he even know Hay had written *The Bread-Winners*? Whichever, Keenan left no doubt whom he was characterizing and where they lived. It is clear that Keenan was intent on writing a parable describing a moral paradigm diametrically opposed to *The Bread-Winners*.

Archibald Hilliard, the main character (John Hay), is a prima donna, a brilliant investor, and an inside trader of railroad stock. What really burned Hay's hide was Keenan's intimation that Aaron Grindstone's daughter is awkward with large feet and that Hilliard had to convince himself that he loved Eleanor (Clara) regardless of her father's millions. Grindstone, "rules his daughter as he rules his railway. He holds the string of the house-latch and pulls only when millions knock."[57] Of course, Keenan pointed out that when dancing, Eleanor was as tall as Hilliard (in real life taller). Keenan was as unjust as Hay had been, claiming that Grindstone had bought property around a church, driving it out of town with fumes from his forges.

Hilliard finds himself in emotional turmoil, as to just what are his motives in wanting to marry Eleanor. "He had hoped and tried to make himself believe that sentiment, and the sacred fire that lights the altar of love were urging him toward Eleanor. But the more he saw her, the clearer her nature revealed itself, the more certainly he saw that she was not the ideal of his youth."[58] But Hilliard would not make the matrimonial decision; Grindstone would. The cunning dad would make sure that Hilliard's

---

[56] Granville Hicks, "The Conversion of John Hay," *The New Republic* (June 10, 1931) 100-101.

[57] Henry Keenan. *The Money-Makers: A Social Parable* (New York: D. Appleton, 1885) 51.

[58] Ibid., 168.

choice to marry his daughter would not be tainted by money. To Hilliard he disclosed,

> My daughter has been engaged for some time to a gentleman whom I approve very heartily. By her own consent, he has been paying her the attentions of an accepted suitor. I hold that engagement sacred. Should she break it and marry anyone else, she will take that step in defiance of my wishes, and she will not get a penny of my fortune. Here is the deposition I have made of my estate in the contingency of my daughter's disobedience. He rose, and going to a drawer in the tray of his traveling-chest, drew out a paper and proffered it to Hilliard.[59]

The catastrophe in the Keenan narrative would not be a bridge, but a conflagration consuming a Grindstone property. In a theater, 200 lost their lives and another 100 were crippled for life. The local newspaper, *The Eagle,* pointed out plainly that the disaster was due to the "defective construction of the building, and that the blame must be put either on the ignorance of the architect, or the parsimony of the owner."[60] The testimony of the architect indicted Grindstone. He produced a letter written by himself to Grindstone, protesting "against the substitution of iron for masonry in the supporting walls and buttresses, and a firm assertion of the architect's purpose to give up the work; for if built as Mr. Grindstone ordered, the building would be unsafe, and most ultimately crush out the lateral walls, or give way under a simultaneous movement of a multitude."[61] Keenan must have known of the contention between Stone and Joseph Tomlinson as he wrote his novel, a task which demanded little literary imagination. Grindstone was found guilty, and shot himself to death "ten minutes after he heard a newsboy shout the verdict."[62]

It was almost as if Keenan was taking revenge against both Stone and Hay or insanely jealous of Hay. The author described dailies and journals as being of one accord: "Only the naked fact that the deaths of 200 people were laid at the door of one man's parsimony and greed."[63] The tragic denouement of the novel could not have been crueler. The author left no doubt as to the identity of Aaron Grindstone. He "was found in the bath, cold and dead—an expression of serenity on his care-worn face. He was

---

[59] Ibid., 176-177.

[60] Ibid., 266.

[61] Ibid., 273.

[62] Ibid.

[63] Ibid., 279.

stretched at full length in the crimson water, and a pistol was found at the bottom of the tub when the body was taken out."[64] The muckraking author concluded with his final condemnation; the Old Testament Prophet Amos could not have been blunter.

> Money, the priest at the altar of that worship whose creed is inequality, and whose ideal is self. It is at best a tediously-threshed sheaf, this tale of money-making; and, but for the romantic enchantments of the millionaire magician, we couldn't be expected to read it - could we? Why, after all, shouldn't he get who can, and he keep who has? Isn't the story old as Naboth's vineyard; old, indeed, as Tubal's crusade in search of equal part-age? It was not only when God was wroth, that to punish Israel he gave the people a king![65]

The novel actually did more damage to Hay than it did to Stone because the latter was now dead. Aaron Grindstone had trumped Hilliard (Hay) by calling his bluff and ferreting out his motives. In the novel Hilliard breaks off the engagement because he discovered that his declaration of love was not for a woman but for mammon. This may have hurt all the more since Hay for much of his married life had carried on at least an affair of the affections with Lizzy Cameron.

The malicious attack on both himself and his father-in-law threw Hay into panic mode. To no avail he beseeched the publisher, Appleton, to withdraw the book. He bought up every copy he could in New York, Boston, Philadelphia, and elsewhere. (Wouldn't this make the publisher print further copies and make *The Money-Makers* a best seller?) He was further disturbed when he read a *Cleveland Journal Review* claiming *The Money-Makers* was, "much better in all its parts than *The Bread-Winners.*"[66] This may be because truth is stranger than fiction, and Keenan came closer to the truth than had Hay. The "Robber Barons" gave birth to the muckrakers, and no era had ever outdone the likes of Ida Tarbell, Upton Sinclair and Lincoln Steffens than did the "Gilded Age." The only solution for Hay was to move to Washington D.C. and build a house adjacent to that of his friend Henry Adams. Within two years after Stone's death, the houses were complete, and the relocation accomplished.

---

[64] Ibid., 336.

[65] Ibid., 337.

[66] Taliaferro, 231-232.

Lizzy Cameron

The most piercing sword of *The Money-Makers* was that art imitated life. According to John Hay's most recent biographer, John Taliaferro, Hay's affections were turned to another woman. Though the author does not accuse Hay of sexual infidelity, there was a long dalliance of infatuation with sexual overtones, and much sexually sublimated language. Lizzy Cameron was beautiful, lithe, slim waisted, charismatic, and flirtatious. Her husband, Don Cameron, was 24 years older than she, and seemed not to mind or be jealous of her multiple liaisons with other men. In a letter to his sister-in-law Flora, December 8, 1879, Hay remarked that Lizzy "Was looking far more beautiful than ever."[67] With Clara not on his Paris trip in November of 1886, John responded to Lizzy's dinner invitation: "Your invitation is seductive to a cookless wanderer."[68] Lizzy wrote Henry Adams from Paris in May of 1887 where she was frequently escorted by John Hay that they had gone "in a lover loge to a ballet. I actually felt wicked and improper. He did too, for he felt obliged to follow up the precedent and tell me how much he loved me."[69] He referred to her as his "tantalizing goddess."[70]

John Hay was more than simply enchanted with Lizzy Cameron; he was slavishly fixated. Again, art imitated life. He had written in *The Bread-Winners*, concerning Arthur Farnham, "He gave himself up to that duplex act to which all un-avowed lovers are prone — the simultaneous secret worship of one woman, and open devotion to another."[71] Thus the novel became a confessional.

After reading Thayer's biography of John Hay, Lizzy wrote Henry Adams: "I think Mr. Thayer makes (Hay) more a decided vigorous character than he really was - to me he seemed timid, unself-asserting, and almost feminine in the delicacy of his intuitions and in his kindness."[72] In other words, John Hay was everything that Don Cameron was not. Lizzy Cameron died in 1944 at the age of 83.

---

[67] Ibid., 188-189.
[68] Ibid., 255.
[69] Ibid., 269-270.
[70] Ibid., 277.
[71] Hay. *The Bread-Winners*, 176.
[72] Taliaferro, 551.

Enduring Relationship with Stone

Stone probably never knew of his son-in-law's infidelity, but he had some awareness of his condescending attitude. The father-in-law was sufficiently secure in his own accomplishments that he was not unduly threatened. And even more importantly, his daughter had married up, not in money, but in manners, relationships, and intelligent discourse with the more refined of society. Stone was proud of the fact that he had enabled this transition and Hay did respect his father-in-law, perhaps at times even being awed by him. Stone had been unshaken at least outwardly while losing millions in the 1873 panic. Hay was fully aware that he had learned and grown from his father-in-law. When he left Cleveland for Washington he carried with him a business swagger, a self-possession of common sense, the ability to make wise economic choices, and to profitably invest inherited money. He wrote to Whitelaw Reid with both jest and seriousness, "The moral is, buy real estate and don't speckylate." To be sure, Hay was convinced that Stone was a better man than Daniel Drew, and most of the other Wall Street shysters.

Hay respected his father-in-law; he really perceived him as the Rock of Gibraltar. During the 1877 strike he wrote, "I do not refer to the anxiety, etc., for you are not a man who would be over anxious ever in a general panic; but you would have been very much disgusted in anger I am sure."[73] Hay did not cower before Stone, and in many ways thought himself to be his equal, if not superior, and he had maybe even greater respect for Julia Stone. To Reid he wrote concerning his engagement to Clara, "There will be an internecine war before Mrs. Stone consents to give up her daughter - wherein I sympathize with her. Before many centuries I shall win."[74]

The Enduring Legacy of Hay and the Stone Family

John really did love Clara. In the spring of 1873 he wrote, "My mind and spirit are yours as well as my heart and life. I love you, your soul and body, your goodness and your beauty. You are my inspiration and reward. I worship you."[75] More importantly, Clara represented the literary aspirations, artistic abilities and intellectual endowments necessary for Hay's

---

[73] Thayer, Vol. II, 6.

[74] Ibid., Vol. I, 351.

[75] Patricia O'Toole. *The Five of Hearts: An Intimate Portrait of Henry Adams and His Friends: 1880-1918* (New York: Ballantine Books, 1990) 49.

compatibility with any female. In June of 1868, Clara had written an academic exercise before graduating from Cleveland Academy: "It is a very old idea that literary women cannot make good housekeepers. It seems natural to believe that a woman cannot attend to belle letters and write poetry (for proficiency in these take a great deal of time) and also find opportunity for her household duties."[76]

Patricia O'Toole argues that Hay did not marry Clara for her father's money, but really was attracted to her physically as well as other qualities already mentioned. Hay sent a picture of his wife to at least one friend, if not others, and was not embarrassed by her stoutness.

To a degree Hay was always out of sorts in Cleveland. Though growing in cultural aspirations, the city was a far cry from the centers of power and intellectual stimulation offered by New York and Washington D.C. As his father-in-law's health worsened, so did his. He became depressed with an almost constant headache and severe insomnia. He consulted with a Philadelphia doctor who recommended a summer trip to Europe, which Hay took accompanied by his brother. But returning to Cleveland meant the return of even more severe symptoms. He wrote John Nicolay "The other day I had the most ridiculous attack I have ever had — I thought I was dead for half an hour.... I feel rickety yet I have been trying my best to get to work again with very indifferent success."[77] It is possible Hay was so empathetic with his father-in-law that he was experiencing a psychological condition known as transference.

Not only was Clara attractive to John, she was attractive to others. Clarence King, explorer and philanderer, a close friend of the couple hinted that he desired a wife like Clara who was "Calm and grand... the best of the 19th century."[78] The Stone household had produced two exceptional women. King would later write to Hay that Clara's repose and tranquility would make even "Nirvana seem fidgety." To Hay he said "Only one in a million times does Providence pour out 'the full cup' for man to drink and 'for you it has.'"[79]

Tyler Dennett wrote, "With the Western Reserve as a state of mind Hay had some kinship, but, in general, his idea of the good life and the prevailing ideals of the territorial Western Reserve - which had no theory or place

---

[76] Ibid., 52.
[77] Ibid., 59.
[78] Ibid.
[79] Ibid., 92.

for a leisure class - impinged on one another acutely. He was a spiritual outlander."[80] Dennett was only partially correct. The Western Reserve was far more than a state of mind and Hay other than an outlander; Cleveland and its environs represented relationships and experiences that forever transformed him. First, he inherited money which allowed him to fulfill his fondest dreams of becoming a "gentleman scholar." He hobnobbed with the literary illuminati of his age: Henry Adams, Henry James and William Dean Howells. He leisurely traveled throughout Europe absorbing its architecture, art and culture, he fulfilled the richest appointments that American politics had to offer: Ambassador to England (Court of St. James,) a U. S. minister to Paris, and Secretary of State under McKinley and Roosevelt. In the year that he died, he summed up his life:

> I know death is the common lot, and what is universal ought not to be deemed a misfortune; and yet - instead of confronting it with dignity and philosophy, I cling instinctively to life and the things of life, as eagerly as if I had not had my chance at happiness & gained nearly all the great prizes.[81]

Second, the Cleveland-Stone inheritance was far more than money affording opportunity for bigger and better things. He would forever be incarnationally identified with the Western Reserve. When he spoke before the Pioneer's Association for the Western Reserve, he romanticized,

> Those populous Western cities...those towering chimneys, with their streaming banners of smoke, at once the monuments and the temples of a worship whose divinity is labor, and whose purpose is the progress of man upon earth; the strange network of iron roads carrying the land in every direction, along which rush those mightiest of men, those engines which had given unimaginable development to commerce and communication, and that still more mysterious system of metallic cobwebs in the air...[82]

Hay was not just blowing smoke. He had developed a respect and genuine affection for the inhabitants and industrial prowess of those who strove to transform themselves and the communities in which they lived. Thus, Hay chose (or Clara chose for him) to be buried in the Stone family plot in

---

[80] Dennett, 103.

[81] Taliaferro, "Hay quote" (preceding Table of Contents.)

[82] Frederick Cople Gaher, "Industrialism and the American Aristocrat: A Social Study of John Hay and His Novel, *The Bread-Winners,*" *Journal of the Illinois State Historical Society*, Vol. 65, (Spring 1972) 89-90.

Cleveland's Lake View Cemetery, one of the world's most stately and serene settings, rather than in Washington, D.C. where he had rubbed elbows with the powerful. Because of his world wide fame, his grave is some 15 feet removed from his wife Clara and their son Adelbert and far more unique and imposing. It is graced with a six-foot high "warrior-angel" sculpted by Augustus Saint-Gaudens. The truth is, John Hay was a divided man, not at home in either the industrial or aristocratic world or maybe at home in both. Frederick Cople Gaher states, "His tenuous attachment to both elites (industrial and leisured wealth) made him a member as well as an observer of two enclaves and deepened his perspective of them."[83]

Hay was a rich man in so many ways, and this enrichment endeared him to almost all persons he ever met. He was a descendant of Amasa Stone, almost as much as his own children. Stone had full confidence in his son-in-law. On January 4, 1883, he wrote Hay, "Should I be taken away, you are the one to be at the helm."[84] Hay would captain the Stone ship only for a brief period. That task would eventually fall on the other son-in-law, Samuel Mather.

---

[83] Ibid., 93.
[84] Taliaferro, 220.

Chapter 7

The Wreck

Stephen Peet

Though many excellent articles have been written on the Ashtabula Bridge disaster examining reasons for the bridge's collapse, attaching guilt to various individuals, excavation of the wreckage site, etc. no full-length book narrative has been written on the event except one. It is not surprising that Rev. Stephen Peet, the Congregational pastor in Ashtabula and probably the most educated person in the village of 2,500, would try to recall and preserve for posterity the horror of those few hours. As a minister and pastor of the Congregational Church, Peet was strategically prepared and in position for an unexpected and extraordinary event like that which no other pastor had ever experienced. He had been trained in anthropology, and was secretary of the American Anthropological Association, as well as being editor of the quarterly *American Antiquities*.[1] Peet's over 200-page account written in the succeeding year of 1877 and published in 1878, brimming with homiletical and vivid prose, is melodramatic and hyperbolic, and does not serve as an analytical description of what happened on the night of December 29, 1876.

> The parched lips were sealed forever; the stifled breath could no longer send forth a cry or groan; the carnival of death had at last silenced all of its victims; the slaughter complete; "blood and fire and vapor of smoke."

---

[1] Thomas E. Corts, ed. *Bliss and Tragedy: The Ashtabula Railway-Bridge Accident of 1876 and the Loss of P. P. Bliss* (Birmingham, AL: Samford University, 2003) 122-124..

(Kings James apocalyptic imagery) The flames leaped and danced, and floated high above their heads, and death was exalted in all its forces. The canopy of darkness arched the snow-covered valley, while the fiery billows rolled between. All that man could do was stand and look upon the scene appalled.[2]

Peet made no attempt to place the event within the cultural context of America's centennial year or trace the confluence of factors and individuals, and how they coalesced. He gave almost no attention to reasons for the bridge's collapse and the manner in which it was constructed. In spite of Peet's inability to place historical and psychological distance between himself and the event, his monograph is not without merit. Without Peet's book, many details of the deadliest train bridge disaster in the history of the United States would be lost forever. As Peet came to the end of his narrative, he was not bereft of systemic and theological analysis: "The haste to get rich and the desire to make men serve the purpose of money – getting, and the control over many by the enrichment of a few, will destroy the sense of accountability and blind men, so that they run profanely into the very place that God has the hidings of his power, but the result is they do not know how to control the lightnings and to control the storms and they are appalled at the calamity that their own temerity has brought down."[3] The gist of this statement suggests that if Peet did not know Amasa Stone, he knew of him, and intentionally indicted him for the disaster.

Founding and Development of Ashtabula

Ashtabula County was organized in 1811, including the township of Ashtabula with six other organized towns. In that same year, mail service was established between Ashtabula and Buffalo and the carrier was allowed 12 days for the round trip and "fourteen days when the mud was deep and the water was high."[4] Ashtabula was incorporated as a town in 1831 and established at least 3 firsts: the first to run its own street car railway, the first to adopt a city manager plan, and the first to offer a school for postal workers.[5] Ashtabula would not be so boastful about another

---

[2] Stephen D. Peet. *The Ashtabula Disaster* (Chicago: J.S. Goodman-Louis Lloyd and Company, 1877) 41.

[3] Ibid., 205.

[4] Moina W. Large. *History of Ashtabula County* Vol. I (Topeka, Indiana: Historical Publishing Company, 1934) 104.

[5] Ibid., 169.

uniqueness. In 1847 the Presbyterian Church installed John Ingersoll, father of Robert Ingersoll, as their pastor. The latter became America's best-known agnostic. Ashtabula was a very religious community, boasting of 26 churches by 1900. Several of these churches had ethnic origins including Italian Catholics, Finnish Lutherans, Finnish Congregationalists and Swedish Congregationalists, leading us to conclude that Ashtabula had a decidedly Scandinavian population. Immigrants were attracted to the low but steady paying jobs offered by the harbor.

For the most part, the citizens of Ashtabula were ardent abolitionists and quite vocal about it. The town was an ideal "transition station" for the Underground Railroad with train service and port of exit for escape to Canada. An 1850 editor of the *Ashtabula Sentinel* stated its contempt for the fugitive slave law enacted that year. A judge was paid $5 if he freed a slave and $10 if he sent the slave back to his owner. The local anti-slavery committee resolved that rather than catching slaves, the citizens would "feed and protect them." There was no attempt to camouflage their obligation to escaping slaves. The newspaper stated that the "Underground Railroad in this section of the state was doing a fair business today…"[6] Concerning the transitory blacks, "We learned that they met with no difficulty in finding food, shelter and necessary assistance in their cause."[7] The editorial in the *Sentinel* further boasted that "no fugitive slave can be taken from the soil of Ashtabula County back to slavery."[8] The large frame house of William Hubbard was a perfect stop for hiding in the basement and slipping out under the cover of darkness to board a ship to freedom. William lost his physician brother, George Hubbard, in the wreck.[9]

Ashtabula Bridge History

The first bridge over the Ashtabula River was built in 1852, a Howe Truss made completely of wood and painted white. "A watchman was kept constantly at the bridge to warn people off and to look out for fire."[10] In 1865, Stone's all iron bridge replaced the all wood bridge. The watchman was done away with. The first bridge was 180 feet long, but the second

---

[6] Ibid., 444.

[7] Ibid.

[8] Ibid.

[9] This house still stands in Ashtabula today and serves as a museum to the Underground Railroad. It is listed on the National Register of Historic Places.

[10] Large, 181.

shortened to Stone's 154 feet by filling in the bridge abutments. In 1872, a "roundhouse" for 8 locomotives was built in Ashtabula. In the same year, the Lake Shore ceased to run freight through Ashtabula on Sunday. Maybe this is the place to note that after Charles Paine and his family escaped the Chicago fire with only the clothes on their back, a $5,000 collection was presented to Paine, which he declined. The money was given to the Cleveland Bible Society, which in turn placed a rack and Bible in each coach owned by the Lake Shore.[11] These two simultaneous acts of piety may have been pure coincidence, but neither would ward off the impending catastrophe.

We know that Stone's bridge or at least the plans for it was begun in 1863. The engineers who reported to the Ohio Legislature found an 1863 design of the bridge signed by Joseph Tomlinson. The 1864 annual report of the Directors of the Cleveland, Painesville, and Ashtabula RR Company, indicated $87,508.03 had been spent on the Ashtabula and Conneaut bridges.[12] As the reader will learn in Chapters 9 and 10, a lengthy amount of time passed between the beginning of the work and the bridge's completion. Correcting the errors in the bridge and stabilizing it so it would stand on its own without underpinning and carry railroad traffic took almost exactly a year from July 1865 to July 1866.[13] The bridge failed to stand on its own twice. After making the corrections described in Chapters 9 and 10, the bridge finally stood after removing the scaffolding a third time.

In Stone's defense he had a lot on his plate as he attempted to correct the defects in the bridge. He was emotionally sabotaged because his only son was drowned in June of 1865 (discussed in Chapter 11). He was also building one of the largest railroad depots in the world: Cleveland's Union Station at 603 feet long, and 108 feet wide, opened on November 10, 1866.[14] Its size would not be surpassed in the United States until Vanderbilt's Grand Central Station was built in 1871. He was also building the Cleveland Academy in 1865 (discussed in Chapter One). Stone was at his wit's end, was unpleasant to be around, and was impatient with A.L.

---

[11] Dave McLellan and Bill Warrick. *The Lake Shore and Michigan Southern Railway* (Polo, Illinois: Transportation Trails, 1989) 62.

[12] *Annual Report of the Directors of the Cleveland, Painesville and Ashtabula Railroad Company for the year ending December 31, 1864.* (Cleveland: Fairbanks, Benedict and Company, 1865) 6.

[13] "The Bridge," *The Cleveland Leader* (January 8, 1877).

[14] Tom Barensfeld, "Cleveland's Union Railroad Station," unidentified newspaper, courtesy of Case Western Reserve University archives.

Rogers to whom he had entrusted the responsibility for erecting the bridge. The trial and error of the trial-and-error method was overwhelming.

In 1872, the Lake Shore extended a spur line to the Ashtabula harbor and in 1873, the harbor began to import iron ore and export coal and coke, becoming one of the most important ports on Lake Erie. The town was a critical stop for one of the busiest railroad trips in the United States, New York to Chicago. Between 1870 and 1880, the population of Ashtabula more than doubled from 1,999 to 4,445.

Because of the spur line, William Vanderbilt paid a visit to Ashtabula in 1872. John P. Manning, local agent for the Lake Shore, hosted him, the whole time attempting to explain the town's strategic importance to the New York Central's Vice President. But Vanderbilt was unimpressed and did not perceive the need for the spur to the Ashtabula Harbor. When he was awakened from a nap at a stop between a "cat tail swamp on one side and against a bank of dirt on the other," he snapped disgustingly "Why did you bring me down here? Let's go back."[15] There is no record that Vanderbilt ever returned, not even to investigate the failure of a bridge which essentially belonged to him. The bridge was owned by the New York Central of which William became President, six days after the accident when his father died in New York.

The Snow Belt

It was one of the most scenic train rides in America. As a passenger rode over the Ashtabula River Gorge, the view was unobstructed because the train sat on the top chords of the bridge. No braces, rods, or rails obscured the plush green of summer, the red, brown and gold hues of fall, the bright resurrection of buds and leaves in spring, and even the winter wonderland that was plenteous in the Northeast Ohio snow belt. The depth and flow of water in the creek varied with the heavy rains of April and May and the long dry spells of August and September. Ohio may be America's most beautiful "treed" state, with its array of maples, oaks, poplars, ashes and sycamores in full display throughout the year. Even when stripped of their foliage in winter, the ever-constant ice and snow glisten in the shafts of sunlight, or drape themselves over the hillsides under constantly cloudy winter skies rendering an aura of mystery and assurance that winter even in Ohio always comes to an end.

---

[15] Large, 211.

But winter does come with a vengeance to northeast Ohio. It was a blizzard, a whiteout with almost zero visibility. Winds gusting up to fifty miles per hour picked up tons of snow from the warmer waters of Lake Erie and dropped drifts several feet high on the tracks of the Lake Shore and Michigan Southern Railroad running from Buffalo to Cleveland. With temperatures hovering around 15 degrees, wind chill (a term not used then) would have made the air feel like it was below zero. "Snow Belt" characterized this area along Lake Erie more than any other location within the United States; such weather, both then and now, brought life to a standstill.

If only Dan McGuire, engineer for the 32-ton Socrates, and Gustavus D. "Pap" Folsom, engineer for the slightly larger Columbia, had decided to wait until morning. No evidence exists that they telegraphed Charles Paine, the Lake Shore General Superintendent in Cleveland, for advice under such brutal conditions. By its very nature technology is not defined by caution, not American technology anyway. The very essence of technology is the attempt to defy or transcend nature. To implement efficiency in the face of obstacles, whether those challenges consist of gravity, darkness, climbing up or tunneling through mountains, uphill or upstream travel, as well as inclement weather is a victory for humankind. To not stare down or overcome such challenges is to admit failure. For railroad travel in 1876, failure was not an option unless nature and nature's God decided otherwise.

The Pacific Express, the New York Central's fastest train, already two hours behind schedule, pulled out of the Erie station about 5:00 p.m., December 29, 1876. The lead locomotives, Socrates and Columbia, pulled the train while an additional two locomotives pushed the train until it picked up sufficient speed to leave the station yard. What should have been a 45-minute run became a two-hour lumbering, sloshing, slipping ride over tracks covered with up to two feet of snow. It was treacherous going. At 7:28 p.m., the train arrived at the Ashtabula Bridge, a 154-foot span, which stood about 75 feet above the Ashtabula River, not much more than a two-foot-deep stream, originating from tributaries in the hill country to the east around the Pennsylvania state line, and emptying into the Ashtabula Harbor.

Ashtabula was a picturesque sleepy village of some 2,500 people, ill prepared for the worst railroad bridge disaster in the history of the United States. The village would have been ideal to transplant to a model train display during the Christmas holidays in the twenty-first century. Some read, others played cards, simply talked, reveled in remembering the happy holidays recently spent with eastern seaboard family, particularly in Boston where the train had originated. Others prepared for bed. But at least a few of the some 150 passengers stared anxiously out of the windows into the blinding storm.

The Collapse of the Bridge

The train was running on the left-hand track, the south track in that the train was moving east to west. Upon approaching the bridge, Dan McGuire slowed the train down to approximately 15 mph., slower than normal given the treacherous conditions, to avoid sliding into the station just beyond the western end of the bridge. When he was about 50 feet from the west end abutment of the bridge, located about one half mile from the center of Ashtabula, he heard a loud crack. Only the two locomotives with their tenders and the two forty-foot-long express cars would have been on the bridge at the time of the initial break. Within a couple of seconds, the rear of the Socrates began to tilt down, and upon looking behind him, McGuire noted that the headlamp of the Columbia was shining up at about a 45-degree angle. By instinct, the veteran engineer gave his engine full throttle, barely able to roll up on the west end abutment, pulling his tender car behind him. (The tender held the coal for heating the water in the boiler of the locomotive, producing steam.) Obviously, the coupling between the tender of the Socrates and Columbia would break, and the Columbia would plunge into the gorge, pulling the rest of the train down with it. The train fell to the left, south, pushing the bridge to the right, north.

In addition to the two locomotives and their tenders, the train consisted of 11 cars: two express cars with packages and freight, two baggage cars filled with luggage, two-day passenger cars, a car for passengers who smoked, a "drawing room car," (also referred to as a parlor car) more elaborately and comfortably furnished than a regular passenger car, and three sleepers named the "Palatine New York", the "City of Buffalo," and the "Osceo," also referred to as the Louisville sleeper. The two regular passenger cars had straight back seats; the parlor car boasted an interior panel with walnut and cedar, running hot water, and some 20 plushy-cushioned arm chairs. Because Cornelius Vanderbilt disliked George Pullman, an animosity fueled by the vanity and pride of competition, he contracted with Webster Wagner for his sleepers.

Stone and Vanderbilt intersected once again in the production of Wagner sleeping cars. Wagner would not have survived without Vanderbilt and according to railroad scholar John White, the Wagner Sleeping Car Company was a subsidiary of the New York Central. Vanderbilt provided the production money for the original cars in return for a 75% interest. John White claims "The Commodore made it clear that he would have no cars operating over his line which did not pay him a major share of the revenue, and that if Wagner did not like the terms, he could take his cars elsewhere."[16] The officers of the New York Central were the officers of the sleeping car company which would have included Stone. This is not to say that the Wagner cars were more dangerous than Pullman cars. It does suggest the irony that luxury equated to catastrophe. It is also ironic that in 1882, Wagner was incinerated in one of his own sleeping cars when trains of the New York Central and Hudson Railroad collided.[17] His body was so charred that it was barely identifiable.

By 1880 a Wagner sleeper boasted of "an interior of solid mahogany relieved with inlaid woodwork upon the section panels. The ceilings are decorated oak, and the seats are upholstered in scarlet plush. There are two windows to each section with spring-roller curtains; and also projecting dust-guards on the outside to intercept cinders."[18] The elaborate decorations included cushioning for both seats and beds, lacquered panels and cloth drapes; all were combustible, a perfect fuel for the conflagration about to take place. This description would have likely been accurate for an 1876 sleeper.

The Columbia fell upside down on the first express car; this was fortunate since no one was in the express car. Less fortunate were the 40 travelers in the second passenger coach which was struck by the front end of the smoker car. All 40 were almost instantly killed, or died within a few minutes from the impact. Five of the sixteen passengers on the smoker car were killed. Only one person survived the parlor car because it was crushed by the "City of Buffalo" sleeper. This sleeper landed on its front end, hurtling everyone in that direction. The 21 on that car were crushed or burned to death. Almost everyone on the "Palatine New York" sleeper

---

[16] John H. White, Jr. *The American Railroad Passenger Car* (Baltimore: The Johns Hopkins University Press, 1978) 233.

[17] "Meeting a Terrible Fate – Nine Persons Crushed and Burned in a Collision – A Train Crashing into the Rear of the Atlantic Express – Nine, Perhaps Twelve Victims Caught in the Burning Cars – State Senator Wagner Among the Dead – Narrow Escape of Many – Others – Terrible Scene at the Wreck." *The New York Times*, January 14, 1882.

[18] White, 239.

survived, because the car landed flat on its wheels. The three passengers on the "Osceo," the last sleeper on the train were lost, with only the brakeman, A.L. Stone, surviving.[19]

Reaction of Both Ashtabula Inhabitants and Survivors

Almost everyone in the village heard the crash and then a series of crashes. Henry Apthorp's wife exclaimed to her husband, "My God, Henry, No. 5 has gone off the bridge."[20] Apthorp was one of the first to arrive at the scene. Others quickly came: the engineer of the pump house, a saloon keeper from one of the hotels, and the foreman of the fire engine "Lake Erie." McGuire pulled the whistle on his locomotive with two sharp blasts. The bell at the fire house pealed through the village. Given the weather conditions and six-foot drifts of snow, the short distance to the wreck was difficult to traverse. Villagers who had sufficient strength attempted to walk to the site, but in the twenty or so minutes of interlude, the rapidly engulfing fire made reaching passengers almost hopeless. Peet described the scene: "The call for buckets went up from below. One old man, seventy-six years old, was in the midst of the wreck, chopping for dear life and calling for buckets at the same time. His son, arriving late, plunged into the midst of the fire and began to work like one made desperate with despair."[21] Those who took pails formed a bucket brigade, but the effort was almost as futile as spitting in Hell.

Pandemonium broke out; most who were not instantly killed were disoriented, stunned, and even delirious. Unfortunately, the inhabitants of Ashtabula were almost as disoriented as those on the train. This is not to say that many of the Ashtabula citizens did not exhibit exceptional courage and emergency measures. Ashtabula Fire Department volunteer Fred W. Blakeslee waded down into the icy water, and pulled survivors through windows. Michael Tinley made his way across the Ashtabula Creek, and chopped holes in the sides of cars until he drove the axe through his foot and had to be removed himself. The brakeman Stone ran to the telegraph

---

[19] Charles A. Burnham. "The Ashtabula Horror" in Corts, 85-86.

[20] Stephen D. Peet. *The Ashtabula Disaster* (Chicago: J. S. Goodman-Louis Lloyd and Company, 1877) 20. For quoting Peet, I have utilized the Gutenberg e-book. The page numbers are vastly different than the original book. https://www.gutenberg.org/files/47359/47359-h/47359-h.htm.

[21] Ibid., 24.

station to warn any approaching train that there now existed a chasm with no bridge.

The survivors assisted one another. H. L. Brewster pulled a fellow passenger to shore. A passenger named Marion Shepard assisted others; she was observed by C. E. Torres dipping her handkerchief in the water and washing away the blood from the face of a wounded man. Shepard was referred to as the "angel" of the night. She was travelling from Albany, New York to Ripon, Wisconsin to visit her future husband Frank Hamilton. She rescued several passengers from the Palatine, the car she was on, dragging them some thirty feet up the bank. She broke windows with planks in order to extract passengers. Though wounded in the back, she assisted doctors throughout the night.[22] She left us the most extensive first-person account:

> The passengers were grouped about the car in twos, fours, and even larger parties. Some were lunching, some were chatting, and quite a number were playing cards. The bell-rope snapped in two, one piece flying against one of the lamp glasses, smashing it, and knocking the burning candle to the floor. Then the cars ahead of us went bump, bump, bump, as if the wheels were jumping over ties. Until the bumping sensation was felt, everyone thought the glass globe had been broken by an explosion. Several jumped up, and some seized the tops of the seats to steady themselves. Suddenly there was an awful crash. I can't describe the noise. There were all sorts of sounds. I could hear, above all, a sharp, ringing sound, as if all the glass in the train was being shattered in pieces. Someone cried out, "We're going down!" At that moment all the lights in the car went out. It was utter darkness. I stood up in the centre of the aisle. I knew that something awful was happening, and having some experience in railroad accidents, I braced myself as best I knew how. I felt the car floor sinking under my feet. The sensation of falling was very apparent. I thought of a great many things, and I made up my mind I was going to be killed. For the first few seconds we seemed to be dropping in silence. I could hear the other passengers breathing. Then suddenly the car was filled with flying splinters and dust, and we seemed to be breathing some heavy substance. For a moment, I was almost suffocated. We went down, down. Oh, it was awful! It seemed to me we had been falling two minutes. The berths were slipping from their fastenings and falling upon the passengers. We heard an awful crash. As the sound died away there were heavy groans all around us. It was as dark as the grave. I was thrown down. Just how I fell is more than I can say. A

---

[22] https://www.engineeringtragedy.com/marion-shepard

gentleman had fallen across me, but we were both on our feet in a moment. Everyone alive was scrambling and struggling to get out. I heard someone say, "Hurry out; the car will be on fire in a minute!" Another man shouted, "The water is coming in, and we will be drowned." The car seemed lying partly on one side. In the scramble a man caught hold of me and cried out, "Help me; don't leave me!" A woman, from one corner of the car, cried, "Help me save my husband!" He was caught under a berth and some seats. I was feeling around in the dark, trying to release him, when someone at the other end of the car said they were all right and would help the man out. I groped along to the door crawling over the heating arrangement in getting to it. While I was getting out at the door, others were crawling out the windows. On the left the cars were on fire. On the right a pile of rubbish, as high as I could see, barred escape. In front of me were some cars standing on end, or in a sloping position. I followed a man who was trying to scale the pile of debris. I got up to a coach which was resting on one edge of the roof. The side was so slippery and icy I could not walk on it, and so I crawled over it. The car was dark inside, and oh, what heart rending groans issued from it! It seemed filled with people who were dying. Two men, a Mr. White, of Chicago, and a Mr. Tyler, of St. Louis, helped me down from the end of the car. Then I was in snow up to my knees. Mr. Tyler was badly gashed about the face, and was covered with blood. This stain on my sleeve was blood from his wound. Right under our feet lay a man, his head down in a hole and his legs under the corner of a car. He asked help, and Mr. White and Mr. Tyler released his legs somehow, and some other men carried him away. It was storming terribly. The wind was blowing a perfect gale. By this time, the scene was lighted up by the burning cars. The abutments looked as high as Niagara. Away above us, I could see a crowd of spectators. Down in the wreck there was perfect panic. Some were so badly frightened and panic stricken that they had to be dragged out of the car to prevent them from burning up. Before we got out of the chasm, the whole train was in a blaze. The locomotive, the cars, and the bridge were mixed up in one indistinguishable mass. From the burning heap came shrieks and the most piteous cries for help. I could hear far above me the clangor of bells, alarming the citizens. We climbed up the deep side of the gorge, floundering in snow two feet deep. They took us to an engine house, where there was a big furnace fire. The wounded were brought in and laid on the floor. They were injured in every conceivable way. Some had their legs broken; some had gashed and bleeding faces, and some were so horribly crushed they seemed to be dying.[23]

---

[23] Corts, 123-124.

Most of the remaining passengers trapped inside the cars were immediately engulfed in flames. Sperm whale oil spilled out of the candle holders, carrying the flame with them. Coal "Baker heaters" tumbled through the cars, igniting ticking, drapery, clothes, and lapping up varnish. The burning flesh could be smelled all the way into the village and would leave 42 people unidentifiable, some of them completely vaporized. The blaze lighting up the sky could be seen for miles. The volunteer fire department was thrown into confusion; the fire hose was never hooked up to the pump house; the horse drawn steam engine was pulled down to the disaster site, but never put into use. Some declared that the Lake Shore had given specific orders not to put water on the fire. A hand engine stood next to the pumphouse, but it too stood idle. Screams, cries, and groans, could be heard from the perishing victims, while horror and shock mortified those who so intensely wanted to help, but were beaten back by the engulfing flames. Some just stood within the eerie no-man's land that separating the freezing cold from incinerating heat, transfixed by a horrifying nightmare staggering the imagination.

Charles Leek

All disasters produce heroes and no one exemplified this truth more than Charles Leek, the assistant telegraph operator for Ashtabula. He stayed at his station for fifty straight hours, sending out messages and answering the steady barrage of questions that came to his switchboard. At the time of the accident, Leek, a black man, was 26 years old. His father, John L. Leek, was born into slavery, and his son Charles had been freed by his master. When Charles was three years old, his father moved from Rhode Island to Ashtabula, Ohio, and opened a restaurant next to the Lake Shore Depot. At the age of nineteen he began work for John Manning, the telegraph operator, as a messenger boy. Within two months, Charles had mastered the telegraph switchboard, and became a full-time operator. Thus, he became the first African-American telegraph operator in America. He then took telegraph operator jobs several places, but for some reason just before the wreck, returned to Ashtabula. This move was providential in that when John Manning was promoted and moved to head of telegraph operations at the Ashtabula Harbor, Charles Leek was tagged to take Manning's post and was thus promoted to Head of Telegraph operation at the Ashtabula Depot. This was another first for an African American.

Leek also became known as a talented musician and master of the violin. He conducted Leek's orchestra for some twenty years. When fifty years old, he left the telegraph office and took over his father's restaurant business. Unfortunately, he died at age 51, from a massive heart attack.[24]

After the Coroner's Jury's verdict, Leek wrote a scathing letter, criticizing its condemnation of the Fire Department. It also gives us a perspective on the emotions and crises that the Ashtabula citizens experienced on that night.

To the Editor of the Herald:

I have read carefully the verdict rendered by the coroner's jury, especially the seventh paragraph; and to say that I am shocked at the action taken by said jury, in saying that those persons who were first to arrive at the scene of the disaster were responsible for the burning of the wreck, would but faintly express the feelings of those persons. Such a verdict was never known in the history of any country. If any man possessing an average amount of intelligence will for a moment imagine himself present that night within a few minutes after train No. 5 went down, and hearing the cries of help from the wounded, would he not start at once and try to lend a helping hand? Or would he turn a deaf ear to such entreaties and go and look for hose, buckets, etc., to put out the fire, allowing those persons to perish?

My opinion of such a person who would leave them at that critical moment is this, that his room would be far preferable to his company in this world. I think the people should take this late consideration. It being a very stormy night, wind blowing furiously and the snow very deep, we all worked at a great disadvantage. The coroner's jury in their verdict condemn the action of those persons who were first to arrive at the scene, and who aided much in caring for the wounded. These, then, are the persons who are held responsible for the burning of the wreck; this seems a very cruel verdict. Let us look into the matter. There were but few persons at the wreck when I arrived, and I know we all worked like men. I thank Almighty God that so many lives were saved, and had we turned our attention to putting out the fire instead of assisting those who needed our assistance at that moment, I think that in the sight of God we would have been guilty of a great crime. I will make the assertion that not one of us knew the hose in the Lake Erie engine house would fit the fire plug at the pump house, and I

---

[24] Information provided by Leonard Brown, executive producer of documentary film, "Engineering Tragedy: The Ashtabula Train Disaster."

ask any sensible person if then was any time to find out. The fire fanned by such a fierce breeze spread with amazing rapidity, which goes to show very clearly that had we gone after hose in such a terrible snow storm, that by the time we returned the fire would have made such headway that it would have been utterly impossible to have saved those persons who were taken out and as a result they would have been roasted alive. If, on the other hand, everything had been in readiness at the pump house, when the train fell, so that a stream of water could have played upon the fire at a moment's notice, then, in all probability, it could have been put out very easily. But as there were no hose kept at the pump house and none nearer than Lake Erie engine-house, no person possessing a heart could turn his head from the terrible scene and go up the bank after hose when more lives could be saved below. This is a terrible charge against those persons who know they did their duty. They did everything in their power to save human life under the circumstances then existing and this is their thanks. According to the Jury's decision it would have been much better, perhaps, had we remained away, for in that case it may be possible that we would not have been blamed or censured so severely.

C. B. Leek[25]

Medical Response

The survivors were dragged up the steep banks. The one flight of wooden steps that went to the top of the west end abutment remained intact, but was useless since the totality of the bridge now lay on the river floor. The Eagle Hotel was turned into a hospital. Stephen Peet described the scene. "It was a horrid place. A dirty barroom. Rooms which had never known a carpet, but whose floors were soon covered with snow and water, little bedrooms just large enough to hold a bed and a washstand, without carpets, with snow and water; beds that consisted of filthy sheets and miserable straw ticks, it was a house forbidding in every respect."[26]

At 1 a.m., a train load of six physicians, several rescue workers, General Superintendent of the Lake Shore Charles Paine and Chief Engineer Charles Collins, arrived from Cleveland. But by this time, the holocaust

---

[25] Letter of C. B. Leek to the editor of the Herald, "The Ashtabula Verdict," March 12, 1877 in "Bridge Disaster Scrapbook" given by Thomas Corts to the Ashtabula County Historical Society at the Jenny Munger Gregory Museum in Geneva on the Lake, Ohio.

[26] Peet, 58.

had completed its destruction and salvific attempts fell under the proverbial assessment, "a day late and a dollar short." E. W. Richards, who was
appointed coroner (there was no official coroner in town), recalled, "Mr.
Paine, superintendent, was on the spot in a few hours after the accident and
was unremitting in his efforts to bring order out of the chaos that prevailed.
I have never known a more tempestuous night than that of the 29th of December last."[27] The surgeons from the Homeopathic College in Cleveland
spent the night dressing wounds, setting limbs, removing a gilded slither
of wood from the collarbone of Alfred S. Parslow, and amputating at least
one arm.

Five doctors resided in Ashtabula at the time of the accident: Ephraim
King, Freeman Case, Henry Fricker, O. H. Moss, and a Dr. Eames. Of the
five, King seems to have offered the most exemplary service. He tirelessly
worked throughout the night, and for weeks visited the wounded housed
throughout Ashtabula. He was an ardent Methodist, served as Mayor of
Ashtabula, and owned a drugstore. Before moving to Ashtabula, King
taught at the Erie Street Medical College in Cleveland. On December 29,
1876, King was 46 year old and worked himself to exhaustion.[28]

As Charles Collins looked out over the wreckage, he wept uncontrol-
lably. The Chief Engineer was as helpless as everyone else. The ensuing
nightmares overwhelmed him. Whether he died by suicide or murder (explored in Chapter 13) he was in an extreme state of despair and depression.

Identifying the Dead

What came next was even more sickening and macabre than what had
happened during the wreck, identifying persons by examining body parts,
clothing, jewelry, and whatever other accessories or belongings would ac-
company someone taking a journey. Scores of letters came to Ashtabula
describing "hair and eye color, clothing, jewelry, possessions, birth marks
and scars."[29] Edward Rivenburg from Spencertown, New York sent a mis-
spelled letter to the Ashtabula postmaster, attempting to find his brother,
William W. Forbes, who ended up not being on the train. Nonetheless, his
brother described him as having

---

[27] Barbara J. Hamilton. "Who's Who? Identifying Victims of the Disaster."
Letter from E. W. Richards to Samuel Bruner, October 29, 1877, Corts, 27.
[28] Courtesy of Leonard Brown.
[29] Ibid., 23.

light brown mustash and chin whiskers. His chin whiskers shorter than his mustash. Has blew eyes, has a mark over his right eye and a mark between his lower lip and chin which had been cut. and has black culered clothes and had flat feet. His heels was smashed down in his feet when he was small. The hair on his head was light brown. He said when he started that he was agoing to have his mustash and whiskers cut off and had coarse boots on and well worn was bought last summer had a pair of buckskin gloves and was all pine pitch, ahandling pine timber, and a black common hat.[30]

Dr. George Hubbard's mother from Rapid City, Iowa described socks that her son had been given at Christmas. A portion of his leg was found with sock intact, and that was the only part of the son's anatomy shipped back to Iowa. Samuel Bruner from Bethlehem, Pennsylvania had his son, daughter-in-law, and two grandchildren with him on the train. The fire so completely consumed the family that only the son's gold watch could be identified. Of another, only an arm could be found, and later identified by a gold ring. John T. Wilson from Winchester, Massachusetts described his brother William as wearing "fine men's underclothing, bosom shirts with gold studs and collar buttons, no buttons other than gold on his linen shirts. Sometimes he wore a jet sleeve decorated with a monogrammed 'W.'"[31] All the description was in vain as William Wilson lost his life in the wreck, but his body, clothing, and belongings were never found. But for others, details were helpful. For example, a letter described a diamond ring worn by Robert Steindal. All that could be found of Robert was an arm, with a diamond ring on a finger.[32]

Several bodies were reduced to black lumps; bodies were cut in two. One full sized woman could be identified by her torso, missing all of its appendages.[33]

The freight house was converted into a morgue, thirty-six bodies were arranged, in boxes in a double line along the sides; a few had been taken out with their bodies uninjured, except as they had died from the breath of the fire. These were placed by themselves upon the floor and from their very attitude showed how awful had been their death. They were mostly men. There they lay, with limbs distorted, with hands uplifted, with averted

---

[30] Ibid.
[31] Ibid., 29.
[32] Ibid., 30-31.
[33] Peet, 68.

faces and with all the agonized and awful shapes which death by fire must produce.[34]

For days volunteers raked, hoed, and shoveled the bottom of the stream while they stood in icy water. The careful excavation found a bone with a chain wrapped around it, a cap which had belonged to "Marvin", a string that had graced a child's hair, a key belonging to E. P. Rogers, a coat owned by Mr. J. Rick, silver buttons from Hoyd Russit's coat, a chain worn by Minnie Mixer, a pocketbook belonging to a Mr. Smith from Toronto. All of these were duly catalogued and sent to relatives or whomever could identify them with plausibility, often based on remembrances of the person. Fraudulent claims were made, but almost always quickly assessed as false. E. W. Richards oversaw this mammoth task while at the same time chairing the Coroner's Jury, which met six times per week. As late as three weeks following the incident, Richards wrote to an inquirer, January 17, 1877: "Went down to the depot and reviewed the remains of part of a body found in the debris this day-viz: the lower part of the backbone, and some of the intestines – all burnt to a crisp – and brought them to the undertaker for burial."[35]

Thieves Descend

While the wreck brought out the best in humanity, it also revealed the worst. Thieves descended on the wreckage collecting money, jewelry, watches, wallets, coins, and whatever else was deemed valuable. Scavengers demonstrated no decency or respect for human dignity. As Alfred S. Parslow was lying in the Eagle Hotel with a long piece of cornice driven through his collar bone, his pocket was relieved of $300. A man from Hartford had the boots stolen off his feet; a wallet was found belonging to an identified body, but with $7,000 missing.

---

[34] Ibid., 35.

[35] Charles A. Burnham, "The Ashtabula Horror," Corts, 78-99, unnumbered page.

No story could top that of a young man who was headed to California and lost both his mother and sister in the wreck.[36] As he lay in the icy waters with broken ribs and a deep gash on his forehead, he had the presence of mind to detect thieves. He hid a valuable watch inside his shirt, a purse in his vest, and another in the pocket of his "pantaloons." As someone assisted him to the hotel, he was so weak he fainted, but not before he felt a hand reach into his shirt. Upon regaining consciousness, he discovered that his ticket to California and both purses were missing; somehow the thief had failed to extract the watch. Of course, dead bodies were much easier to ravage than living. From one deceased person was stolen a diamond pin, a commander's badge, a "Sir Knight's pin," as well as other jewelry. Some of the thieves were apprehended and stolen property recovered.[37]

Who Was Aboard the Train?

We will never know exactly how many were aboard the train; the best estimates were between 150 and 200. The "manifest," i.e., passenger list, would have been checked and persons accounted for in Buffalo because several trains converged on that city. But seemingly head counts were not made for subsequent stops between Buffalo and Ashtabula. Because the passenger list was uncertain the number of dead cannot be stated with exactitude. Charles Paine, General Superintendent for the Lake Shore, stated in early February 1877 that there were 80 fatalities and 69 survivors. Both of these numbers would grow over the next several months, and even years, until more recent scholars list 78 survivors and 98 fatalities.[38] Astonishingly, of the remains of the deceased passengers, 48 could not be identified. In other words, if 48 persons had been cremated at the same time in the same fire, the destruction could not have been more complete. Much of the wreckage was nothing other than a crematorium.

---

[36] The young man was Edward Trueworthy, whose father Thomas Trueworthy also survived the wreck. Edward's sister collected newspaper articles after the accident, creating a scrapbook which was sold on eBay in 2004 and purchased by Thomas Corts. It is now in the possession of the Ashtabula Historical Society. eBay: "Ashtabula, OH train disaster documents 1876-7. Item number: 2216889925. Courtesy of David Tobias.

[37] Darrel E. Hamilton. "Almost the Perfect Disaster," Corts, 7-8.

[38] The book edited by Corts lists all who lost their lives, and whether their remains were identifiable. 140-145.

Recollections of Survivors

The survivors would gradually be able to describe the terror that seized them and the miraculous intervention which spared their lives. Alfred Parslow from Chicago recalled, "As soon as I could collect my thoughts, I went to work to extricate myself but how I did it, I'm unclear. I only know that I found myself out of the car and into the fragments of ice and floating pieces of the wreck. The screams of the dying and crushed broke upon my ears and were the most pitiful sounds that I have ever heard. I managed to reach unbroken ice and from there I climbed up the height."[39] George White, travelling on the sleeper Palatine, remembered "The scene within was horrible, heart-rending… indescribable. It was enough to unnerve the bravest. There were maimed and bruised men, women, and children, all held down by cruel timbers. They were in different stages of delirium and excitement. Some were screaming, some were groaning, and others praying."[40]

H.A. White from the sleeper City of Buffalo told his story:

The first thought that came into my mind was that I was dead; that it was no use for me to stir or try to help myself. I waited in that position until I heard two more crashes come, when all was quiet; I then tried to see if I could not raise what was on and around me and succeeded. I opened my eyes and the first thing I saw, was the glass in the top of the door that opened into the saloon in the rear end of the car. I struck that immediately with my hand and thrust my head through it. I spoke then. Up to this time there was not a shriek or voice heard in the car that I was in…all had been stilled.[41]

Peet described the experience of Foster Swift and his wife:

Mrs. Swift retained her senses and presence of mind. She was badly injured at the time but did not realize it. When the accident occurred, there was a terrible crash. The bell rope snapped like the report of a pistol and the lights were extinguished. As the cars went down there was no noise. Her husband was hurled across the aisle and held down senseless. She was wedged in between two seats, and extricated herself. She spoke to her husband but he made no reply and she thought he was dead. The agony of her mind at the moment was fearful to contemplate. She finally with the

---

[39] Peet, 51.
[40] Ibid., 52.
[41] Ibid.

aid of Mr. White got him out.[42]

J. E. Burchell, an attorney from Chicago and passenger on the train, described the horror of the night but not without misperceptions:

> The first thing I heard was a cracking in the front part of the car, and then the same cracking in the rear. Then came another cracking in the front louder than the first and then came a sickening oscillation and a sudden sinking, and I was thrown stunned from my seat. I heard the cracking, and splintering and smashing around me. The iron work bent and twisted like snakes, and everything took horrid shapes. I heard a lady scream in anguish, "Oh! Help me!" Then I heard the cry of fire. Someone broke a window and I pushed out the lady who had screamed I think her name was Mrs. Bingham....
>
> The cracking of the flames, the whistling wind, the screaming of the hurt, made a pandemonium of that little valley, and the water of the freezing creek was red with blood or black with the flying cinders. I did not then know that any lives had been lost. All had escaped alive, (he was wrong) though all were bruised or injured. The fire stole swiftly around the wreck, and in a few moments the cars were all in flames. The ruins covered the whole space between the two piers, the cars jammed in or locked together. One engine lay in the creek, smashed to pieces, (again he was wrong)the ruins breathing steam and fire....
>
> I did not go back to the wreck, but from the engine-house door I could see into the ravine, and the fearful scene it presented. The sight was sickening. The whole wreck was then on fire, and from out the frozen valley came great bursts of flame. There were crowds of men there, the fire beat them back, and they could do nothing. The wounded were lying around in the snow, or were laid on stretchers or taken on the backs of men and carried up the bluff. The spectacle was frightful but those who had gone to assist worked steadily and well in spite of the intense heat.[43]

Finding Closure

The citizens of Ashtabula would never completely find closure and healing for the event of December 29, 1876. For months they answered inquiries and retrieved belongings. Many local families housed victims

---

[42] Ibid., 55.
[43] Whittle, 290-296.

of the wreck for days until they could be united with their family or returned to their homes. The whole of the village was involved in providing consolation for the survivors and solace to those who had lost loved ones and felt helpless because of being hundreds of miles from home in a village they had never known existed. Scores of reporters descended upon Ashtabula trying to provide grisly details for their hometown newspapers.

A burial plot for the remains of the dead was prepared at Chestnut Grove Cemetery and a funeral was held Friday, Jan 19, 1877. Many of the city's buildings were draped in black and the whole of Ashtabula came to a standstill. Though the main service took place in the Methodist Episcopal Church, the largest sanctuary in town, all of the churches participated, including the Baptist, Presbyterian, Episcopal, and Congregational. The mile long procession included sleds carrying the clergy, friends and family of the deceased, and the members of the City Council. On foot were the Saint Joseph Temperance Society, Ashtabula Band, Ashtabula Light Guard, Ashtabula Light Artillery, relatives of the dead, curious spectators, members of the press from all over the country, and almost all Ashtabula inhabitants.[44]

The Ducro Funeral Home had made a painstaking attempt (without the benefit of DNA) to identify body parts and came to the conclusion, but not full confidence, that they had the remains of 22 persons. The remains were placed in 19 coffins and buried in 19 graves adjacent to one another. Three coffins with body parts were left at the depot in hopes that the remains would be identified by friends or relatives. Finding none, they too were placed in graves at the same site several days later. John Ducro, owner of the funeral home, assisted in shipping body parts to families.[45]

In 1892 T. W. McCreary, one of Ashtabula's leading business men, initiated a capital campaign for the purpose of building a suitable monument to those who lost their lives.[46] On Memorial Day 1895, some 5,000 people gathered to witness the unveiling of a 35-foot cenotaph at the grave site in Chestnut Grove Cemetery. The lead locomotive for the ill-fated train, the Socrates, was on hand as a most visible reminder of the disaster. Though both the Socrates and Columbia continued in service for many years, no one can identify when and where they were eventually scrapped. The 200-pound bell which sounded the alarm for the tragic event today sits

---

[44] Darrell Hamilton, 15.

[45] Barbara Hamilton, 32. The Ducro Funeral Home is still in business today.

[46] Ibid., 18.

in front of the Ashtabula Fire Station on Main Street. The monument lists the names of forty people, some of whose remains were never found. The inscription reads

TO THE MEMORY OF THE

UNRECOGNIZED DEAD

OF THE ASHTABULA TRAIN DISASTER,

WHOSE REMAINS ARE BURIED HERE.[47]

As the years rolled by survivors died, and claims were made as to the oldest person still living who had been on the fateful train. In 1941 Harold Bennet, then 101 years old, claimed to have been a "newsboy" on the train. Bennet's story did not bear scrutiny, as he would have been 36 years old at the time of the disaster, far too aged for a "newsboy." He claimed to have been in the Ashtabula hospital for 11 months, but there was no Ashtabula Hospital in 1876.[48] In 1960, Effie Neely died at the age of 101 while living in Troy Township, Geauga County, Ohio. Neely was eight years old the night she escaped from the train. The last surviving employee was in all likelihood James F. Hunt, who at 19 years old served as fireman on the Socrates.[49]

On Saturday, December 30, the citizens of Ashtabula chose eight citizens to serve as a Coroner's Jury for the purpose of investigating the accident. Other than the choice of Edward W. Richards, who was Justice of the Peace and appointed coroner, because Ashtabula had no official coroner, it is unclear as to the process by which the other men were chosen. Interestingly, Ashtabula citizens did not believe that any woman could offer any meaningful insight. The men faithfully met for 68 days, rendering a verdict on March 7.

Photography was in its infancy. The first vivid photographs were produced only about 15 years earlier, when Matthew Brady astounded the world with his pictures of slain soldiers from the battlefields of the Civil War. Fred Blakeslee, a member of the Ashtabula Fire Department, had opened up a photography shop on Main Street. The next morning, he

---

[47]   https://www.google.com/search?q=Ashtabula+Monument+epitaph+tbm+i ssch+source=iu+icx=1&fir=81Sw5wAsk0z

[48]   Corts, 151-152.

[49]   Darrel Hamilton, 19.

hauled his cumbersome equipment down into the gorge. Because of the cold, he had to intermittently thaw out his "wet plates" in a nearby switchman's shanty.

One of his photographs became a classic; the destruction of the train was so complete that replicating it for a movie set or for some other purpose would be impossible. (Though Leonard Brown in his documentary "Engineering Tragedy: The Ashtabula Train Disaster," with the skill of a master cinematographer has created believable authenticity.) The tangled iron, melted train equipment, broken and bent I-beams, and thousands if not millions of scraps impossible to identify gave the appearance of utter destruction. Other than the Columbia lying on its back, and the void between the colossal abutments, one would have been hard-pressed to identify the remains as a train and bridge, fully intact and serviceable only about 12 hours earlier. Everything of which the train was constructed was gone except that which was made of metal. Blakeslee had frozen the wreck in time, a gift to every serious researcher and even persons with a casual interest in the aftermath of an event that radically changed present and succeeding generations.[50]

For Henry Adams, the cruel fate of the ninety-eight victims was inevitable. It was the demonic determinism written into the very nature of technology and modernity.

> "Every day nature violently revolted, causing so-called accidents with enormous destruction of property and life, while plainly laughing at man, who helplessly groaned and shrieked and shuddered, but never for a single instant could stop. The railways alone approached the carnage of war; automobiles and fire-arms ravaged society, until an earthquake became almost a nervous relaxation."[51]

Even though Adams had exaggerated, he was prophetically accurate in foreseeing technological advancement careening towards destruction. The Ashtabula Bridge Disaster constantly reminds us that we all live in the irony that while humanity has unlimited potential for good, it often acts as its greatest enemy.

---

[50] Blakeslee is not the only person who took photographs. J. M. Green also took pictures, but seemingly sometime after Blakeslee. Email: Barbara Hamilton to Darius Salter, July 25, 2022.

[51] Brown, *The Last Aristocrat*, 368.

The best known painting of Amasa Stone age unknown
Courtesy of Case Western Reserve University Archives

Amasa Stone sometime in his 40s
Courtesy of Case Western Reserve University Archives

Amasa Stone sometime in his 50s
Courtesy of Leonard Brown

Julia Ann Gleason Stone
Courtesy of Case Western Reserve University Archives

Adelbert "Del" Stone - only son of Amasa Stone who died at age 20
Courtesy of Case Western Reserve University Archives

Clara Stone Hay - oldest daughter of Amasa Stone - Public domain

Flora Stone Mather - youngest daughter of Amasa Stone - Public domain

Andros Boyden Stone - younger brother of Amasa Stone, Jr. and primary business partner - Courtesy of Western Reserve Historical Society

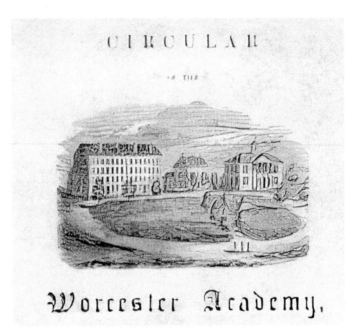

Circular of the Worcester Academy attended by Amasa Stone, Jr. - Public domain

Amasa Stone's boyhood home on the family's Charlton, MA farm - Courtesy of Amy Hiatala - Archivist at Old Sturbridge Village, Sturbridge, MA

John Milton Hay - Clara's husband & man of world fame - b/w photo
portrait, 1897 - Public domain

Samuel Mather
Samuel Livingston Mather, Flora's husband & wealthiest man in Ohio,
photo 1908 - Courtesy of the Cleveland Public Library

William Howe-genius brother-in-law of Amasa Stone, Jr.
- Courtesy of the American Society of Engineers

Amasa Stone's mansion at 1255 Euclid Avenue - Courtesy of Western
Reserve Historical Society

Charles Collins - Chief Engineer of the Lake Shore & Michigan Southern
R.R. - Public domain

Mary Collins - Wife of Charles Collins - Public domain

Joseph Tomlinson - a gentleman and expert draftsman whose role
in the bridge will always remain controversial - Public domain

Harmon House in Ashtabula that still stands today - Home of Mary
Collins' parents - Courtesy of the Ashtabula County Historical Society

Charles Leek - who was at his telegraph post for 55 continuous hours - Public domain

Gustavus "Pap" Folsom - Engineer of the Columbia - Public domain

G.W. Knapp - Chief of the Ashtabula Fire Department
who consumed too much alcohol - Public domain

H.P. Hepburn - Ashtabula's Mayor & also employee
of the Lake Shore & Michigan Southern R.R. - Public domain

The famous Blakeslee photo taken on the morning after
the wreck - December 30, 1876 - Courtesy of Leonard Brown

Depiction of the wreck site by *Harper's Weekly*

Fred Blakeslee - who took early morning photographs on
December 30, 1876 - Courtesy of the Ashtabula County Historical Society

The Ashtabula Bridge - with a train on it before disaster -
Courtesy of the late Charles Burnham.

J. M. Green photo of wreck clean-up - Courtesy of the late
Charles Burnham

The west end abutment of the bridge after the wreck - Courtesy of the
Ashtabula County Historical Society

The lead locomotive Socrates that continued in service for years -
Courtesy of the Ashtabula County Historical Society

The Columbia - The second locomotive turned upside down after the
wreck - was later put into service - Courtesy of the Ashtabula County
Historical Society

The replacement wooden Howe Truss Bridge after failure of the iron
bridge - Courtesy of David Tobias

Ashtabula Depot and Freight House - Courtesy of Leonard Brown

Marion Shepard "Angel of the Night" - Public domain

Phillip Paul Bliss - Public domain

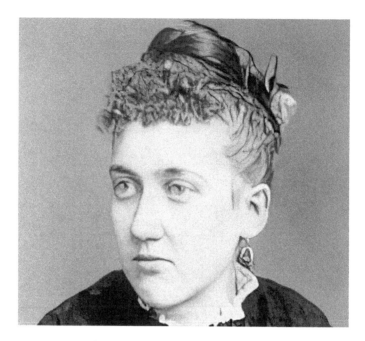

Lucy Bliss - wife of P. P. Bliss - Public domain

The pumphouse at the west end of the bridge had no firehose that fit the hydrant - Courtesy of David Tobias

Ashtabula around 1879 - Courtesy of Library of Congress

Abraham Lincoln funeral procession when it came through Cleveland in
May 1865 - Courtesy of the Cleveland Public Library

168

Downtown Cleveland in 1861 - A well-developed city - Public domain

The John Hay house next to Amasa Stone's house on Euclid Avenue -
Courtesy of the Cleveland Public Library

The Mather Mansion- Courtesy of the Cleveland Public Library

Amasa Stone Chapel on the campus of Case Western Reserve University -
Courtesy of Case Western Reserve University Archives

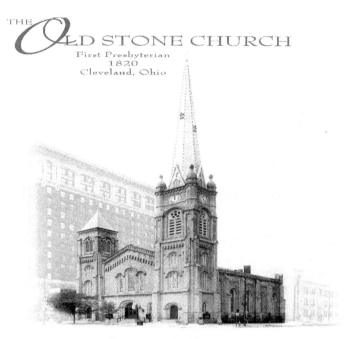

Courtesy of Old Stone Church

The LaFarge windows dedicated to Amasa Stone
at the rear of the Old Stone Church - Public domain

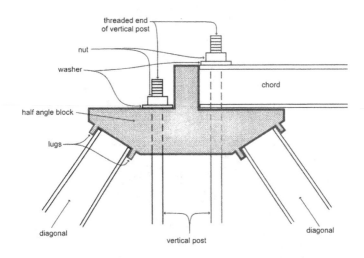

An ideal graphic of an angle block with braces in their vertical position-held in place by lugs - This connection was never implemented - Courtesy of Wikipedia

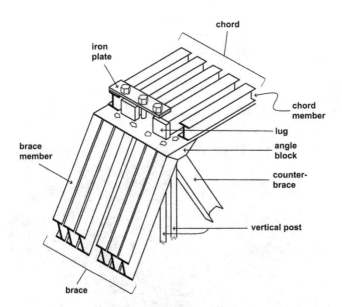

An imagined graphic of an angle iron and the bracing for the Ashtabula Bridge - Courtesy of the Wikimedia Commons Original by Charles MacDonald - Redrawn by Tim Evanson

Leonard Case Jr. - Public domain
Nemesis of Amasa Stone - for whom Case Western Reserve
Univesity is named.

Monument in Chestnut Grove Cemetery, Ashtabula, Ohio, in
memory of those who perished on the evening of December 29, 1876

Grave of Amasa Stone, Jr. - Lake View Cemetery
- Courtesy of Cleveland Public Library

Present day bridge over Ashtabula Creek - Courtesy of Leonard Brown

Chapter 8

The Loss of Lucy and Philip Paul Bliss

Moody and Mass Evangelism

The preeminent fact now remembered about the Ashtabula train disaster is that gospel songwriter and evangelist Philip Paul Bliss was aboard the train and lost his life. However, this knowledge would belong only to church musicians, historians, and a few pastors. Even though they have vaguely and faintly heard that Bliss was killed in a train wreck, hardly any of them would know that it took place in 1876 at Ashtabula, Ohio.

If one subscribes to Malcolm Gladwell's thesis in his book *Outliers* that "success" is largely circumstantial, that it is due to time, place and an innate aptitude in keeping with available opportunities, then Bliss was born into the right family at the right time and place.[1] The single most predominant factor within American post-civil war Protestantism was revivalism, and revivalism was big business. No less than the rise of other industries during the Gilded Age, religion would make use of technology, the innovation of means to increase the yield. Rather than simple word of mouth invitation, or a gradual increase in attendance because of perceived other-worldly phenomena of which newspapers would finally take note, there would now be advanced organizing and marketing no less than for a P.T. Barnum circus. The reason Dwight Moody used Barnum's "Hippodrome" in New York City was more than just the need for a building which would house a gathering of thousands. It would also create the entertainment venue associated with the structure. Moody biographer Bruce Evensen writes of Moody and Barnum:

Both men were Yankee farm boys who had grown up in the Connecticut

---

[1] Malcolm Gladwell. *Outliers* (New York: Little, Brown & Company, 2008).

River valley, without their fathers. It had thrown both of them back on their own resources, which proved considerable. Each became an eager supersalesman — Moody in ministry, Barnum in the big top — but before their names became known, Moody sold shoes and Barnum Bibles. Each had chosen an enterprise that suited his "natural inclination and temperament."[2]

Those who performed in an evangelistic campaign as a soloist or a conductor of a choir, as well as other musical groups whether they be vocal or instrumental, became almost as important and well known as the preacher. Thus, history would retain the names of Moody and Sankey, Sunday and Rhodeheaver, as well as the more recent Graham and Shea.

Before the Civil War, no singer was associated with evangelist Charles Finney or any other well-known preacher. The appeal of gospel singing was part of a much more sweeping change within American religion. As I have written elsewhere, "Moody was the prototype of a new breed of professional evangelists. He gave himself to the 'business of conducting revivals,' Everything was carefully planned in advance, nothing was left to chance."[3] Not that Moody would take shortcuts, or even short circuit God's power. As Bernard Weisberger has written of Moody, "His own schedule during a revival was a fearful thing to contemplate: up before dawn for study, two or three prayer meetings, lectures or training sessions with workers, the large evening meeting, and more often than not, a talk in the inquiry room afterwards."[4]

The Development of a Gospel Song Evangelist

In 1876, no one's "gospel music" star was rising faster than that of Philip Paul Bliss. In temperament, gifts and training, Bliss fit the prevailing theological ethos as exemplified by Moody and the lesser-known Daniel Whittle. Both, as well as every other "successful" evangelist in the latter half of the nineteenth century, lacked formal theological education. When Moody was asked about his theology, he responded he did not have one.

---

[2] Bruce J. Evensen. *God's Man for the Gilded Age: D. L. Moody and the Rise of Modern Mass Evangelism* (Oxford: Oxford University Press, 2003) 93.

[3] Darius Salter. *American Evangelism: Its Theology and Practice* (Grand Rapids: Baker Books, 1996) 101. "business of conducting revivals" borrowed from Winthrop Hudson.

[4] Bernard Weisberger. *They Gathered at the River* (Boston: Little Brown, 1958) 210.

The transcendent, sovereign, fire-breathing God of Jonathan Edwards had been traded in for an immanent loving God who employed captains of life boats, grabbing as many as possible from perishing. God was just waiting to pull them in. Moody "translated the Gospel into the literary style of the contemporary magazines which were deeply tinged with Victorian sentimentality."[5] Moody's services were characterized by spirited singing, fundamentalist interpretation of the Bible, warm vernacular preaching, and a noticeable absence of involvement in social and political issues. The folksy ambience of hearth and home filled Moody's tabernacles, and Bliss and Sankey were just the kind of people to gather around the piano and provide appropriate music.

Philip Paul was born to Isaac and Lydia Bliss July 9, 1838, in Clearfield County, Pennsylvania. John Bliss, Isaac's father, had been a not-so-successful itinerant preacher. He, as well as the rest of his family, loved church music and was constantly singing. When he died in January, 1864, Philip Paul wrote, "Pa Bliss died, the best man I ever knew. He lived in continual communion with his Savior, always happy, always trusting, always singing." Philip also remembered his grandfather as "always a poor man, but early in the morning and after the toil of the day in the evening, sitting on the porch of his humble home, his voice would be heard in song, and I can almost hear him now sing 'Upon the other side, come to that happy land, come, come away.'"[6]

The piety of the John Bliss home was not lost on his son Isaac, who raised his five children in a home of strict religiosity. Philip's wife Lucy remembered that she had been rebuked by his father for "laughing on Sunday." But in spite of being puritanical and a hard scrabble poor farmer, Isaac was a jovial man who like his father was "constantly singing, praying, and reading the scriptures." According to Daniel Whittle, with whom Bliss would associate himself as an evangelist, his father's influence, along with his "mother's daily lessons, and contact with the grand scenery around his home, the mountains, valleys, forests, and streams of which were ever dear to him, made up for the most part, the influences that were brought to bear upon his first ten years of life." Most unique was

---

[5] H. Shelton Smith, et al. *American Christianity: An Historical Interpretation with Representative Documents* Vol. II (New York: Charles Scribner's Son, 1963) 321.

[6] D. W. Whittle, ed. *Memoirs of P. P. Bliss* (New York: A. S. Barnes & Company, 1877) 16.

his genetic musical inheritance. Philip Paul "early developed a passion for music, and would sit and listen with delight to his father's singing, when but a child, and very early sang with him. He would readily catch up a line, and whistle it or play it upon some rude musical instrument of his own manufacture."[7]

But the musical atmosphere of the home did not preempt the daily reminders of poverty due to the father's lack of common-sense ability to earn a living, and the family's peripatetic lifestyle. By the time Philip was ten, the Bliss family had lived in four different places in Pennsylvania and Ohio. The family was so impoverished that the son left home at age 11, and according to Bliss scholar Thomas Corts, "From age eleven to sixteen, his independent existence was disciplined by work on farms, as a cook in logging camps, as a log-cutter, and sawmill laborer, work that yielded as much as nine dollars per month with board, and board was his greatest need."[8]

Bliss never resented his hard lot in life, but rather was grateful for it. In a letter to his mother on February 19, 1871 he wrote, "For whatever of small advantages, small houses, plain living, threadbare, patched clothing, back woods society and unpleasant recollections of my childhood, I have to cherish these precious thoughts - my parents prayed for me, even before I knew the meaning of prayer, and they consecrated me to the Lord and His service."[9] This same attitude and work ethic characterized him when he enrolled in Susquehanna Collegiate Institute in Towanda, Pennsylvania, where he wrote a friend "I'm a kind of chore-boy, but I am not ashamed of it. I saw wood, bring water, sweep rooms at so much a piece, and am resolved to earn every penny I possibly can honorably."[10]

Throughout the toil of his adolescent years, Bliss was developing a handsome physique and visage, which would be noted for the rest of his life. F. W. Root, with whom Bliss worked in Chicago, commented, "If ever a man seemed fashioned by the Divine hand for a special and exalted work, that man was P. P. Bliss. He had a splendid physique, a handsome face, and a dignified striking presence."[11] Fellow evangelist Whittle described him as of "large frame finely proportioned, a frank, open face with

---

[7] Ibid., 17.
[8] Corts, "The Loss of Bliss in Ashtabula," 42.
[9] Corts, 44.
[10] Whittle, 232.
[11] Ibid, 36.

fine large expressive eyes and always bright and cheerful, full of the kindest feeling, wit and good humor, with a devout Christian character, and an
unsullied moral reputation."[12] Another wrote, "His personal appearance
and bearing were such as to attract and win respect and friendship wherever he went. Nature had lavished on him a profusion of charms. Not Saul
or David was more eminent among his fellow men, for fine physique and
manly beauty."[13] In the estimation of another, Bliss's nobility of presence,
"Never failed to produce an impression on the passer-by."[14]

Conversion and Marriage to Lucy Young

Bliss's life was full of providential circumstances and events, none
more so than finding God and finding a wife. When the family lived at Elk
Run, Pennsylvania, Bliss went forward to commit his life to Christ at the
age of 13 in a revival held by a Baptist preacher. He was later baptized by
a Christian minister in a running creek. The Christian Church, also known
as Disciples of Christ, was founded by Thomas and Alexander Campbell,
Bliss and his wife joined the Presbyterian Church immediately after marriage and he would later be the director of music for the First Congregational Church in Chicago. Perhaps his mongrel beginnings portended
Bliss's transcendence over denominationalism and sectarianism, a characteristic of all the populist and popular evangelists of the late nineteenth
century. Bliss did join a local Baptist church after his baptism, but this was
camouflaged to the extent that, "many thought him to be a Methodist."

In 1857, the 19-year-old Bliss attended a musical convention in Rome,
Pennsylvania, where he boarded with the O. F. Young family. He and his
younger sister resided with the family during the winter of 1858, when
he fell in love with Lucy, the oldest daughter. They married June 1, 1859.
For the rest of 1859, Bliss was employed by his father-in-law, "picking
up stones, repairing fences, planting, hoeing, and shelling corn, chopping
logs, mowing hay and oats, and doing other similar chores."[15] But the
Young family had much more to offer than manual labor; all the members
seemed to be musically gifted. Lucy taught her husband to play the organ.

---

[12] Ibid., 21.

[13] Ibid., 301.

[14] Ibid., 307.

[15] Bob J. Neil. *Philip P. Bliss (1838-1876): Gospel Hymn Composer and
Compiler* (unpublished Ph.D. dissertation, New Orleans Baptist Theological
Seminary, 1977) 25.

Lucy's brother O. W. Young wrote several hymns. Bliss's father-in-law bought him a "melodeum" (a reed organ similar to an accordion) and with the instrument and his horse Old Fanny, "he was in business as a professional music teacher."[16]

Lucy was not an unseen partner, remaining in the background of her husband's career. She often accompanied him, singing in a quartet, playing an instrument, or rendering a solo in his evangelistic services. Bliss wrote in a letter to his father-in-law March 5, 1866, "Again thank you for a sweet gentle being, who has been vastly too good for me and one of whom I was in every way so early unworthy, yet, without whose existence life would seem a blank! Ah, She *is* all the world to me."[17] In 1874, Bliss wrote one of his former teachers, "As she, (Mrs. Bliss) is not to read this, I must say she is an extraordinary woman. You don't know many women of such unselfish devotion, sublime faith, and childlike trust. She lives so near the Lord that I ought to be a good man. Humanly speaking my life would have been a failure without her. God bless her."[18]

Beginning of Professional Musical Career

Bliss first came to the attention of the musical world by writing secular songs and singing in a patriotic quartet, "The Yankee Boys," and forty-four secular songs published by Bliss have been identified. Most of them consist of sentimental and nostalgic lyrics similar to his contemporary Stephen Foster, who wrote "My Old Kentucky Home." The following words from the Bliss song "The Heart Makes the Home," are exemplary.

'Tis the heart makes the home,
 Ever brightly to bloom;
Tis the heart, tis the heart makes the home,
Through the dark weary day,

Or in Joy's milder ray;
Tis the heart, tis the heart makes the home,
Though humble and poor,
Is my cot on the moor,
The love-light brightly beams thro' the gloom;

---

[16] Corts, 47.
[17] Neil, 23.
[18] Corts, 63.

Though storms gather round,
Purest joys here are found,
And I turn, fondly turn to my home.[19]

The Yankee Boys brought Bliss into contact with the Chicago music publishing firm of "Root and Cady," who seemingly for a brief time sponsored the quartet. As invitations became infrequent, the group was disbanded, and Bliss was hired by George Root to write articles and songs, as well as conduct musical conventions around the Chicago area. Beginning in 1870, he became choir leader at the First Congregational Church of Chicago and Superintendent of the "Sabbath School," a position he held for over three years. The Chicago sojourn brought him into contact with the two most strategic and career defining people in his life, Dwight Moody and Daniel Whittle.

Encountering Moody and Whittle

Moody had not yet gained the worldwide fame that began with his mass crusades in England. He started his ministry by gathering hundreds of children for the Sunday services at the YMCA in Chicago, a city already experiencing the problems of rapid industrial growth. By 1869 Moody was preaching in the streets, and when the crowd gathered to sufficient size, he invited them inside a theater for a more structured service. Philip and Lucy Bliss, "just happened" to spot him and followed along. Moody picked Bliss out of the crowd as he tried to boisterously and vibrantly help the rather lame musical segment of the service from his seat in the congregation. Though Bliss would never become Moody's official song evangelist, as did Ira Sankey, Bliss often assisted Moody in his Chicago services, and the two of them would be in contact for the rest of Bliss's life.

Bliss did become the song leader for the lesser known, but quite effective evangelist Daniel Whittle. Moody introduced Bliss to Whittle, and the two of them traveled together to a Sunday School convention where Whittle was one of the featured speakers. He often went by the name of Major Whittle because of his service to the Union during the Civil War. Whittle, born in 1840, was severely wounded at the Battle of Vicksburg in the summer of 1863. After having been sent home, he found himself making a patriotic speech in Chicago. He recalled:

I, a boy of twenty-one, was put forward to speak with Bishop Simpson

---

[19] Ibid., 96-97.

on the platform behind me waiting to give his address. I was weak from my wound, and felt foolish at being in such a position. Directly in front of me, in the centre of the hall, a sturdy young man jumped to his feet and cried: "Give him three cheers!" I recognized the face of Mr. Moody as he led the cheering with great earnestness. This manifestation of sympathy nerved me for the few words that followed, and I have often thought it was a specimen of what his courage, faith, and example have been to me all through his life. When I told him sometime afterward of how much good his sympathy had done me that night, and how vividly I remembered his earnest determined look as he led the crowd, I was rewarded by his reply: "I took you into my heart that night and you have been there ever since!"[20]

Whittle later worked for the Elgin Watch Company in Elgin, Illinois, just west of Chicago. As he associated with Moody, he gradually sensed he was "called" into full-time Christian ministry. Moody and Whittle were of a kindred spirit and the former encouraged the latter in evangelistic preaching.

"Hold the Fort"

Whittle often told the story that inspired Bliss's first and best-known gospel song, "Hold the Fort." As W. T. Sherman was "marching to the sea," his supply lines had been cut off in an area 20 miles from Kennesaw Mountain in Georgia. The Union soldiers thought their cause to be lost until they saw a white signal flag from the top of the mountain which was translated, "hold on," "hold out," or "hold fast," but in Whittle's telling, the words became "hold the fort." Immediately upon arriving back home in Chicago, Bliss wrote the lyrics to "Hold the Fort."

Ho, my Comrades, see the signal, waving in the sky! Reinforcements now appearing, victory is nigh.

Refrain: "Hold the fort, for I am coming." Jesus signals still; Wave the answer back to Heaven, "By Thy grace we will."

See the mighty host advancing, Satan leading on; Mighty ones around us falling, courage almost gone!

---

[20] William R. Moody. *D. L. Moody* (New York: Garland Publishing, 1988) 499. Bishop Matthew Simpson was no doubt giving his "War Speech" which he gave multiple times. As I have argued elsewhere, this speech was the most important proclamation for the Union cause during the Civil War. Simpson was a personal friend of Lincoln and preached his funeral at Springfield, IL.

See the glorious banner waving! Hear the trumpet blow!

In our Leader's Name we triumph, over every foe.

Fierce and long the battle rages, but our help is near; Onward comes our great Commander, cheer, my comrades, cheer![21]

Not only did "Hold the Fort," become the most popular "gospel song" in England and America, it also became a colloquial expression. According to *Harper's Magazine*, "Probably no modern hymn has been more widely sung in England and America than the one just named ('Hold the Fort.')."[22] Corts claims, "Not limited to English, 'Hold the Fort' was translated into nearly all the European languages, into Chinese, and the native language of India and it was also heard in Swedish and in Zulu."[23] The song was also used at political rallies, prohibition causes, suffrage conventions, as well as by the labor movement. It was included in a labor songbook as recently as the 1950s.

But as is often the case, poetry failed to represent the facts. There was no fort on Kennesaw Mountain, and Sherman's frontal assault ended in tremendous loss. Sherman biographer Robert L. O'Connell writes:

> The attack came on June 27, aimed at the southern spurs of Kennesaw. It began badly and then degenerated. The temperature climbed to over a hundred. The rebels' fire was accurate and brutal. Eventually, a few blue-coats got within feet of the gray entrenchments, but most had been pinned down or simply shot. By the midafternoon, around three thousand Union soldiers had been killed or wounded, nearly four times Confederate casualties. At this point, Sherman resorted to one of his best military traits: he recognized he was beat and cut his losses.[24]

Evolution and Characteristics of Gospel Music

Though the term "gospel music" was used as early as the seventeenth century, the term "gospel song" or "gospel hymn" was not popularized until two books were edited by Bliss, *Gospel Songs* 1874 and *Gospel Hymns and Sacred Songs* 1875, the latter coedited with Sankey. Music scholars

---

[21] library.timelesstruths.org

[22] Corts, 52.

[23] Ibid.

[24] Robert L. O'Connell. *Fierce Patriot: The Tangled Lives of William Tecumseh Sherman* (New York: Random House, 2015) 143.

Harry Eskew and James Downey state, "Gospel hymns are generally subjective or hortatory, and are often addressed to one's fellow man, and center upon a single theme which emphasized through repetition of individual phrases or refrain following each stanza."[25] I would add that Gospel songs are more experiential, testimonial, individualistic, and less theological and doctrinal than the hymns of Charles Wesley or Isaac Watts.

At least some of the above characteristics are true of Bliss's hymns. Each of the four stanzas of "I Will Sing of my Redeemer" begins with the personal pronoun "I." All five petitions in "Break Thou the Bread of Life" utilize the word "me;" this characteristic was quite intentional on Bliss's part. He was evidently criticized by at least some at the First Congregational Church in Chicago because of the sentimentalism and simplicity of his song selection and thus resigned. "I know not what they'll do. More showy music was demanded and I resigned. I must insist on playing music for devotion and public worship. And I can have no sympathy with operatic or fancy music for Sunday."[26]

However, there were exceptions to the rule; one of the most objective theological and Christological hymns ever written is Bliss's "Hallelujah! What a Savior."

> "Man of Sorrows!" what a name
> For the Son of God, who came
> Ruined sinners to reclaim.
> Hallelujah! What a Savior![27]

In total, 303 sacred songs have been identified as having been written or scored by Bliss, many of them for which he wrote both the words and music. For the song which became the most popular in all of Bliss's hymnody he wrote only the music. His friend Horatio Spafford, according to legend, identified at least the approximate spot in the Atlantic Ocean where his four children had drowned, and penned the words to "It is Well with My Soul." It may have been well with his soul, but in the opinion of some, may not have been well with his mind. He and his wife moved to Jerusalem, waiting for Jesus to return. Who could blame him after having lost all of his children?

---

[25] Quoted in Timothy Kalil, "P. P. Bliss and Late Nineteenth-Century Urban Revivalism" in Corts, 104.

[26] Corts, 55.

[27] *Sing to the Lord* (Kansas City, MO: Lillenas Publishing Company, 1993) 253.

"Whittle and Bliss" that was almost "Moody and Bliss"

While Moody and Sankey held campaigns in the large cities of Boston, New York, Philadelphia, and Chicago, the Whittle/Bliss team went to smaller cities such as Jackson and Kalamazoo, Michigan, Peoria and Rockford, Illinois, and Minneapolis. They then ventured down south to Lexington, Kentucky; Nashville, Tennessee and Mobile, Alabama. Whittle recorded of the last stop, "Mr. Bliss's singing was greatly enjoyed by the colored people, and he is even more moved by their wild and plaintive melodies. When he had been singing the song of his composing, "Father I am Tired" they would be broken down in uncontrollable emotion."[28]

Whittle was a prolific song writer himself, having composed the beloved hymns "I Know Whom I Have Believed" and "Moment by Moment." But, unlike other evangelists we have already referenced, the singing evangelist Bliss would become more popular than the preaching evangelist Whittle. The world would have hardly heard of Whittle outside of his association with Bliss and the resulting *Memoirs of P. P. Bliss* which he edited in 1877, and is still the best single source for Bliss biographical information.

Moody biographer J. C. Pollock suggests that had Moody met Bliss earlier, the latter would have been his song evangelist rather than Sankey. In 1869 Moody lamented, "To think that such a singer as Bliss should have been around here for the last four years and we not known him." In Pollock's assessment, Sankey was Moody's second choice. When Moody received his invitation to London in August of 1872, a trip that would make him famous, he invited Bliss to be his song evangelist. "Bliss regretfully declined because he could not afford, with a young family, to abandon professional engagements. There was no alternative but Sankey the amateur."[29] The truth is, Bliss was more talented and charismatic than Sankey. It may have been Sankey's voice, but it was mostly Bliss's music that was sung at the Bristol campaign. From Liverpool, Moody wrote Whittle regarding Bliss:

> "I am delighted with his music. I do not think he has got his equal on earth at this time & raised up of God to do the work that he is doing, his hymns are already sung around the world, we are using them to do a great work

---

[28] Whittle, 67.

[29] J. C. Pollock. *Moody: A Biographical Portrait of the Pacesetter in Modern Mass Evangelism* (New York: The MacMillan Company, 1963) 102.

in this country. To let you know something of how they are liked we are selling of the penny book 250,000 per month they are going all over the United Kingdom & God is using them as much as he did Wesley's hymns in the days of the Wesleys. Tell Bliss to keep at it, God bless him is my earnest prayer."[30]

The Great American Campaigns

In 1875-1876, Moody held perhaps his greatest American campaigns, the first in Philadelphia, the second in New York, and the third in his home-town of Chicago. He began his Philadelphia effort in November 1875, and finished in January 1876. The department store magnate John Wanamaker secured the largest Philadelphia building, the old Pennsylvania Railroad freight depot, in which 10,000 chairs could be placed. Wanamaker adver-tised Moody's arrival with his own merchandising philosophy, "I owe my success to newspapers."[31] Night after night 12,000 squeezed into the de-pot with thousands turned away at the door. Newspaper man John Farney was not only impressed with the crowds, but even more impressed with the planning committee which had been as "Systematic as an astronomer and that this new evangelism was not content alone to trust some higher power."[32] Nothing more exemplified evangelism having gone mainstream than the night President Grant, his cabinet, and the Justices of the United States Supreme Court showed up.[33]

It was then on to New York City's Hippodrome where P. T. Barnum had staged the "greatest show on earth," and Moody himself would be no less prominent as reported by *Harper's Weekly* March 11, 1876. Barnum rented Moody the huge building, in which almost 15,000 could be packed, for $1,500/week, the cost underwritten by the tycoons Cornelius Vanderbilt and J. Pierpont Morgan. In spite of the numerical success, disillusionment began to creep over Moody as the newspapers day after day touted his celebrity status, and Moody began to fear that he was the center of at-tention rather than Christ. And not all newspaper editors were favorable, including Charles Dana of *The Tribune*, who claimed that the attenders at Moody's meeting were already religious, and that the evangelist was do-ing little to combat the corruption which included "drunkenness, dishon-

---

[30] Ibid., 144.
[31] Evensen, 83.
[32] Ibid., 85.
[33] Ibid., 86.

esty, wrongdoing, crime and evil."[34] Nonetheless, with additional services being added throughout the day, the crowd swelled to 30,000. The event was overwhelming for Moody, and in particular, the police who attempted to get through the door, with one officer almost trampled to death, having to be hospitalized.

For two weeks Moody preached and Sankey sang. With Vanderbilt in the crowd, Moody preached on Matthew 16:19, "Lay not up for yourselves treasures on earth, where moth and rust corrupt." One reporter described Moody's delivery as "a great radiating center," and "sermons seemed thrown off by a centrifugal force of their own, rather than by mental effort."[35] Moody then took the summer off and began his campaign in Chicago, his home town. Beginning the effort in a depressed state, the evangelist began to unravel and, in his desperation convinced Bliss to return to Chicago, which would lead to the song evangelist's death.

The Chicago Campaign and Pressure on Bliss

In the second week of August 1876, the Chicago city fathers including pastors, newspaper men, and wealthy individuals, headed by George Armour and Marshall Field, began to build a tabernacle at the corner of Monroe and Franklin Streets. "Moody Tabernacle" would seat upwards of 8,000 people. Fliers and posters were sent out and put up by the thousands. Some people scoffed that if Moody was sent by God, why were so many posters needed? When the newspapers touted Moody and his workers as a power from "on High," one reader accused the press of making an idol of an "ignoramus."[36] On the first night, 10,000 had crammed into the Tabernacle and another 10,000 were not able to get in. *The Chicago Times* noted that Moody knew "how to talk to customers whether they are looking for dry goods or religion."[37] Moody was a lifelong salesman who lived by the 11th commandment, "Let there be advertising."

While Moody fully resonated with the publicity, he continued to wrestle with his celebrity persona. The events in his personal life leaped out of his control. At the beginning of the campaign his epileptic brother died. He rushed back to Northfield, Massachusetts for Samuel's funeral, who was a professed Unitarian. The evangelist's daughter caught scarlet fever, the

---

[34] Ibid., 113.
[35] Ibid., 117.
[36] Ibid., 139.
[37] Ibid., 145.

leading cause of death among Chicago's children. Moody in spite of the masses, or maybe because of the masses, began to feel very lonely.

During the summer of 1876 Bliss had worked with Moody in eleven services across Massachusetts, New Hampshire, and Vermont. During the Chicago meetings Moody often used Bliss as song evangelist during the day, preserving Sankey's strength and voice for the evening. Bliss terribly missed his children back in Rome with their grandparents, and was determined to leave Chicago for Christmas with his family under the condition that he would return, but according to Whittle's diary Bliss had an "almost unaccountable aversion to the idea."[38] Bliss left Chicago on December 15, hoping he would not return, at least in the near future.

The Last Christmas and Road to Death

The ten days at Rome with Lucy, the boys and Lucy's parents were filled with all the joy and wonder that Christmas could provide a family in love with one another and with God. Bliss played Santa Claus and distributed both homemade and purchased toys to the members of his family, hoping all the time that Moody and Whittle would forget about him. But then the dreaded telegram came with the news that his singing had already been advertised in the newspapers for end of the year services. Lucy and Philip said goodbye to the children on the morning of December 28 with plans to be in Chicago on the evening of the 29th. A carriage transported Philip and Lucy to a train station in Waverly, New York, and the two boarded the train headed for Buffalo, checking their trunk all the way through to Chicago. But the train broke down in Hornellsville (now called Hornell), only about 20 miles above Waverly, where they spent the night in the Osborne Hotel.[39] Perhaps Whittle and Bliss were tired of being the side show for Moody and Sankey. Whittle recorded in his diary "We would rather go almost anywhere else than Chicago and shrank from following Mr. Moody there."[40]

Christian theology revels in repeating narratives when someone, because of whatever reason, just misses getting on a doomed plane or train, but there is no theodicy (justification of the ways of God) sufficient for a Christian cut off in the prime of his ministry, by missing the right train, and getting on the wrong one because of some freak mechanical failure.

---

[38] Ibid., 152.
[39] Corts, 68.
[40] Ibid., 67.

Several did testify to premonitions about boarding the Pacific 5 Express headed to Chicago. They either did not begin the trip or got off at some point before Ashtabula. Mrs. J. M. Richards ended her journey at Rochester because she feared going further. The most remarkable story was told by a Miss Hazen who was to be married in Pittsburg on January 17, 1877, to a man who lived in Chicago. But the trip included spending Christmas at Buffalo and New Year's in Cleveland. Her journey was to begin in Buffalo on the train destined for destruction. She was staying with her aunt who "had a dream of danger by going on that particular train, and so strongly did the dream impress her, that when they reached the depot, 'Auntie' positively refused to go on board the train. This Aunt had been a slave of her parents who raised Miss Hazen when her parents had prematurely died."[41]

On the morning of December 29, Philip Paul and Lucy caught a branch line of the Erie Railroad to Buffalo, New York. There they boarded the westbound Pacific Express, owned by the Lake Shore and Michigan Southern Railroad, a "crack train" scheduled to arrive in Chicago on Saturday morning, December 30. ("Crack" meant that it was the pride of the fleet.) Rolling west along Lake Erie, the train encountered a worsening snow storm. By the time the train arrived in Erie, Pennsylvania, it was two and a half hours late. The train had been scheduled to arrive in Ashtabula, Ohio at 5:15 PM, but it was now about that time when the train left Erie. The snow was so deep that two "pusher" locomotives were employed to enable the train's departure from the Erie station. Seemingly, the Columbia had been added in Dunkirk, New York, about 70 miles east of Erie, Pennsylvania.[42]

The weather continued to deteriorate on a trip that should have taken no more than an hour for the fifty miles between Erie, Pennsylvania and Ashtabula. For two hours the train plowed through snow drifts, a fight between man's technology and nature's elements. What Philip Paul and Lucy were doing during the train's final minutes is left mostly to conjecture. One surviving passenger reported the hymn writer in a parlor car, "with work spread out on his lap."[43]

---

[41] "Warned in Dreams: People Who Were Saved From Ashtabula by Presentiment" from the *Chicago Daily Tribune*, n.d. The Ashtabula County Historical Society scrapbook.

[42] Whittle, 293.

[43] Corts., 68.

Nothing of the husband and wife's bodily remains or earthly posses-sions were ever found, not "a bone nor a button." The most persistent story is that Philip Paul got out of the train alone by crawling through a window, but returned to save his wife and children, all of them being consumed by the engulfing fire. Of course, the two children were not on the train, and even Whittle who arrived on the scene two days after the disaster questioned the narrative. The Bliss story seems to have been confused with Charles Brunner, who perished in the wreck along with his wife and two children.[44] Whittle showed a picture of Bliss to as many of the 60 or so survivors whom he could access, and found only one lady who could identify him.

Funeral Services and Legend

Expecting Philip Paul Bliss's arrival, Moody had said the "revival" will be well served by the return of Whittle and Bliss but just how he planned to utilize them was not entirely clear. The news of the disaster reached Chicago about 9:00 p.m., due to the time difference, two and one-half hours after the wreck. But Whittle and Moody did not learn of the wreck until picking up the Saturday morning newspaper. Whittle was "crushed" and Moody "inconsolable with grief."

Marshall Field donated a thousand yards of "mourning crepe" to wrap around the interior of the Tabernacle, and when Moody tried to speak that night, "it was only in broken sobs and in a low voice that was tremulous with grief."[45] Some noted that prior to the accident Moody had seemed "somber," and Moody himself confessed to a "certain sense of foreboding."[46] Some would identify these emotions as signs, omens, or portents, but the truth is, they work best in hindsight as recollections or reflections, not as prophecy or premonition. No one warned Bliss not to get on a train attempting to fight its way through a blinding snowstorm.

A memorial service was held in the Chicago Tabernacle on January 5, with 10,000 on the inside, and another 10,000 on the outside. A collection of $10,000 was to go to the two Bliss children. In that Cornelius Vanderbilt had passed away on January 4, 1877, it was left to Henry Ward Beecher to juxtapose the lives of the two men:

---

[44] Ibid., 70.

[45] Evensen, 155.

[46] Ibid., 153.

Here is a man almost unknown except as a sweet singer in Israel. His life suddenly ceased; a few papers noted it, he held no such place in the world as Vanderbilt. It is not right that I should compare these men except to say that the latter was the vaster in the lower range of the mind in strength of character, and yet it seems to be that Mr. Bliss has done a far grander work — He opened the door through which 10,000 souls have seen the other life; has made a heaven not of brass, but transparent; has caused joy to blossom, and made the name of Jesus Christ to be effulgent, and brought something of the very spirit of heavenly grace upon the earth; has made little children understand the glory of the Savior's love and service, and his whole life has been put into the work of sanctifying the dispositions of men. . . .[47]

A memorial service was held at Rome, July 10, 1877 to dedicate a 22 ft high cenotaph, erected in honor of the Blisses, at which both Moody and Sankey were present. The front side reads, "Erected by the Sunday Schools of the United States and Great Britain in response to the invitation of D.L. Moody as a monument to Philip P. Bliss, author of 'Hold the Fort' and other gospel songs."[48] The other side reads:

Sacred to the memory of Philip P. Bliss, born in Clearfield County, Pa., July 9, 1838, and Lucy Young Bliss, his wife, born in Rome, Bradford County, Pa., March 14, 1841, who left their home in this village Thursday morning, Dec. 28, 1876, for Chicago, and met their death at Ashtabula, Ohio, Friday evening December 29, 1876, by the falling of a bridge, by which a train of cars was broken into fragments and consumed by fire, some eighty persons being killed. It is believed that the bodies of Mr. and Mrs. Bliss were consumed to ashes, since nothing recognizable as belonging to their earthly tabernacle has ever been discovered.[49]

In 1895, as narrated in the previous chapter, the citizens of Ashtabula erected a 35-foot-high obelisk in Chestnut Grove Cemetery. Around and under it are 19 coffins of unidentified bodies, and an additional four boxes of "body parts and scraps of clothing." Whether the ashes of Philip Paul and Lucy Bliss are contained in the boxes is not known. Along with the 40 or so victims believed to be buried at the location, the monument has inscribed, "Philip Paul Bliss and wife."

---

[47] Neil, 38.

[48] Ibid., 40.

[49] Ibid., 40-42.

The two Bliss sons, Philip Paul and George Goodwin, each had successful careers. Philip Paul born in 1872 was aptly named in that he became a composer of music. "His total compositions included about 200 instructive piano pieces, many operettas, (both words and music) scores of secular cantatas, courses, about 100 songs, a graded course for piano in four volumes, and other works for piano, organ, violin, and cello." He died at the relatively young age of 61 in Oswego, New York leaving behind his widow Lena (Major) who died in 1940. The younger brother, George Goodwin born in 1874, attended Princeton, became an accountant, and died the same year as his brother at the age of 59. Strangely, his wife Mary (Belcher) died the same year as her sister-in-law by marriage, Lena (Major) in 1940. Neither couple had any children, and there are no living descendants. A disappointing discovery for a genealogist or historian.[50]

In 1811, Methodist Episcopal Bishop Francis Asbury wrote to one of his lay preachers, Jacob Gruber "We congregate annually three or four million! Campmeetings! The battle axe and weapon of war, it will break down walls of wickedness, part of hell, superstition, false doctrine."[51] Asbury had plenty of reason for optimism; the Methodist Episcopal Church had grown seven times as fast as the American population between 1800 and 1810. Revivalism in the form of camp meetings, protracted scheduled services at local churches and ultimately city-wide campaigns would provide the growing edge for American evangelism in the nineteenth century. Without formal liturgy, this form of worship incorporated vernacular and populist elements that were the most evangelistically effective, not necessarily the most theological and biblical. Democratic, participatory, affective and emphasizing the altar call as its most critical sacrament, revivalistic worship demanded a compatible and facilitating music. A new market was born and ushered in what might be called the "golden age of gospel hymn writing," defined by the likes of Julia Ward Howe, Fanny Crosby, George Duffield, Charlotte Elliott, Frances Havergal, Ira Sankey and Philip Bliss.

The variety of Bliss's gospel music was far-ranging, but he seemed to have been most adept at writing songs that were invitational and evangelistic such as "Almost Persuaded," "Ho! Everyone Who Thirsteth," "Let the Lower Lights Be Burning," and "Only Believe." In the most comprehensive rhetorical and musical analysis of the work of Philip Paul Bliss,

---

[50] Neil, 29-32.
[51] Salter, *America's Bishop,* 307.

David Smucker states, "Bliss was the only major figure to combine the roles of soloist, teacher, composer, compiler, and poet during the peak years of the gospel music tradition. His training in singing schools, musical conventions, and musical normals during the 1850s, his successful contributions to popular songs in the 1860s and his vital evangelical spirituality prepared Bliss to assume his role as a representative figure."[52]

When Bliss died at the age of 38, he was approaching his most creative years. He had before him half of his life that no doubt would have produced sacred songs which were even more profound and prolific than those written in the few short years that had been given him. We can conclude without doubt that Philip Paul Bliss made the best of the years that God granted.

---

[52] David Smucker. *Philip Paul Bliss and the Musical, Cultural, and Religious Sources of the Gospel Music Tradition in the United States, 1850-1876* (unpublished Ph.D. dissertation, Boston University, 1981) 371.

Chapter 9

The Coroner's Jury Investigation

Uniqueness of the Coroner's Jury

Like most small towns, in particular those lacking a hospital, Ashtabula was without an official coroner. Edward (E.W.) Richards, a Justice of the Peace, would fulfill that role. Other than his occupation within the community, the reason for his appointment remains a mystery. He had no official training as a coroner, nor any formal education in medicine. He immediately chose six persons to serve with him, no women included, to act as a "Coroner's Jury." The other members were H. L. Morrison, T. D. Faulkner, Edward G. Pierce, George W. Dickinson, Henry H. Perry and F. A. Pettibone. All of them were merchants and other business types, well known in the community. According to railroad accident historian R. John Brockmann, this particular arrangement was a common practice for any community experiencing a serious technological tragedy within its vicinity.[1] Richards explained the factors that led to his position:

> I would respectfully call your attention to the circumstances under which I assumed the charge together with an explanation defining my cause of action therein. The office of county coroner was vacant and being called upon the evening of the 30th, by our very respectable citizens to act as such, I accepted the trust, although with many scruples on account of inexperience in such matters as well as also in the progress not informed about it until 10:00 at night and the scene of the accident was not visited until the morning of the 31st about 36 hours after the falling of the bridge. We found the officers of the railroad, the mayor on the ground with a police

---

[1] R. John Brockmann. *Twisted Rails, Sunken Ships: The Rhetoric of Nineteenth Century Steamboat and Railroad Accident Investigation Reports, 1833-1879* (Amityville, New York: Baywood Publishing Company, Inc., 2004).

force. The Express Company was in charge of its own property and the U.S. Mail agent looking after its interest - the R.R. Co. also had the most efficient men there with Supt. Paine at their head and then even afterwards I found myself working in harmony with them all and notwithstanding that the latter was the most abused loss of affliction on record. I will bear my testimony to their manly conduct and worth all through the trying circumstances in which they were placed. Under this state of things and having full confidence in their management, I did not change their program, for they were doing the work better than I could.[2]

After examining steamboat and railroad accident investigations throughout the nineteenth century with careful rhetorical analysis, Brockmann noted the changing social and scientific factors affecting Coroners' reports. Changes in corporate management and technology took place throughout the century. Obviously, the American frontier had left eastern Pennsylvania and ventured some five hundred miles to the west. Nothing is more telling as to the rising sophistication and influence of the old Northwest than the fact that four of the first five United States presidents came from Virginia, and eight of nine through twenty-nine came from Ohio, by far more than any other state.[3] Amasa Stone was representative of both the industrial and political power shifting from the east coast to west of the Allegheny Mountains.

The committee's lack of expertise exposed it to severe criticism. The *Railroad Gazette* condemned the committee members as being "Proverbial for their stupidity, and when the subject to be investigated involves some of the most abstruse facts involved... the average coroner and his jury are as helpless and imbecile as so many children would be in dealing with the facts or drawing conclusions thereof."[4] Yet, the *Gazette* admitted that railroad companies cannot be trusted to investigate their own accidents because they were likely to "remove all evidence including the cause of accidents, usually without reference to finding out the cause for their occurrence." Thus, "The only courts which represent the interests of the public, at least as far as investigation into the causes for the purpose of preventing similar accidents are concerned are coroner's juries..."[5]

---

[2] Copy of original courtesy of David Tobias.

[3] They are William Henry Harrison, 9; Ulysses Grant, 18; Rutherford Hayes, 19; James Garfield, 20; Benjamin Harrison, 23; William McKinley, 25; William Taft, 27; and Warren G. Harding, 29.

[4] Brockmann, 216.

[5] Ibid., 217.

The jury convened on Saturday, December 30, the day immediately after the wreck, and met each day for ten weeks, excluding Sundays. The committee's investigation was so thorough that even the *Gazette* had to admit, "We doubt whether there has ever been a more thorough investigation of a railroad accident in this country... due to the fact that the accident was so interesting to experts in bridge building... and more wonderful than all, it has presented its conclusions in a calm, clear, judicial statement not unworthy of a judge on the bench, and certainly not to be expected of a jury, not to say a coroner's jury."[6]

Railroad Employees

Dan McGuire, engineer of the Socrates, was the first employee of the Lake Shore and Michigan Southern to testify. He recalled leaving Erie about 5 p.m. There were two locomotives on the train. "I could not tell what other cars there were of my own knowledge.... I was of two car lengths of being across the bridge. An average car length should be forty feet, I should think." He then said that he thought "the bridge gave way somewhere near the center or this side of the center." After stopping his locomotive, McGuire jumped off and ran back to the crossing and ascertained that the whole train was off, and down with the bridge. "I went down where Engineer Folsom lay or sat up next to the abutment..." By that time, "There was a fire in the head car, blazing up pretty light, but I could not tell how far it had advanced on the car; there was a fire in another car ahead but it had not broken out through the top." McGuire further stated that he remained "until all of the persons were taken out alive; when I left the wreck was all in aflame, but I could not say how near it was to being all consumed."[7] The most critical part of his testimony was his perception that no part of the train came off the tracks while the Socrates was on the bridge.

> I heard the bridge snap, as I supposed; I pulled my throttle out and looked back and saw the engine Columbia sink; then the Columbia's draw bar broke; my engine ran up maybe 150 feet from the abutment before I got her stopped, before hearing that cracking which I spoke of, I think there was no part of the train off the track. I did not feel any jar or anything

---

[6] Ibid., 220.

[7] "The Ashtabula Accident," unidentified newspaper, courtesy of Carrie Wimer, Genealogy/Local History Archivist at the Ashtabula, Ohio Public Library.

which would indicate that any car or any part of a car was off the track.[8]

On Friday, January 26, James Doran, Track Master of the Lake Shore Road, was called to the stand. He testified that the bridge was made up of a double track, but only one of them was used until "The whole road was double tracked which was about four or five years ago." He examined the track east of the bridge the morning of December 30 and was confident the train had not run off the track.[9]

George Reid, Supervisor of bridges for the Lake Shore, stated that he had inspected the bridge in September and believed that "no defect" could have escaped his attention. The bridge in his estimation was strong, having never needed an extraordinary repair, though he had discovered "loose bolts twice in inspecting the bridge." A.L. Rogers, who had supervised the assembling and raising of the bridge eleven years earlier, claimed that he had inspected the bridge in October and "found it in good condition."[10] As Roger's present job description was unclear, but seemingly he was still employed as a carpenter for the Lake Shore.

Robert McIntyre identified himself as having been a conductor and brakeman for the Lake Shore Road for six years. Looking out of the window of his house, he saw the head lamp of the Socrates and thought the train was stuck in the snow. Then he heard someone say the train had run off the bridge. After going down the steps to the bottom of the ravine, he went to the north side of the train. "When I first went down there, I got on the tender of the locomotive that was down, and from there I got over

---

[8] Ibid.

[9] "The Disaster! What Has Been Said and Done at the Coroner's Inquest," *Ashtabula Weekly Telegraph* (February 2, 1877).

[10] Both the Reid and Rogers testimonies are found in the Leverich album. In 1877, Gabriel Leverich made a presentation to the American Society of Civil Engineers on the Ashtabula Train Disaster. For his presentation he prepared an album of newspaper clippings. Some of his sources are poorly identified or not at all identified but no less authentic. The pages are not numbered, and I have utilized articles that are impossible or difficult to access elsewhere. The album is in the possession of Linda Hall Library, Kansas City, Missouri. The Linda Hall Library received the holdings of the Engineering Society Library in New York City in 1995. The holdings filled 18,000 linear feet of shelves, making the Linda Hall Library one of the "foremost engineering libraries" in the world. Even before the acquisition, it was "the largest privately funded science and technology library in the United States." See Bruce Bradley, "A Library of First Resort for Science, Engineering, and Technology: The Linda Hall Library. "*Proceedings of the IATUL Conferences.* Paper 13. https://docs.lib.purdue.edu/latul/1995/papers/13.

to a baggage or express car, I think. This first car was partly on its side." McIntyre then walked along on the lower side. "I know it was quite steep as I had to hang on to the upper edge to keep from sliding off. I saw this Mr. Reid trying to get out a man whom I thought to be the conductor; and I saw a lady on the end of one of the cars which had caught fire, with one of the timbers lying across her below the knee and I tried to get her out but I could not."[11] He would not be the only person to meet with futility that evening.

On January 20, Albert Congdon identified himself as the "Master Mechanist" of the Lake Shore when the bridge was built. He oversaw the fabrication of all the parts. The I-beams were rolled at the Cleveland Rolling Mill and the angle blocks were cast by the Lake Shore shop foundry in Cleveland. It is assumed that all parts would have passed Congdon's inspection before being shipped to Ashtabula, but he did not explicitly make that claim. He did make three confessions: he did not have sufficient material to make the bridge as Tomlinson had designed it, and he did not designate where parts (braces, chords, angle blocks, tension rods, etc.) were to be placed before shipping them. Also, he did not calculate (did he know how?) the "strength of tension of compression members." He left the impression that he said little to nothing to Tomlinson about these issues as he did not consider himself a "competent bridge man."[12]

Upon the parts of the bridge arriving at Ashtabula, Rogers informed Congdon that Stone "had given him orders to erect it, but he did not know how. I asked him why he did not go and tell Mr. Stone so, and he said he did not like to. I then told him as much as I knew."[13] This causes us to ask, was Rogers intimidated by Stone, or simply afraid he would lose his job?

Charles Paine's testimony was perfunctory. He did say, "Mr. Collins never expressed any distrust of the safety of the bridge to him, and, on the contrary, said to him that its only fault in his opinion, was its great surplus of strength."[14] (Collins would have been more correct if he had said "great surplus of materials.") He said nothing of his own knowledge of the bridge, or even his impressions after having arrived on the scene a few hours after the wreck. He presented the jurors with a list of 72 adults

---

[11] Unidentified newspaper article courtesy of Carrie Wimer.

[12] "Cleveland, Ohio, January 21," *New York Tribune* (January 22, 1877) Leverich album.

[13] Ibid.

[14] Ibid.

and 8 children supposed to have been lost, and 65 identified survivors. He further said, "This list includes the names of those known to have been on the train, both passengers and employees, many of whom have not been identified by any articles found among the remains." The conductor listed "128 adult passengers, 6 train hands, 5 sleeping car hands, 3 express men, 2 baggage men, and 1 newspaper boy." He disavowed having given "an order not to put water on the fire, and he knew of no official having given such an order."[15]

The following from George Carpenter, who was then employed by the King Iron Bridge Company in Cleveland, recalls the actual raising of the bridge. At that time, he worked for the Lake Shore shop in Cleveland and remembered that when the braces were sent to Ashtabula, they had no marks on them to show where they belonged. He also remembered putting the webs of the diagonal I-beams horizontal as Tomlinson had designed them. "In erecting the bridge, we experienced considerable trouble in the top chord, the members of which were too long. I applied to Mr. Congdon to know what I should do and he came down here with Mr. Stone. They did not give me any advice while they were here but afterwards told me to send the angle blocks to Cleveland, to have the lugs planed off. (These were the top lugs on the angle blocks and were not planed off but planed down or thinned.) This was done and we had no further trouble."[16] He then recalled after the bridge was erected, Mr. Stone inspected it, and told him that it was well done.

Congdon was not as confident as Stone, and told Carpenter the main braces would have to be changed or the "damned thing would go in the river." When the braces were turned from horizontal to vertical, the lugs on the angle blocks had to be chipped off or planed down. (These were the lugs on the bottom of the angle blocks as explained below.) Carpenter, in his testimony, condemned the bridge. "I did not consider this system of lateral bracing in this bridge good for anything. I have been a locomotive engineer...I had no practical experience in bridge building until I worked on this one. I never knew of their ever building any other wrought iron bridge at the Lake Shore shops."[17] (Lugs were protrusions from the angle

---

[15] *New York Tribune* (February 7, 1877) Leverich album.

[16] "What Has Been Said and Done at the Coroner's Inquest," *Ashtabula Weekly Telegraph* (February 9, 1877).

[17] Ibid.

blocks to hold attached braces in place and should not be confused with the lug nuts on an automobile wheel.)

There were two sets of lugs on the angle blocks; the top lugs held the I-beams for the chords in place. Or to put it another way, the I-beams buttressed up against the lugs which were 6 inches high, 3 inches wide, and 1 11/16 inches thick.[18] The lugs on the bottoms of the angle blocks were positioned to receive the compression braces horizontally. When the beams were inserted between the two lugs, one of the lugs was on the top of the web and one on the bottom. They were of no use when the compression braces were turned to the vertical position. The angle blocks were returned to the Lake Shore shop and the lugs on the bottom of the angle blocks completely sheared off. Thus, the compression braces were held in place only by friction, provided by the weight of the bridge and the strength and tightness of the tension rods. Vibration of the bridge when a train passed over it, caused shifting of the braces. This would be the primary problem of the bridge.

The Fire Department Controversy

The jury swore in G.A. Knapp, Chief of the Ashtabula Fire Department, who was questioned as to why no water was placed on the fire. The questioning would address the most controversial decision made in the minutes immediately following the wreck, and his conflicting testimony would do nothing to render comprehension and clarify the decision, if not reveal outright deceit and incompetence. Knapp testified that the Lake Shore station agent, A. A. Strong, ordered, "We don't want water, but aid for the wounded."[19] After making that point, Knapp became even more confused and thus confusing; he did not know that there was any valve called a fireplug at the pumphouse. He found the door to the pumphouse locked; he began to stretch the hose down the hill toward the burning train. Who gave orders to bring the "steamer" (a small engine on two wheels pulled by two horses) down to the river is not clear, but nonetheless, it was standing next to the pumphouse but never produced a drop of water.[20] In hindsight, Knapp believed the fire could have been put out, and bodies saved for recognition by attaching a hose from the Lake Erie engine house

---

[18] I calculated the dimensions from information on page 81, figure 5, in Charles MacDonald's report discussed in Chapter 10.

[19] Unidentified newspaper article courtesy of Carrie Wimer.

[20] See "The Fire and Water Controversy" in Corts, 173-177.

onto the fireplug early in the evening, or by using the steamer at a later hour.

The jury continued to investigate why water was not put on the fire. On Monday, February 5, Harvey Tilden, hydraulic engineer in charge of all the water works for Ashtabula, testified the main pump of the firehouse was the best in the business for the city's fire hydrants. One was located at the pumphouse at the west end of the bridge. "No hose was provided for these hydrants, because they had bad luck with several kinds of hose and were testing hose in view of supplying the whole road at the time of the accident."[21] In other words, there was no hose which fit the hydrant at the west end of the bridge.

Tilden's testimony was verified by Knapp, who when he was recalled to the stand, back-pedaled from his previous statement that the central office in Cleveland had ordered that no water should be put on the fire. He claimed in his previous testimony he had tried the available hose and it fit; he now said, "There was no fire hose at or in the pumphouse, and that the hose upon which he tried to cap was a section of rubber hose left by the Fire Department for trial and did not fit either the cap or the hose in use by the Fire Department at Ashtabula."[22] James A. Manning, foreman of the fire engine Lake Erie also had to reverse himself by saying that

> There were fire hose at the pump house fitting the hydrant, and also that he turned off the cap of the fire-plug on the hose at the fire engine house and found they would not fit, and was afterwards obliged to admit that they did so fit, desires to say in explanation with the approval of the jury that there was no fire hose at or in the pump-house, and that the hose upon which he tried the cap was a section of rubber hose left with the Fire Department for trial, and which did not fit either the cap or the hose in use by the Fire Department at Ashtabula.[23]

Mayor Hepburn told the jury he had requested, "A road be made to the pump house," presumably, from the stairs that went down to the river bed on the west end of the bridge. It was his opinion "if water had been thrown when he first got there, it would have scalded to death any who were still

---

[21] "What Has Been Said and Done at the Coroner's Inquest," *Ashtabula Weekly Telegraph* (February 9, 1877).

[22] Ibid.

[23] Ibid.

alive in the wreck and save no lives…"[24] He then disavowed giving orders "concerning throwing of water that night, and knew nothing of any orders given that water should not be thrown."[25] Not only was Hepburn mayor of Ashtabula, he was employed as Assistant Civil Engineer of the Franklin Division of the Lake Shore Railroad. Serving in both positions may have compromised Hepburn's testimony. Being desirous that his name be cleared from this controversy, he ordered Knapp, who served under his appointment, to write a letter denying that he gave orders not to put water on the fire. Whether Hepburn dictated the following letter or Knapp wrote it is not known.

Ashtabula, Jan. 1st, 1877

This is to certify that I received no orders of any kind, either directly or indirectly, from H. P. Hepburn, Mayor of Ashtabula, at the burning of the wreck of train No. 5, or Pacific Express, which was precipitated with the iron bridge into Ashtabula River on the evening of the 29th of December, 1876; nor did I hold any conversation with him of any kind that evening; nor have I any knowledge of his giving any orders to any foreman of the different fire companies that evening.

G. A. Knapp

Chief Engineer of Fire Dep't.[26]

The bottom line in all of the above is that the Ashtabula Fire Chief, the Ashtabula Fire Department and the citizens of Ashtabula were totally unprepared for a fire at the Ashtabula River Bridge. Knapp had the reputation within the community of being incompetent and turning to the bottle too often. On that night he was a man not in command of himself or anyone else. As the Ashtabula Fire Chief, he was anything but a leader. The jurors would later conclude that the first people to arrive at the scene of the disaster "who seemed to have been so overwhelmed by the fearful calamity that they lost all presence of mind and failed to use the means at hand, consisting of the steam pump in the pumping house and the fire engine *Lake*

---

[24] "What Has Been Said and Done at the Coroner's Inquest," *Ashtabula Weekly Telegraph* (January 25, 1877).

[25] Ibid.

[26] "The Disaster! Unhappy Opening of the New Year at Ashtabula Particulars, Full Particulars, Incidents, Acc., of the Accident, Official Report of the Coroner's Inquest," Unidentified newspaper. Courtesy of Cary Wimer.

Erie and its hose..."[27] But as we discovered in Chapter 7, Charles Leek perceived this assessment as too harsh and defended the fire department as well as the Ashtabula citizens.

Tomlinson Testimony

Joseph Tomlinson's part in the catastrophe is so unique that we need a brief biographical sketch. In that he disagreed with Stone and in the eyes of many proved to be right, Tomlinson is often painted as the hero and Stone the villain. But a closer look will demonstrate that the matter is not that simple.

Tomlinson was born in Ruskinton, Lincolnshire, England, June 26, 1816, and passed away on May 10, 1906, in Rapid City, Iowa. He was one of fourteen children, four girls born before him. The arrival of a boy made the family so happy that the church bells pealed throughout the village. As a boy he was considered dull, and his parents removed him from school and apprenticed him to a cabinet maker. During his seven years of apprenticeship, he demonstrated himself adept at mathematics, draftsmanship, and just about anything that required mechanical attention. At some point he attended a "mechanics institute," where or for how long is unclear. At the age of 24, Tomlinson emigrated to America, landing in New Milford, Connecticut. Here he experienced a portent of the Ashtabula Disaster. After inspecting a railroad bridge, he remarked to his pastor, Noah Porter, who eventually served as President of Yale, that upon inspection of a railway bridge he noticed that miscalculations were made regarding its resistance and weight. The bridge later collapsed.

Tomlinson himself experienced a bridge failure. He utilized Nova Scotia iron on a bridge over the River St. John at Grand Falls, New Brunswick. After great fanfare and celebration of its completion, the bridge collapsed the next day. However, he built many bridges that lasted for decades, if not a century. The first bridge he ever built was at Pittsfield, Connecticut in 1844, and was still standing at the time of his death, over sixty years later. He showed remarkable powers of concentration and endurance by

---

[27] Brockmann, 224.

working three days without sleep to remedy a "quicksand" problem at Whitehall, Connecticut.[28]

After the fallout with Stone, Tomlinson moved to Kansas City in 1867 and designed the first bridge ever built over the Missouri River. The bridge was built and located in Kansas City, yet it was named the Hannibal Bridge. The Hannibal Bridge is a case in point where fact needs to be separated from fiction. Tomlinson was not the designer of the Hannibal Bridge; the bridge was designed and built by Octave Chanute, its chief engineer. In the 200 plus page book on the bridge, Tomlinson is mentioned four times, and Chanute referred to 84 times. Tomlinson was first hired as a draftsman for the bridge, very similar to the job which he held for Stone: "At the same time a set of plans for the fixed spans was prepared by Mr. Tomlinson under the direction of the chief engineer (Chanute) which were to be adapted only if on a fair examination they were found to be preferable to those submitted by outside parties."[29] We presume that Tomlinson beat out the competition. After his drafting work, Tomlinson was superintendent of the day shift,[30] and was afterwards appointed superintendent of the night shift.[31] On October 1, 1867, Tomlinson was hired as "the superintendent of the superstructure." He oversaw the erection of the bridge because he knew better than anyone else where the exact parts were supposed to go, and how they were to be attached. This no doubt would have been true for the Ashtabula Bridge if he had remained an employee of Amasa Stone.

After the Kansas City stint, Tomlinson moved back to Canada where he supervised the construction of lighthouses. He pioneered the use of petroleum, replacing the seal oil that had been used for the main lamps of lighthouses. Shortly before retirement, he was sent by the Canadian government to build a bridge across the Fraser River in England.[32]

By all accounts, Tomlinson was a man of imposing physique and commanding appearance, standing about six feet. His face was handsome, graced by a neatly trimmed beard and a towering bald forehead. He en-

---

[28] William P. Anderson and J. A. L. Waddell "Memoir of Joseph Tomlinson," *Transactions of the Canadian Society of Engineers* Vol. 19, 321-325. Additional information taken from "Home of a Great Bridge Builder," *The Cedar Rapids Weekly Gazette* (March 11, 1963).

[29] Octave Chanute and George Morrison. *The Kansas City Bridge* (New York: De Van Nostrum, 1876) 116.

[30] Ibid., 87.

[31] Ibid., 91.

[32] *The Cedar Rapids Weekly Gazette.*

joyed a long retirement of some twenty years as a gentleman farmer of 800 acres just outside of Cedar Rapids, Iowa. Cleveland, rather than being a bitter memory, must have inured itself to him because he and his second wife are buried in Woodland Cemetery in Cleveland. Fellow engineers stated of him: "Mr. Tomlinson's principal characteristics as an engineer were extreme thoroughness in detail, neatness of execution, great persistency and absolute honesty. He did personally a great deal of drafting on his various designs, and very few draftsmen have ever equaled him in neatness of execution and fineness of work."[33]

Tomlinson first appeared before the Coroner's Jury on January 11, testifying that he had made the drawings for the iron Howe truss bridge over Ashtabula Creek, but "never approved of a wrought iron Howe truss bridge over a large span." He further claimed that the main braces were not made as large as originally designed, and it was his intention they should be strengthened.[34] Comparing Tomlinson's testimony before the Coroner's Jury and the Ohio Legislature, it seems he was referring to diagonal bracing on the sides of the bridge, consisting of compression braces and counter braces crisscrossing each other. There were also two other kinds of bracing on the bridge: lateral bracing was on both the top and bottom of the bridge connecting the top chords on each side and bottom chords on each side. In other words, the lateral bracing connected the top chords with one another and the bottom chords to each other. Sway bracing connected the top chord on one side with the bottom chord on the other side, and the bottom chord on one side with the top chord on the other side. All these sets of bracing crisscrossed with one another and should have been securely fastened but were not, which was another primary fallacy of the bridge. But Tomlinson did not make this clear.

Tomlinson informed the jury, "Forging and the fitting of the ironwork was done in the Lake Shore shops in Cleveland according to his plans and pattern."[35] (Again, there seems to be a contradiction in that he protested the use of iron for the bridge, but he oversaw the prefabrication of its parts. But as we will discover later in the Legislature hearing, he did not protest until it was too late.) He then claimed Stone had altered his plans so that the braces were defective, but did not say exactly how Stone had changed

---

[33] Anderson and Waddell.

[34] "Ashtabula, January 11," *New York Times* (January 12, 1877) in Leverich album.

[35] *Ashtabula Weekly Telegraph*. No date, Courtesy of Carrie Wimer.

his design. When asked exactly what the differences were between Stone and himself, he could not be specific. "This is very hard to answer. My opinion was that he did not like to have anyone around him who had an opinion of his own."[36] Tomlinson recalled that at this point in the conversation, his relationship to the company (the Lake Shore) was "severed." Tomlinson went on to say he did not think a better set of rods or chords were ever put into a bridge.

After examining the wreckage on January 15, Tomlinson told the Coroner's Jury some of the braces had slipped out of place before they were last painted, some of them as much as three inches.[37] (He was speaking of the diagonal bracing.) He said the braces should have been fastened to the angle blocks, and if so, the wreck could not have happened. He did emphasize that the braces needed to be strengthened, which Stone eventually did by adding two additional braces for each panel. One newspaper concluded that Tomlinson had suggested there was improper inspection of the bridge, but Tomlinson did not exactly make this accusation, at least at the Coroner's hearing.

The closest we can come to an exact wording was given by *The Cleveland Herald* on January 16, 1877.

> He entered the employment of Mr. Stone for the sole purpose of making the drawings for this bridge. He objected to building a Howe truss bridge in iron. He did not at the time mention this fact to Mr. Stone, his reason being that he considered Mr. Stone's judgment superior to his own. He thinks a good bridge can be made of iron, but it is unnecessarily heavy and much more expensive. He thinks Mr. Stone intended to have a first-class bridge without regard to expense. The only defect he knew of at the time it was being built was that the rolled beams for braces and chords were not made as large as the model. He mentioned this fact to Mr. Stone, who, on several occasions, would become impatient and tell him to attend to his own business. He does not think that this defect could have caused the accident.[38]

The problem with Tomlinson's testimony was that many of his insights came in hindsight. He continued to give specifications to Albert Congdon

---

[36] Unidentified newspaper article, Courtesy of Carrie Wimer.

[37] "Cleveland, Ohio, January 16," *New York Tribune* (January 17, 1877) in Leverich album.

[38] "Ashtabula," *The Cleveland Herald* (January 16, 1877). From Ashtabula Historical Society scrapbook.

for manufacturing the parts, but yet believed that iron was too heavy for the bridge's design. He did not let this be known to Stone. According to Congdon, Tomlinson designed the angle blocks. Tomlinson was supposed to have inspected the angle blocks before they were shipped. Whether he did or not is unknown. The angle blocks were tailor made for the diagonal braces to be laterally attached to them. This was done by casting small lugs or wedges on the bottom of the angle blocks. In order to make room for turning the braces to the vertical position and add additional bracing as we have already noted, these lugs were sheared off. Obviously, this arrangement was a compromise of the stability of the braces and thus the bridge. No wonder that Tomlinson discovered while examining the wrecked bridge that six braces were out of place.

Expert Witnesses

On Tuesday January 30, Job Abbott, a civil engineer and Vice President of the Wrought Iron Bridge Company of Canton, Ohio was sworn in. After examining the wreckage, he believed the tension members to be of ample strength, in fact heavier than needed. (The tension members were the 2-inch round rods connecting the bottom chords with the top chords at each angle block.) By the direction the bridge had fallen, the I-beams forming the main braces of the second and third panel from the west end in the south base, was the place where the bridge failed. (This was correct according to MacDonald and Gasparini.) He then made three contradictory statements. First, "The bridge had become gradually weakened by use, the form of construction being such that many of the compression members were strained beyond their safe working capacity." Second, "The iron bridge should not deteriorate by use, provided it not be strained above a safe working capacity." Third, "The devolvement (increasing weight) of the locomotive might have caused the destruction of the bridge, but does not think that such was the case."[39] He too, made suggestions as to how

---

[39] Unidentified newspaper article, Courtesy of Carrie Wimer.

the bridge may have failed, but backed off from asserting that they were sufficient cause for the bridge's collapse.[40]

On February 6, the jury examined B. Morton, a New Haven, Connecticut civil engineer, who only provided more obfuscation. The braces should have been fastened to one another where they crossed. Yet, according to Morton, "The factor of safety was amply sufficient." By examining drawings of the bridge, he observed the lateral braces of the bridge were not attached to the chords at the proper places, but "did not think that the bridge should have deteriorated with use by proper care." And while slight displacement of the main braces would weaken the bridge, "Such displacement should have been noticed by the inspector and remedied immediately." Morton had intentionally, or unintentionally, placed two huge question marks in his testimony: Was the bridge given "proper care?" And, was it carefully inspected, at least to the extent that the inspector would notice this "displacement?"[41]

On February 10, Charles Hilton, engineer of the New York Central and Hudson River Railroads was called to the stand. His hedging and being of little help may have been due to his position within the New York Central. For every defect he pointed out, he followed with the disclaimer that the mistake would not materially affect the bridge. The lower braces were not connected at the "best points," but that was not a problem because of the stiffness of the lower chords. (We assume that Hilton meant the bottom lateral braces, which were attached to every other angle block. The X of the lateral braces covered two panels rather than one, making it insufficient.) Though the braces were a little off from the angle blocks because additional braces had been added he "would not consider those defects would constitute a fatal element of weakness in the bridge." The bottom line was that Hilton would not give a "theory concerning the failure, because the greatest strains were borne by members different from those towards which indications point as ones which did fail." In other words, the braces that failed were under less strain than the ones that did not, but

---

[40] According to David Simmons, Abbott's "company was a national manufacturer of wrought iron bridges, among the nation's largest at the time. He therefore had an overwhelming vested interest in advocating the strength and reliability of iron bridges at a time when the nation was still in the process of shifting away from wooden bridges." Email from David Simmons to Darius Salter, February 1, 2022.

[41] *New York Tribune* (January 31, 1877). Leverich album.

Hilton gave no explanation as to why. It was difficult to pin liability on anyone or identify any specific defect from Hilton's deposition.[42]

E. N. Beebout, civil engineer of the Canton Wrought Iron Bridge Company, condemned the bridge by citing "the system of lower lateral bracing employed as insufficient," and thought "this bridge was not as strong to resist lateral motion as it would have been had the track been on the lower chord instead of the upper one." He then referred to the problematic construction of the bridge as had been reported by A. L. Rogers "that the upper chords of the bridge were too long and they were shortened; the bridge when put up, immediately settled down almost a straight line (losing its camber); it was then raised up twice and wedges put in it, and at length it stood firm and strong."[43] He did not give a specific reason as to why the bridge ceased to be firm and strong causing its failure.

On January 12, A. Gottlieb was sworn in and identified himself as an engineer of the Keystone Bridge Company of Pittsburgh. He proceeded to argue that the dead weight of the bridge should be no more than 2,200 lbs. per linear foot, when the actual weight of the bridge was from 3,000 to 3,200 lbs. per linear foot. How Gottlieb arrived at these amounts from a strain sheet, were as unclear to the jurors as they are to us. He estimated that the locomotives with their tenders were 55 tons each; in other words, combined with the dead weight of the bridge, they were too heavy, but this was not the conclusion of any other engineer who examined the bridge.[44]

Albert Howland, a civil engineer from Boston, was at this time 31 years old and had graduated in 1871 from the Massachusetts Institute of Technology having studied mathematics and civil engineering. His mathematical ability was so exemplary that he was asked to teach mathematics at MIT, but did not accept the offer. A confirmed bachelor all of his life, he was "regarded as an encyclopedic source of accurate knowledge." Seem-

---

[42] "What Has Been Said and Done at the Coroner's Inquest," *Ashtabula Weekly Telegraph* (February 16, 1877).

[43] "Cleveland, Ohio, January 25, 1877," *New York Tribune* (January 26, 1877). Leverich album. David Simmons claims that E. N. Beebout and Job Abbot are the same persons.

[44] "Ashtabula: Continuation of the Coroner's Inquest," *The Cleveland Herald*, (January 13, 1877) Ashtabula Historical Society scrapbook.

ingly, the examination of the Ashtabula Bridge disaster was the high point of his career.[45]

Howland and A. Gotlieb were the only engineers to appear before the Coroner's Jury in Ashtabula and to give a report to the Joint Legislative Committee in Columbus. Since we do not have an adequate newspaper report from Howland's testimony in Ashtabula and assuming that he would not have contradicted himself or given different information before the two committees as they concurrently occurred, we need to access the 8-page report given in Columbus. After having spent two weeks examining the bridge, he gave the most adequate description of the bridge and its setting found anywhere.

Howland stated he saw an actual plan of the bridge made by Tomlinson, but "in several important particulars, the plan and structure do not correspond,"[46] but was not specific on the points of difference. The angle blocks were quite substantial, "33 and 3/4s inches long and 25 inches wide at base where they bear against the chord, the metal being ¾ inches thick."[47] He described the diagonal braces as "6 inch I-beams, 21 feet 6 inches long with planed ends, and placed with web in plane of truss."[48] More critically Howland stated, "The only projections on the face are lugs (generally in the form of an angle to receive the corner of the braces), one-half inch thick, one-half inch deep, and one and one-half inch long, each way; a large part of them have been entirely, and others partly chipped off."[49] They were the lugs on the bottom of the angle blocks to which we have already referred.

Howland observed that there was little correspondence between the "web" and the weight of a beam. (The web is the space or distance between the flanges of the I-beams.) The struts were insufficient for the compressing members and almost in every case, strained above the safe limit, and in very many cases, to many times that limit.[50] The braces were not bolted to the angle blocks; thus, there was nothing but friction to keep them in place.

---

[45] "Albert H. Howland," *Journal of the Boston Society of Engineers* Vol. 2 (Boston: 1915) 131-132.

[46] Howland's testimony from *"Report of the Joint Committee Concerning the Ashtabula Bridge Disaster Under Joint Resolution of the General Assembly"* (Columbus: Nevins and Myers, State Printers, 1877).

[47] Ibid., 37.

[48] Ibid., 37-38.

[49] Ibid., 37.

[50] Ibid., 44.

"The bearing of the braces on the angle blocks showed displacement in the planes of the trusses, varying in amount from half an inch to one- and one-half inches in twenty to thirty cases – one beam was half off the angle block. In some cases, there was lateral displacement of about one inch."[51]

In Howland's opinion, the train should have not been on the top chord because "the panels were long and the beams were shallow and had no surplus metal."[52] (There were 14 panels, and no one else suggested that the panels were too long.) A strong wind was more likely to sway a top chord than a bottom chord. In that the floor beams were made out of wood, Howland believed "the elasticity of wood and iron are so vastly different that they are not suited to resist conjointly the same force acting upon them in the same manner....How this combination of wood and iron would act in transferring a lateral force to the abutments, would be entirely a matter of conjecture, but, that little should be risked on the chance in its acting harmoniously and efficiently, seems plain."[53] This was a risk Amasa Stone had taken, but possibly never occurred to him.

The following was critical in Howland's argument and straightforward enough for a layman to understand:

> The members of the braces are plain I-beams, without a hole for a bolt or rivet, *fastened* to nothing; not connected to each other except at the inter-section of the mains and counters by a yoke that could hardly fail to wear loose; depending upon each other for lateral support; that neither should be required to give; with nothing to prevent them from bending sidewise par-allel to each other. Very slight inequalities in length, such as are indicated by the angle blocks, would give one of the beams a tendency to bend in one direction or the other by the uneven distribution of the strains which would result. This same effect would be produced by the unevenness of the bear-ing faces of the angle blocks, which were partly chipped instead of planed, a defect in itself. Probably half of the braces had their bearing area reduced by chipping the corners of the flanges where the vertical rod interfered; in many the half of one flange was cut away; I noticed one whose bearing area was reduced to one flange and half the web....[54]

Howland recommended in conclusion:

In view of the appalling nature of this disaster, it seems fitting to state here

---

[51] Ibid., 45.
[52] Ibid.
[53] Ibid., 46.
[54] Ibid., 44-45.

that I think there is urgent need of thorough and complete experiments on the forms and qualities of iron now used in important structures, especially for members subject to compression, such experiments as have already been inaugurated with the encouragement, to some extent, of the National government.[55]

Due Diligence

In spite of being unfairly maligned by the press and those who thought themselves more sophisticated and informed than the bumpkins in Ashtabula, the Coroner's Jury had given due diligence to the task. Without technical knowledge, they had utilized and imported expert witnesses and, on the scene, eye witnesses to provide a comprehensive and coherent understanding of the events which had taken place on December 29, 1876. The immediacy of the investigation, both in time and proximity, rendered a judgement that was both fair and honest without being overly accusatory, and yet, not retreating from holding all persons accountable who played any part in the drama that had quickly begun, and for all practical purposes, had been over in about three hours.

It was difficult to be objective about an event which had literally engulfed every inhabitant of Ashtabula. To see clearly through the confusion, grief, horror, fear, and just about every other emotion known to human existence was a daunting task. To be at the same time sympathetic and judicial was a challenge. The investigation demonstrated that E. W. Richards, the foreman for the inquest, was a person of uncommon intelligence and leadership abilities. He precisely stated the delicate charge facing the members of the jury:

> Without power under the law to compel the attendance of unwilling witnesses, nor authority to enforce the production of papers, it will be seen that the jury labored under serious difficulties in the prosecution of what they deemed their duty, and we desire to return our thanks to the officers of the railway company who so cheerfully aided us in securing the attendance of witnesses, and also to those gentlemen, who, as civil engineers and experts of high standing in their profession, traveled long distances and spent days and weeks in making a thorough and critical examination of the wrecked bridge at the joint request of the jury and railway company. The written reports and testimony of these gentlemen have been of great service to us and will well repay careful perusal, as on them is based our

---

[55] Ibid., 49.

verdict in regard to the bridge.[56]

The jury examined 54 witnesses: 21 employees of the Lake Shore and Michigan Southern Railroad, both those on the ground and on the train, 10 Ashtabula citizens, 9 members of the Fire Department, and six survivors of the accident.[57] Most impressively the jury had examined eight expert witnesses, attempting to comprehend the accident through the eyes of professionals not employed by the Lake Shore, seven civil engineers, plus Joseph Tomlinson. Many of them came from great distances, such as Albert Howland from Boston and Charles Hilton from New York City. Brockmann emphasizes that the Ashtabula Coroner's Jury excelled because of the expert witnesses.[58] These civil engineers came on their own time and expense because of a vested interest. The companies no doubt picked up the tab with the exception of Tomlinson, who billed the jury for his travel expenses.

> The experts were drawn to investigate this disaster, provoking responses from as far away as Boston because this bridge design was thought to be the future of bridge building since it was created wholly from iron. Moreover, the bridge's Howe Truss design represented the current status quo in bridge design…and in dozens of bridges already created for railroads clear across the country: "The failure of the Ashtabula Bridge not only alarmed the general public, but also shook the blind confidence of railroad companies in their existing bridges."[59]

Conclusion of the Coroner's Jury

It was the conclusion of E. W. Richards that though the experts had disagreed on some of the particulars, they were uniform in condemning the bridge's design, the manner in which the diagonal bracing was attached to the angle blocks, believing that friction would be sufficient to hold the components in place. In short, there had been too much on-the-spot adaptation and too many critical changes for the bridge to even stand on its own, much less with a train crossing it.

---

[56] Nancy T. Wilcox. *The Ashtabula Bridge Disaster.* Appendix B, 2-3. Unpublished paper. Courtesy of David Tobias.

[57] Ibid.

[58] I am not as confident as Brockmann. I have noted contradictions, hedging and conflict of interest.

[59] Brockmann, 222.

The jury summarized eight conclusions, and two of them are sufficient for us to understand its interpretation of the bridge failure. The third point stated:

> That the fall of the bridge was the result of defects and errors made in designing, constructing, and erecting it; that a great defect, and one which appears in many parts of the structure, was the dependence of every member for its efficient action upon the probability that all or nearly all the others would retain their position and do the duty for which they were designed, instead of giving to each member a positive connection with the rest, which nothing but a direct rupture could sever. The members of each truss were, instead of being fastened together, rested one upon the other, as illustrated by the following particulars; the deficient cross-section of portions of the top chords and some of the main braces, and insufficient lugs or flanges to keep the ends of the main and counter braces from slipping out of place; in the construction of the packing and yokes used in binding together the main and counter braces at the points where they crossed each other in the shimming of the top chords to compensate deficient length of some of their members; in the placing, during the process or erection, of thick beams where the plan required thin ones, and thin ones where it required thick ones. [60]

The fourth point stated:

> That the railway company used and continued to use this bridge for about eleven years, during all which time a careful inspection by a competent bridge engineer could not have failed to discover all these defects. For the neglect of such careful inspection, the railway company alone is responsible.

Of course, these conclusions condemned Charles Collins who was personally responsible to see that the bridge was sufficiently inspected. He did not testify before the Ashtabula Coroner's inquest, though he did appear before the Ohio Legislature Committee. Did the Ashtabula jurors request his presence? Was he too emotionally spent to appear before both commit-

---

[60] Brockmann, 223. The thin braces were the counter braces, and the thick braces were the compression braces. Since Congdon did not designate which parts went where, Rogers may have not been able to differentiate between the two. The compression braces ran toward the middle of the bridge. Then at the middle of the bridge they reversed themselves. Or to put it another way, the compression braces were tilted towards the east for the 7 panels on the west end of the bridge and towards the west on the 7 panels on the east end of the bridge.

tees? Was he afraid of incriminating himself or incriminating Stone? With certainty, he was a very troubled person, not being able to sleep or eat and asking himself, "What could I have done differently?" Quite possibly the jury planned to call him, but by Wednesday, January 17 he was dead.

Conspicuous by his absence was Amasa Stone. The committee received the following note:

> Cleveland O Feb. 8th, 1877
>
> To Whom it May Concern
>
> This is to certify that Mr. Amasa Stone is under my professional care and that he is at present physically unable to leave the city.
>
> Very Respectfully, D.B. Smith, M.D.[61]

Stone may have thought himself above an inquiry conducted by the citizens of Ashtabula, an ad hoc group hastily thrown together without legal authority, and from his perspective had no right to sit in judgement of his many accomplishments, marred by this one failure. Paradoxically, the committee condemned Stone in its summary statement, but did not mention him in its final verdict.

> With the undoubted intention of building a strong, safe, endurable wrought iron bridge upon the Howe Truss plan, he designed the structure, dictated the drawing of the plans, and the erection of the bridge, without the approval of any competent engineer and against the protest of the man who had made the drawings under Mr. Stone's direction, assuming the sole and entire responsibility himself.[62]

The above statement simplifies the complications of assessing responsibility, singling out Stone, and implying that the men surrounding him and critically involved in the bridge's construction were absolved. The committee failed to cross examine the witnesses, which was the main difference between a court hearing supported by law, battled out by attorneys, and overseen by a competent judge. Joseph Tomlinson had designated the parts and specifications for the bridge knowing the bridge was going to be made of wrought iron, but never confronted Stone. Albert Congdon had manufactured parts with "insufficient material," making them smaller and

---

[61] Letter, Courtesy of David Simmons.
[62] Brockman, 224.

lighter than Stone's design and specification called for. After spending a year providing specification for the parts, the number of parts needed and how they were to be positioned, only then did Tomlinson protest that the braces needed to be reinforced, which they eventually were by adding extra braces. There is no question that Stone did not take warmly to his authority and expertise being questioned.

The greatest problem with the hearing and subsequent verdict, is that it was circular, and the defects were noted only in hindsight. The bridge failed because its parts failed, defects that were only noted after the event, not while the bridge was standing and serviceable for 11 years without a single interrupting incident. As we have already seen, three of the civil engineers, after noting what issues with the bridge were problematic, reversed themselves, claiming that the defects should not have been fatal. They may have feared liability issues, or been hesitant to disagree within the guild of Civil Engineers. Though Stone was not a civil engineer, they were well aware of his herculean reputation, and thus treaded lightly. And they may have adopted a position of humility, knowing that all construction in some way is imperfect.

Iron Bridges?

None of the eight conclusions reached by the Coroner's Jury condemned the Ashtabula Bridge because it was made out of iron. Though the engineers who testified before the Coroner's Jury could not predict the future of iron bridges, they were not totally oblivious to the liabilities which engineering historian Henry Petroski summarized a century later.

> The tenacity of steel has added a tensile dimension to structure that brings a release from the dominance of compression as a stabilizing force almost as if from the pull of gravity itself, but the memory of countless iron bridge failures during the 19th Century keeps modern engineers from getting swell-headed over how fast they can build longer and taller structures even today. Indeed, the chronic ailments of iron bridges that evolved with the expansion of the railroads is to this day one of the most discussed and most chronicled chapters in the history of structural engineering.[63]

---

[63] Henry Petroski. *To Engineer Is Human: The Role of Failure in Successful Design* (New York: Vintage Books, 1992) 56-57.

In other words, there was not unanimity on the structural use of iron among civil engineers and bridge experts. F. S. Slataper, Chief Engineer of the Pennsylvania Railroad, opined that the Ashtabula Bridge iron was of good quality, the abutments were sound, and its "atomical structure unimpaired by the jarring of 11 years of use." It was his conviction that the bridge was not wide enough for its height, though he gave no measurements or physical proportions to support his theory. He seemed to be saying the bridge was top heavy. The actual dimensions of the bridge according to descriptions given at the Ohio Legislature hearing were 19 feet 6 inches wide, between outside chords and 20 feet from top chord to bottom chord. In other words, the bridge was almost a perfect square for the entirety of its 154 feet length.[64] Most importantly, Slataper debunked attributing failure to the bridge being made of iron.

> Iron railroad bridges are almost exclusively used in countries much colder than the northern latitudes of the United States - on the Scandinavian peninsula, in Russia and in the north of Germany, and in all cases where properly constructed, climatic influences have had no serious effect upon them. While he conceded that a substantial stone structure was the acme of bridge attainments, yet, owing to expense and other difficulties, a stone bridge was often impracticable; and, next to stone, ranked iron - in economy, safety and stability - and if the latter material was scientifically used, its durability and security could not be called into question.[65]

The jurors did not question the "experts" as to the reliability of iron bridges. In referencing Amasa Stone, E. W. Richards stated, "Iron bridges were then in their infancy, and this one was an experiment which ought never have been tried and trusted to span so broad and so deep a chasm. This experiment has been at a fearful cost of human life and human suffering."[66] According to Petroski, iron bridges were controversial in the latter half of the nineteenth century simply because there were hundreds of failures, but even with the failures, iron was increasingly used. "Despite the apparent risk they came to be regarded as superior to wooden structures in the second half of the nineteenth century because of the increasing availability of iron at decreasing costs, the growing scarcity of timber and its increasing

---

[64] *The Cleveland Herald* (January 11, 1877) No title. Leverich Album.
[65] Ibid.
[66] Brockmann, 224.

cost and the resistance to fire damage, among other reasons."[67] The plain reason it was not addressed is that there was little understanding of the tensile strength of iron.

To this day, there is not unanimity among engineers and railroad historians as to the liability of iron on a cold winter night for a bridge covered with ice and snow and being swayed by fifty-mile per hour gusts of wind with no protection from the elements. According to Stephen Ressler, civil engineer and lecturer on the Ashtabula Bridge disaster, "Iron is 20 times stiffer than wood."[68] According to Ressler, "In the original *wooden* Howe truss, the use of compression members composed of multiple parallel elements was viable, because--even if the parallel elements were of slightly different lengths--they could be effectively clamped to the angle blocks when the tension rods were tightened. But if these parallel elements are made of *iron*, they deform much less under the action of the tension rods--and so a slightly undersized element can remain loose, even after the rods have been tightened."[69] By way of analogy, think of a log house which doesn't have to be perfectly notched in order to receive an inserted log. If it was made of iron, every notch would have to be perfectly calculated in order to receive a perfectly calculated beam. Part of Stone's problem was that he was still working with wood concepts for an iron bridge. According to David Simmons,

> One of the major problems with Stone's design was that he was applying wooden bridge techniques – in which he was widely experienced – to iron construction. It was an awkward attempt that created problems from the get go. But by 1877, virtually no iron bridge was built without direct links or connections between the compression members in the web and the top and bottom chords. That's why the issue kept coming up. Experts were scratching their heads in the 1870s wondering how an iron bridge could possibly stand without positive connections.[70]

To be sure the members of the Coroner's Jury did not become "swell headed;" if anything they erred on the side of charity. E. W. Richards, feeling the full weight of the task before him and his fellow citizens stated, "In entering upon the duty of ascertaining the cause and manner of death of

---

[67] Henry Petroski. "On 19th Century Perceptions of Iron Bridge Failures," *Technology and Culture* Vol. 24, Number 4 (October, 1983) 656.

[68] Zoom meeting with Stephen Ressler, Darius Salter, and Leonard Brown, Thursday, March 17, 2022.

[69] Email, Stephen Ressler to Darius Salter, September 8, 2022.

[70] Email, David Simmons to Darius Salter, February 1, 2022.

the victims of the late railroad disaster at Ashtabula, the jury found themselves at the very outset, embarrassed by the limited powers conferred on coroner's juries by the law."[70] The jurors did not overstep their boundaries, but in being conscientious about the restraints imposed upon them may have not gone far enough or probed deep enough. Prosecuting attorneys, they were not.

---

[70] *Ashtabula News Extra* (Wednesday, March 14, 1877).

Chapter 10

The Ohio Legislature Investigation and Subsequent Engineering Reports

A joint committee appointed by the Ohio Legislature met in Columbus, Ohio, from January 12 to January 30, 1877 to investigate reasons why the Ashtabula Bridge fell. The only exceptions to meeting on the grounds of the State Capitol were the interrogation of Amasa Stone; and Gustavus Folsom. The tone and content of the questioning by legislative members (anyone in the legislature was allowed to question) was quite different from the investigation by the Coroner's Jury. The legislative questioning was more exact, more technical, more accusatory, and far more thorough. For instance, the legislature asked Stone 198 questions and Tomlinson 259 questions. The Committee interviewed five persons, all of them having had some kind of direct responsibility for building the bridge: Amasa Stone, Charles Collins, Albert Congdon, A. L. Rogers and Joseph Tomlinson. They called no Ashtabula citizens, who would have been immediately at the site of the wreck. Why the Committee did not interview Charles Paine, General Superintendent of the Lake Shore, is unclear. Tomlinson's answers were the most complete and Collins' the most incomplete, riddled with the most "I don't knows." All of the witnesses indicted themselves as well as their colleagues.

The exceptions to those who actually worked on the bridge were M. J. Becker, an engineer employed by the Pittsburg, Cincinnati, and St. Louis Railroad, and Gustavus D. Folsom, engineer of the Columbia. Why Becker appeared before the Committee instead of the other engineers, who gave written reports, is unclear. Becker condemned the sleeves or yokes around the crossing compression and counter braces as insufficient. He

emphasized that both the braces and I-beams for the chords were insuf-
ficiently attached to the angle blocks. The lateral braces should have covered each panel rather than every two panels. Becker stated he felt safer
on a bridge with no cast iron parts. In Becker's opinion, "such a bridge as
that at Ashtabula should never have been built, and I am confident that if
that plan had been submitted to a competent board of engineers, before its
erection, that bridge never would have been built."[1]

Because Gustavus D. "Pap" Folsom, engineer on the Columbia, was
recovering from injuries sustained in the wreck, the Legislative Committee met with him at his home in Cleveland. He testified that "he first noticed the bridge giving away when about two-thirds across. It began to
swing toward the south but recoiled and went to the north." The witness
further said, "when the chords on the south side let go, those on the north,
by a very sudden action, drew the bridge in that direction. When I experienced the first sensations of the bridge giving away, I was applying the air
brake very lightly. The engine struck the west abutment, and I think for a
moment was held by the forward engine, while the iron express propped
underneath me, meanwhile the tender swung against the wall; my engine
then dropped down straight and turned over backwards."[2] (If dual tracks,
trains in both the United States and Great Britain ran on the left hand
track.)

Obviously, Folsom was disoriented, but somehow had the presence of
mind to shut off the steam. He remembered that with his right hand he
seized the throttle rachet, and "with the left the air brake caught and hung
there until my engine struck there below. I heard no braking sound until
I heard two distinct crashes behind which I supposed was the bridge and
train striking below, before I fell. I don't think my engine could have made
more than one revolution after I felt the sinking sensation until she struck
the abutment."[3] He identified the Socrates as a 32-ton engine and his own
as a 35-ton.

Two critical aspects of his brief testimony need to be noted for our
purposes. He did not "like" the bridge, one of the main reasons being that
upon crossing it, he always heard a "snapping" as if the joints were set-

---

[1] *Report of the Joint Committee Concerning the Ashtabula Bridge Disaster,
under Joint Resolution of the General Assembly* (Columbus: Nevins & Myers,
State Printers, 1877) 63.

[2] Ibid., 73.

[3] Ibid.

tling together, and he never noticed a snapping sound on any other bridge he crossed. Second, he emphatically disavowed that his engine was off the track before the bridge collapsed.[4]

Amasa Stone

Due to health reasons, Amasa Stone was also interviewed in his home. As to the insufficiency of the lateral braces which crisscrossed between the bottom chords, each "X" covering two panels, Stone claimed the bridge would have been quite safe with no lateral bracing. Obviously, engineers did not agree with this assessment, in that the lateral braces were for the purpose of the bottom chords not bulging out, keeping the bridge in equilibrium and preventing oscillation. Stone continued in his contention that the train had come off the track, rather than there being some defect in the bridge.

> It is very conclusive evidence, to my mind, that the bridge was carried down by the second locomotive in some way leaving the track. The bridge was not strong enough to take a locomotive across off the rails. Had the bridge broken through weakness it would have pulled in the other direction. I understand you to say that the bridge swung to the north?[5]

Stone's rationale for this conviction was convoluted, to say the least, and disagreed with all other reports as to exactly how the bridge fell, the train falling to the south and pushing the bridge to the north. Stone continued:

> Had the bridge broken from its own weakness, it is conclusive to my mind it would have swung to the south. I am convinced, a model test, to the extent of breaking a truss, would show conclusively that that truss would fall to the south and pull the bridge to the south. An engine dropping on the cross-floor beams would tend to deflect them and pull the truss inward — that is, the truss to the north, that the train was passing over on, and when pulled to the north from a vertical line to a small extent it would go down.[6]

Stone admitted he had never built an iron bridge on the Howe truss plan. Even after observing that on his first inspection of the bridge A. L. Rogers had mixed up the braces, he defended his continued use of Rogers by wrongly claiming "He couldn't put the work together different than I

---

[4] Ibid., 72-73.
[5] Ibid., 81.
[6] Ibid.

designed it."[7] Stone denied he was a stockholder in the Cleveland Rolling Mill incorporated under the name of Stone, Chisholm, and Jones, but did identify A. B. Stone as his brother. (That he was not a stockholder was not true according to the letter Stone wrote to Vanderbilt on July 13, 1874, included in Chapter 4.)

Incredibly, Stone stated he never knew the bridge would not bear its own weight, when the falsework (underpinning scaffolding) was removed. Stone additionally claimed he did not know that the bridge had lost its camber when it first settled. In his mind the only trouble with the bridge was the placing of the braces "flat ways instead of vertical." He claimed, contrary to Tomlinson, that they were not designed to go that way. He confessed that a better bridge would have been a stone arch, but "we had determined what funds we had to expend, we were short."[8] The Committee pressed Stone on the lack of connection between the braces and the angle blocks.

> Q: I will ask you, sir, whether it is not the universal practice to provide some means for holding these braces in place?
>
> A: Not that I am aware of.
>
> Q: I will ask you, sir, if the authorities, without exception, don't require some means to be provided for holding these braces in place?
>
> A: They were held by clamp bolts, at intersections. I suppose that would be sufficient to keep them in place.
>
> Question repeated by Mr. Converse.
>
> A: I am not aware of such.
>
> Q: Can you name a single authority who states that it is not necessary to provide other means than the mere pressure, such as lugs, etc.?
>
> A: I never noticed an allusion to it.[9]

On January 31, 1877, Amasa Stone wrote the following letter to the *New York Times*:

---

[7] Ibid., 82.

[8] Ibid., 85.

[9] Ibid., 77.

As to the Ashtabula bridge, it was the last bridge that I designed, and of the 10 to 15 miles of railway bridges that I had designed and erected, a great number of which are now in use, and built 25 to 30 years ago, this one had the heaviest proportions, and was designed as the strongest in all of its parts. Its material was of the best quality, and the work was done in the company's shops by the best workmen; and from the day it was built until it went down it did not show a sign of weakness in any of its parts. Such is the testimony of good bridge men who have been in charge. My theory is, that from some unknown cause the second engine left the track as it struck the bridge, went through its deck, and carried the train with it. No wood or iron bridge is prepared to carry a locomotive across it safely when off the track.[10]

Charles Collins

Charles Collins did not know where to locate the plans and specifications for the bridge. He claimed that during the building of the bridge, he was relieved of his Ashtabula responsibilities because he was building a line for the Cleveland, Painesville, and Ashtabula Railroad Company, from Jamestown to Franklinton.[11] What most indicted Collins was his inexactitude in directions, details, and scheduling for inspecting the bridge. This duty fell to G. M. Reed, the bridge inspector who worked under Collins' supervision. When asked how often Reed was to inspect the bridge, Collins answered, "He has no specified time to make examination. Whenever at such times as he thinks it necessary."[12] Then the accusation from George Converse, a legislator, "Does it not strike you that this is a very loose way to do business, with such vast interests and property and men's lives involved?"[13] Collins responded, "No sir; I thought everything was well taken care of."[14]

Collins would not make a judgement as to the cause of the accident. He did not believe that planing off the lugs and clipping off corners of the braces would have any material effect on the bridge. He understood the angle blocks were so constructed as to receive the braces horizontally

---

[10] "The Ashtabula Bridge: A Letter from Mr. Amasa Stone Concerning It," *New York Times*, January 31, 1877.

[11] *Joint Committee*, 90

[12] Ibid.

[13] Ibid., 91.

[14] Ibid.

rather than vertically, but was not sure that the original plan called for such a placement. (As we will see, the horizontal placement was in Tomlinson's original plan.) When Collins was asked a loaded question, "State to the committee whether the bridge was an experiment or not," Collins answered, "I should think it was, rather, so far as I was concerned."[15]

When asked about the difference in price between an iron bridge and a stone arch, Collins estimated about $25,000.[16] In all, Collins answered, "I don't know" eighteen times, but never more critically than the following exchange:

Q: Did you yourself, as engineer of the road, ever make a thorough inspection of that bridge?

A: So far as looking at it for the safety of the trains.

Q: Did you consider that you had made a thorough inspection? That is the question.

A: To make an analysis of the bridge, I didn't.

Q: I mean, by making a thorough inspection of the road, such as one as would satisfy you in your own mind that that bridge was perfectly safe?

A: Yes, that it was perfectly safe.

Q: I understand that you made such an inspection.

A: Yes.

Q: How long since did you make such an examination?

A: I don't know. It has been some time. I have been on the bridge often and made an examination.

Q: You say you don't know how long it is?

A: No, sir.[17]

---

[15] Ibid., 96.
[16] Ibid., 101.
[17] Ibid., 102.

Probably the most humbling part of the questioning came at the point of confessing that the company kept no hose at the pump house. The pump was used to pump water up to the town's reservoir. Incredibly, Collins stated the engine house was there because "we will want that water for some purpose, and the men put in the waterworks thinking we might want a fountain there at some time."[18] The capability of putting out a fire was only secondary because, "We would have a hose there, probably, if there was no bridge there, for the purpose of wetting the ground and keeping the grass bright - we often do - and if there should be a fire about the building or coal-house, or about the vicinity, it could be used for fire purposes."[19]

Albert Congdon

Albert Congdon as "Master Mechanist" had responsibility for manufacturing and ordering the materials for the parts of the bridge. Congdon claimed that he was the one who, after the bridge was erected, and would not hold its own weight, informed Stone the braces needed to be turned. He explained that the lugs were for the purpose of "keeping the braces in position, but they had to be planed off or planed down when the braces were turned." When Rogers failed to understand the need for camber in the bridge and had the chords shortened because they were in a straight line after the falsework was removed the first time, Congdon informed him, "Mr. Rogers, you don't apparently understand the principle of structure; there is no straight line about this; it's (a) segment circle." At first Stone without seeing the bridge, did not realize the chords were too long in that Rogers was not allowing for camber, and ordered the shortening of the top chords. When Congdon was asked, "by how much," he gave the following answer:

I think, three inches and a quarter. Then they put the top chord in, screwed up the bolts, and when taking the blocks down from under the chords the bridge settled right down; it came down until it was about two and one-half inches concave; and then they didn't take out any more; they stopped. And then Mr. Stone became acquainted with the state of things, and he asked me to go down to Ashtabula with him; we went down and looked at the thing, and then he said that the top chord was too short, and it must be lengthened again to its original length. I lengthened it.[20]

---

[18] Ibid., 103.
[19] Ibid., 105.
[20] Ibid., 110.

Actually, the chords were not lengthened, but plates, also called shims, were placed between the ends of the chords and the lugs on the angle blocks. According to Congdon, Stone did not understand, even after seeing the braces bent, that they needed to be turned to a vertical position. Congdon claimed he had to inform him, "If those braces were turned up edgewise, says I, they would be more than stiff again, and add no more weight to your bridge; you would have more than twice the carrying weight without adding any extra weight to the bridge; and then I spoke to him about the fastening of the end panel, and gave my description how it could be fastened."[21]

Congdon suggested a correction to the bridge that had been mentioned by no one else, at either the Coroner's or Legislative hearing. "My opinion of the bridge – of course you understand that I do not pretend to be an expert, but I do have some judgement about matters – perhaps if the bridge had been well attended to, which there was ample evidence that it was not, and that I should have recommended, when they used a double track, to put in another truss."[22] One assumes the truss would have run the entire length of the center of the bridge. This correction would no doubt have added stability. (We do not know how often two trains were on the bridge at the same time going opposite directions.) Had Congdon suggested to Stone a third truss, he would have, in all likelihood, rejected the idea. Stone, no doubt, wanted to prove that a single truss on each side of an iron bridge would work, just as it did for a wooden bridge. Notice that on the inspection question, Congdon had thrown Collins "under the bus," as we now say, or in this case, "under the bridge."

Congdon, like everyone else, noticed from a previous painting that the braces had changed position and re-emphasized, "I am satisfied in my own mind that had the bridge had been well attended to, as soon as anything was wrong, and showing a disposition to be working out of place, and fastened as it might be fastened easily, the bridge would have been standing now."[23] We do not suppose Collins was sitting in the Legislature which we assume was open to the public, listening to their accusations of dereliction, but he was no doubt reading the newspapers. The guilt feelings were becoming unbearable. The failure of Collins to complete the inspections required by his job description was repeated almost ad nauseum, as exem-

---

[21] Ibid., 112.
[22] Ibid., 114.
[23] Ibid., 115.

plified by the following interchange between Congdon and a Legislature member:

> Q: I understood you to say that you believed that if carefully watched, that it was a good bridge, and it required care taken of it.
>
> A: I said that the bridge had evidence of not having care - not being watched.
>
> Q: Whose duty was it to give it that attention and care that you have reference to, that it didn't have?
>
> A: As far as I know it was in the Chief Engineer's department.
>
> Q: You spoke about some of these braces being moved, and it had been painted since. Do you know how long since it was painted?
>
> A: No sir, I don't.[24]

A. L. Rogers

A. L. Rogers had worked for the Lake Shore Railroad for 20 years, and was 59 at the time of the disaster. Before supervising the erection of the Ashtabula Bridge, his most sophisticated experience had been that of a shop carpenter. He had no exposure to or expertise in working with any kind of iron construction. At the time of his employment by Stone for the erection of the bridge, he was supplying firewood for the locomotives. From our perspective today, and from the perspective of those on the Coroner's Jury and Legislative Committee who questioned him, this occupational transition seems impossible. The manufactured parts of the bridge were not even marked as to their intended placement. Thus, the choosing of what went where became more difficult than the task of a five-year-old child putting together an erector set, which would come with some kind of directions for attaching the parts together.

When Rogers first discovered the upper chord was too long, Stone decided not to shorten the chord, but to "plane off," or diminish the size of the lugs which weakened them They would have been the correct length, if Rogers had retained Tomlinson's 6-inch camber.[25] This means that all the

---

[24] Ibid., 116.
[25] Ibid., 121.

angle blocks were returned to the Lake Shore shop in Cleveland because the lugs were cast with the block as one piece. This was remedied by placing liners, or shims, between the lugs and I-beams of the chords. Throughout the testimonies before the Coroner's Jury and the Legislature hearing, there was confusion as to whether the I-beams forming the chords were cut off at each end of the bridge or the lugs at the end of the I-beams were shaved off. If the latter was done, the bridge was seriously compromised because the I-beams buttressed up against the lugs.

Rogers had no experience with anything needing a "camber," a rise in the middle of the bridge. (In a flatbed semi-truck designed for carrying a heavy load, the bed is higher in the middle so there is less strain on the underpinning, and there is less chance for the bed to go concave, and the load breaking the trailer, or falling through the bed of the truck. Or another way to think about a camber is a very shallow arch compressed in the middle.) When asked if he knew how much camber was designed for the bridge Rogers answered, "I didn't know anything about it. I went to Mr. Congdon to inquire. I knew nothing about the framing of the bridge."[26] At this point, the camber became guess work, Congdon and Stone disagreeing on the amount. Upon removing the scaffolding, also called false work, the bridge was dependent on the tension rods and diagonal compression braces, the latter being deficient in both position and strength. The camber disappeared, and Rogers feared it would collapse if let down further.

After this second failure, fearing another encounter with Stone, Rogers went to see Charles Collins, the person responsible for the operation of the Lake Shore and Michigan Southern Railroad as its "Chief Engineer." Collins disavowed having had anything to "do with the bridge, either design or construction."[27] He told Rogers that the braces would be stronger if they were perpendicular rather than flat, i.e., placed so that the flanges were on the bottom and top rather than the sides, but he gave no reason why. Collins then promised Rogers that he would request Stone to look at the bridge, which he did.

According to Rogers, Collins confirmed what Congdon had already suggested, that the braces needed to be turned to a vertical position. Rogers beseeched Collins to go see Mr. Stone on his behalf, as to what to do about the settling of the bridge because "It was simply because I was afraid that something would be brought to bear, that would cause me to

---

[26] Ibid.
[27] Ibid., 123.

lose my position; that was one reason that I had."[28] The following line of questioning not only explains Rogers' fear of Stone, but gives us some insight into Stone's personality.

Q: How long have you been connected with Mr. Stone?

A: I have known him for over twenty or twenty-one or twenty-two years, and to say that I am acquainted with him to-day I can't.

Q: I will ask you to state whether he would allow Mr. Collins or anybody else to make any opposition to his will or his plan?

A: I don't believe that he would.

Q: Or any suggestions?

A: He might possibly receive a suggestion. From the little that I know of the man, I don't believe he would even receive a suggestion, he might. I can relate a circumstance that might modify that statement, but I wouldn't like to give it in testimony.[29]

Stone and Congdon showed up and decided the braces needed to be turned, and more braces were added. The compression braces consisted of four I-beams, and two were added. But to allow for the two tension rods already in place, the corners of the added I-beams had to be chipped off. This was done, and again the bridge was lowered, and none of the braces bent. This was the third and final time the scaffolding was removed, and the bridge at last proved it could stand on its own. But nothing was done for holding the diagonal braces in position on the angle blocks, because the lugs had been planed off as we have already explained. When the bridge was let down, one of the braces actually slipped because it was 1/16th of an inch too short. Nonetheless, Stone pronounced the bridge ready for use after the third erection. The bottom line was Rogers had never seen a scaled plan (what we would call a blueprint) as it had been drawn up by Tomlinson.

[28] Ibid.

[29] Ibid., 132-133.

Joseph Tomlinson

Ironically, Joseph Tomlinson, out of all of the witnesses, proved himself to be the most knowledgeable about the bridge, and would also make one of the biggest mistakes: identifying the east end of the bridge as the point of failure. Tomlinson was very clear about his intention for the braces to be placed horizontally. (No one on either the Legislative Committee or Coroner's Jury challenged his design.) "Q: In your design, were the braces intended to rest flatways or upon the edge with the weight horizontal or vertical? A: They were intended to lay flat in the first place, and they were changed to stand edge ways."[30]

Tomlinson's solution was to add additional braces, and/or rivet plates to the braces, and connect the braces together (compression and counter) with a yoke and a bolt or rivet where they crossed so they acted as a single unit. They were not bolted, but only tied with a yoke where they crossed. Tomlinson did not hint at the possibility that the vertical position was stronger than the horizontal position. The following interchange is critical to our understanding Tomlinson's mindset.

> Q: The faces of the angle blocks on this bridge were originally intended to receive I-beams, six-inch web and four-and-a-half-inch flange, lying horizontally; now, by turning those bolts and setting the web vertical and the flange horizontally, would there be breadth enough on the face of the angle block to receive them?

> A: There was width enough on the inclined surface of the angle blocks for the six-inch braces in place, but anywhere they interfered with the bolts.[31]

> Q: Did you notice what measures were taken to remedy the deficiency, where the bolts were interfered with?

> A: Yes, a number of the flanges of the braces had been cut away.

> Q: How many of these braces had been cut or chipped, more or less, in order to accommodate the bolts?

---

[30] Ibid., 136.

[31] The bolts screwed through gibbet plates, holding the I-beams down on the angle blocks. An iron gibbet plate was approximately 3 inches wide and 1 inch thick. I am not sure whether Tomlinson meant the bolts or the tension rods because the rods interfered with additional braces, requiring chipping off the corners of the braces.

A: I couldn't tell, but I should think about two in five.

Q: What effect would this chipping away have upon the sustaining power in those braces?

A: Cutting away a corner, slightly, I don't think it would have very great amount, any dangerous effect; but when the flanges are cut away on one side so as to leave only the web and the flange on the opposite side, it would render the braces less stiff; it would very much impair its strength.

Q: I think you spoke in your first examination, you mentioned that upon the face of these angle blocks, there were lugs cast, into which the heads and foot of the braces were introduced to keep them in position. Was there anything on the angle blocks in the modified structure to take the place of these lugs, or, what became of them?

A: The most of them appeared to have been chipped off.

Q: Was there anything to prevent the head or foot in these braces from getting out of place?

A: The greater part of the braces seemed to have no provision made for holding them in place.[32]

Tomlinson also indicted Collins as did Congdon when asked a loaded question (accusatory) "What would you think of an inspector of a bridge who should fail, for any length of time, in making an inspection of the bridge, to notice the fact of these displacements? A: A person who was in the habit of reasoning/thinking strains on a structure, would evidently look at these points the first thing; I should think his observing facilities were not very good if he did not see them."[33]

But evidently, Tomlinson's observing abilities were not very good either. In contrast to everyone who testified, in particular Dan McGuire, the engineer of the lead locomotive, and the later engineering report by MacDonald, he believed the bridge had failed at the east end.

In talking with others, and thinking up the matter after I left here, the conviction was very strong upon my mind that the disaster was caused by some one of the sets of the main braces, near the east end, having got out

---

[32] Ibid., 144-145.
[33] Ibid., 146.

of place, which allowed the structure to fall suddenly; and if that was the cause, that the end braces at the east corner would have slipped off of the angle blocks and struck hard against the masonry, so as to leave an impression. To-day, I went up and examined the masonry at all the corners, and at the south corner of the east abutment I found it had been struck by the braces, which confirms my opinion that the disaster was caused by one of the sets of main braces getting displaced. That would throw everything else out of place in an instant.[34]

A greater problem lay in the fact that the whole time Tomlinson was designing the bridge, he believed an iron bridge on the Howe truss plan to be inadequate, that is, iron was too heavy. Evidently, months went by before Tomlinson ever expressed this opinion to Stone, if he expressed it at all.

Q: Having serious objections to the Howe truss bridge of this length, how came it about that you didn't state those objections to Mr. Stone?

A: Well, I suppose when I first came to Mr. Stone's I looked upon him as a higher authority in bridge matters than myself.

Q: Then Mr. Stone told you just what he wanted done, and how to do it, and you merely drew this plan at his dictation, and in such manner as he directed?

A: Generally, that would be correct; because in carrying out the detail of the plan, there would be, of course, suggestions and consultations between us.

Q: In those conversations did you ever suggest to Mr. Stone that the Howe truss was not the best form of bridge of this span?

A: I think not.

Q: Were you not consulted at all in regard to the plan?

A: No, not in regard to the plan.[35]

Did Tomlinson need the job? Was he afraid of being fired? According to him, Stone had contacted him when he was working elsewhere. This disagrees with Anderson and Waddel, who wrote in their memoir after his

---

[34] Ibid., 151-152.
[35] Ibid., 149.

death, that Tomlinson was already in Cleveland building another structure. Tomlinson remembered having worked with Stone in 1847. When asked, did Stone send for him at the time when he projected building of the bridge, Tomlinson answered, "Yes. He telegraphed me at Freelingham. And how he came to do so, I don't remember. I know I received a telegram from him and a letter; it was in answer to the telegram that I came to Cleveland."[36]

Greater than the conflict of ideas was a conflict of personality. Tomlinson's retiring phlegmatic temperament dictated conflict avoidance, while Stone's choleric decision-making temperament demanded confrontation. Stone was a take-charge, grab the bull by the horns, decisive actor. If it was damned if you do and damned if you don't, Stone did. As a draftsman preferring to work alone, Tomlinson would be classified today as introverted. The rub between the two men was soon apparent, especially to Tomlinson, who was more sensitive to slights and criticism than was Stone, who tended to let them bounce off of him while he charged ahead. Tomlinson recalled, "When I was doing business with Mr. Stone, in my interviews with him about these matters, we never could harmonize. Whenever I made any suggestions, it always led to some discord, and it was in these circumstance that I left the work."[37]

The Verdict

There are many reasons for a person asking questions, especially in a public setting. Questions are often statements of opinion; they demonstrate insight into a critical issue; they display a person's knowledge; they exhibit leadership abilities by taking command of the floor; they position a person for advancement in their profession; they earn social capital, as well as fulfilling all kinds of neurotic tendencies. Those who testified were at the mercy of politicians; exhibiting prowess in obtuse matters may have been more important than obtaining truth. Consider the following line of questioning by Charles Converse to Charles Collins.

Q: Would it not tend to throw the line of compression away from the centre?

---

[36] Ibid., 149.
[37] Ibid., 134-135.

A: I don't consider it would effect it materially.

Q: If the centre line of compression is thrown out of centre of the strut, would it not weaken it very much?

A: Not without it was cut to some great extent.

Q: Suppose it as only one-half an inch?

A: It wouldn't effect it materially.

Q: You think then, that the strut eleven feet long, the line of compression being one-half inch out of the line of the centre iron, would make no difference in its strength as a strut?[38]

Nobody in the world could have definitively answered these questions. Not in 1877 anyway. The Committee did not interview various engineers, as did the Coroner's Jury, but received a report from three engineers who had visited the site on January 18. Whether B. F. Bowen, Thomas Johnson, and John Graham as civil engineers were from the same firm or not, is unclear. As did everyone else, they condemned the lack of attachment of the braces to the angle blocks. Also, almost as everyone else, they noted the lack of dependence of the parts on one another, which led to a diffusion of strength rather than an amalgamation. They suggested a solution:

> If the several groups of beams composing the braces and top chord had each been combined into a single member, by riveting on to their flanges a system of diagonal plates - say three and a half by half inch - running alternately from right to left and from left to right across the entire group, the bridge would have been abundantly safe. This arrangement would have made each group strongest in the lateral direction and weakest in the direction of the webs of the beams; but in this direction the beams offer about five times the resistance that they do laterally. The top chord members could then only deflect in single panel length, and on that account their strength would have been still further increased — twofold. The result would have been that the factors of safety given in the tables would have been increased *five times* for the braces and *ten times* for the chord. They would have been so excessively strong that much of the material might have been omitted.[39]

---

[38] Ibid., 94.
[39] Ibid., 32.

The engineers were suggesting that 3 ½-inch wide and ½ inch thick flat iron plates, be zigzagged across the top of the chords, binding the I-beams together for the entire length of the bridge.

The engineers praised both the workmanship of the bridge and the strength of the material. They concluded by saying,

> We would say that we find nothing in this case to justify the popular apprehension that there may be some inherent defect in iron as a material for bridges. The failure was not due to any defective quality in the iron. It was not owing to the sudden effect of intense cold, for failure occurred by bending, and not by breaking. It was not the result of a weakness gradually developed after the erection of the bridge. It was due simply to the fact that it was not constructed in accordance with certain well established engineering principles. We find no evidence of any weakness which could not have been discovered in the plan and avoided in the construction.[40]

The concluding verdict of the Ohio Legislature included the following:

> The defects in the original construction of the bridge could have been discovered at any time after its erection by careful and analytical inspection, such as the importance of the structure demanded, and thus the sacrifice of life and property prevented.[41]

> The lateral system between the lower chords was defective in this: the struts were placed at every other panel point, and the tie-rods extended across two panels, and instead of being fastened at the ends of the struts, were fastened at alternate panel points, crossing each other at the middle of the strut. The sway braces were too small and too infrequent.[42]

> The lateral system between the upper chords had the same defects as that between the lower chords, with this exception; the floor-beams had small lugs united to them, and they acted as struts.[43]

> No provision was made for holding the members comprising the braces in their places on the angle blocks, and your committee find that many of them were out of place before and at the time the bridge went down. The braces were greatly weakened by imperfect bearings and having their ends

---

[40] Ibid., 33.
[41] Ibid., 4.
[42] Ibid., 5.
[43] Ibid.

chipped off.[44]

There was one weight upon the bridge which has been overlooked, and did not enter into the calculation of the engineers, as an inspection of their statements will show, namely, the snow on the bridge at the time of the accident. The proof shows twenty inches of snow on the ground. It is probable that much of the snow had blown off the bridge; but whatever weight of snow or ice there was on the bridge would still further diminish the factor of safety in both the braces and upper chord.[45]

It would be needless to say that any engineer would be derelict in his duty who did not provide in the construction of a bridge against wind, snow, ice, and the vibration of a rolling load. They are as much to be anticipated and provided against as the law of gravity.[46]

Your committee are of the opinion that a third or centre truss in bridges carrying two tracks would greatly promote safety and security. The material of the bridge was good, and likewise the workmanship, with the exceptions before stated. There was material enough in the bridge and a different disposition of it would have secured five times the strength, and small and comparatively inexpensive additions in the way of diagonals on the braces and upper chord, and a securing of the braces to the brace blocks, would have rendered the bridge secure.[47]

The MacDonald Report

Charles MacDonald was born on January 26, 1837 and educated at Queens University, Kingsland, Ontario and also graduated from Rensselaer Polytechnic Institute at Troy, New York in 1857. His working career included being assistant engineer on construction of the Grand Trunk Railway in Michigan, surveying for the construction of the Philadelphia and Redding Railroad and about 15 years of building bridges in the vicinity of New York City. In 1887 he superintended the erection of the Hawkesburg Bridge in New South Wales. After 1900 he became Vice President of the

---

[44] Ibid.
[45] Ibid.
[46] Ibid., 6.
[47] Ibid.

American Bridge Company and served as president of the American Society of Engineers from 1908 to 1909.[48]

In February 1877, MacDonald gave a presentation on the Ashtabula bridge failure to the American Society of Engineers, meeting in New York City. MacDonald had made a visit to the Ashtabula site a few days after the disaster and made a "partial examination" of the wrecked material. MacDonald said of Stone, "For some time previous, he had been familiar with wooden bridge construction, and is generally considered to be a man of large experience in the requirements of railroad practice and commanded the confidence of a large circle of influential men in his ability to carry out successfully anything he might undertake to do."[49]

MacDonald concluded, after examining the bills of material as verified by both Congdon and Tomlinson, "There appears to have been modification of the original order and as all of these braces were the same length and the erection was entrusted to inexperienced men, it is probable that some of the lighter braces may have been placed nearer the abutments than was originally intended."[50] The I-beams forming the chords were supposed to have gradually increased from 10.35 square inches to 11.85 square inches, but the plan had not been adhered to. In spite of these defects, "The workmanship at splices and connections with angle blocks was very good of its kind, and could not have failed to insure safety at these points."[51] MacDonald commended the floor of the bridge which consisted of "6-inch cross beams, 9.6 square inches area, upon which rested 7 x 14 inch white pine stringers and 3 x 5 oak ties, spaced 2 inches apart."[52] Considering the strength of the flooring and the fact that there was a considerable guard rail for the inside of each track, "Accidents from derailment on such a floor as this are extremely rare."[53]

But on the spot modifications had to be made, as we have already described, so that the braces could be inserted vertically rather than horizontally, in order to add two more braces. Four braces did not possess

---

[48] "Charles MacDonald Engineer, Is Dead", *Brooklyn Daily Eagle* July 9, 1928.

[49] Charles MacDonald. "The Failure of the Ashtabula Bridge: Transactions of the American Society of Civil Engineers" (New York: American Society of Civil Engineers, February 21, 1877) 75.

[50] Ibid., 76.

[51] Ibid., 77.

[52] Ibid., 78.

[53] Ibid.

sufficient strength to hold up the bridge when the falsework was removed. The braces were sent back to the rolling mill, and the ends diagonally cut so that they were jagged in order to fit into the angle blocks. In MacDonald's words,

> This change in the number and position of braces made it necessary to chip off portions of the flanges to prevent them from interfering with the vertical rods, and as the castings had been planed in grooves to suit the first positions, they, too had to be chipped to as nearly a square bearing as possible. All this work was imperfectly done, and the result was, that the braces did not have what would be understood as square bearings.[54]

In examining the wreck, MacDonald noted the exception to straight I-beams occurred at the vertical rod, 2 sections from the west end of the bridge. The I-beams which crossed this point were bent as much as 90 degrees. This was due to what Charles MacDonald and later Dario Gasparini claimed to be the fatal flaw in the bridge. "The cast iron angle block at the top of the second set of braces had the south lug broken off close to the face, and the line of fracture disclosed an air hole extending over one half of the entire section."[55]

Cast iron was exactly that, melted iron poured into a cast or a mold. The problem with cast iron which was not true of wrought iron, strenuously formed by rolling it between heavy rollers, was that iron cast in a mold may leave a bubble, a vacuous inclusion, undetectable because of non-existent radiology. The lugs and angle blocks were cast as one piece, another bad idea. Thus, MacDonald confidently concluded "from the facts stated, there would seem to be little doubt but the failure began in the south truss at the second panel point from the west abutment."[56] (According to Stephen Ressler, this flaw might have not been fatal, if lateral bracing was connected to this particular angle block. But because the lateral bracing covered two panels, this block had been skipped or by-passed.)[57]

MacDonald argued there was nothing wrong with the top chord, and it was only by the fracture in the angle block that the bridge could have failed. "It is certain that the top chords in the center panels must have been strained up to 8,700 lbs. per square inch, thousands of times without

---

[54] Ibid.

[55] Ibid., 80.

[56] Ibid.

[57] Zoom meeting with Stephen Ressler, Darius Salter, and Leonard Brown, Thursday, March 17, 2022.

giving any perceptible sign of weakness, while failure finally took place in the strongest part of the chord probably from a defective detail.[58] This would have been true of the second panel from the west end of the bridge. The defect was not in the chord, but in the angle block. The fissure in the lug had over the years worked its way down through the entirety of the angle block. MacDonald then condemned everyone who had condemned the bridge because of not being informed, or not fully informed, concerning bridge construction and having made "snap judgements." He critically stated, "Judging from the tenor of much that has appeared in the secular press, either as evidence taken under the solemnity of an oath or by way of editorial command, this bridge must have been conceived in sin and born in iniquity."[59] Of course, this statement represented either exaggeration or sarcasm, or possibly both. No one had suggested that Amasa Stone had any evil intentions in building the bridge.

MacDonald confessed that the knowledge of iron bridges was in its infancy. Those who had built iron bridges had built better than they knew, or in other words, the bridge had stood in spite of their ignorance. Gasparini states: "This was specifically true for fatigue crack initiation and subsequent brittle fracture in iron castings, especially at low temperatures. No visual inspection could have detected the extent of the interior flaw or even the likely surface microcrack at the exterior of the lug. The fact that the bridge carried many heavier loads for over 10 years indicates that its static strength was sufficient, despite the difficulties during construction."[60] In MacDonald's perspective, if anyone could have built a bridge made out of iron, it should have been Amasa Stone.

> If then, the state of knowledge at the time has not been under-estimated, the Ashtabula bridge was the result of an honest effort to improve the bridge practice of the country, undertaken by a man whose experience in wooden bridges warranted him in making the attempt. As to his willful neglect of proffered advice, it would be well to suspend judgement until all the facts are brought to light by the proper tribunals. His worst enemies will at least accord to Mr. Stone the possession of common sense.[61]

---

[58] MacDonald, 82.

[59] Ibid.

[60] Email from Dario Gasparini to Darius Salter, January 31, 2022.

[61] MacDonald, 80.

MacDonald then made an unsubstantiated statement, "It (the bridge) passed into the hands of the regular inspection officers of the road, who, from current report, were most careful men, but who had never had more than the ordinary practical experience gained in the handling of wooden bridges."[62] What did MacDonald mean by current report? Current reports did not exist. MacDonald then claimed that not "one railroad in fifty" would incur sufficient time and expense to make the needed inspection of a bridge, at least one that was sufficiently thorough. He claimed the New York Central and the Boston and Albany profited by the fact that the original designers of the riveted work still remained to watch the results of their experiments; but these were the exception.[63] The problem with this statement is that the New York Central did not build the bridge, though at the time of the failure, owned it. Amasa Stone who had built the bridge had no direct responsibility for it at the time of its failure. He was only a director, not an employee. As we argued in Chapter 4, directors took no hands-on responsibility for any operation or material asset by or of the railroad company, though Stone was more informed about the bridge than any other director of the NYC.

MacDonald claimed that there was no current formula for testing the compression members of the bridge other than an unverified and inadequately tested hypothesis referred to as "Gordon's formula." In short, cast iron could not be tested, but there was no reason to discount the use of wrought iron. MacDonald then returned to his unsubstantiated, if not false assumption, that the bridge had been sufficiently inspected but admitted to the need for more adequate inspection. "But the fact exists, that the inherent weakness of this bridge had escaped the notice of a regular inspection which was generally believed to be adequate and the result undoubtedly points to the immediate necessity of a radical change in a system which has shown itself to be utterly unequal to the requirements of the age."[64]

For MacDonald, the problem lay in the fact that iron bridge building was in uncharted territory, and that no one should have to take personal responsibility for the bridge's failure. This was not the conclusion of the two engineers who responded to MacDonald. Edward Philbrook had sent his associate Albert Howland to examine the wreck (which we covered in Chapter 9) and Howland had observed that some of the braces had been

---

[62] Ibid., 83.
[63] Ibid., 84.
[64] Ibid., 85.

removed from their respective angle blocks, and observing that the ends of the braces were not worn, they had never borne their proper portion of the load in supporting the upper chords.

Also responding to MacDonald's paper was Thomas C. Clarke, one of the most authoritative and knowledgeable bridge builders and railroad executives in the world. Scholar Francis E. Griggs claims,

> Thomas C. Clarke was one of the leading engineers in Canada and the United States from 1848 to 1901, working on railroads, waterways, buildings, and primarily bridges. He was a prolific writer with articles not only in American Society of Civil Engineers (ASCE) publications but in British technical journals. He also published many articles in Scribners' and other popular magazines. He was a leader in the formation and running of several major bridge designs and construction companies and worked with many leading bridge engineers of the period. He served ASCE for many years starting as a director in 1890 and was elected President of the Society for 1896-1897. Despite his noteworthy career, he is one of the unknown giants of 19th century civil engineering.[65]

Clarke was much more condemning of the bridge than either MacDonald or Philbrook. He did admit MacDonald was in all likelihood correct that the break took place at the 2nd angle block which was fractured. The bridge, if erected properly and the braces kept in place according to Gordon's formula (which was hypothetical without sufficient scientific proof) should have supported twice the weight under which it actually failed. We can recall that Tomlinson suggested additional braces be added, but in order for the angle blocks to receive 6 braces rather than the 4 for which they were designed, the flanges of some of the braces had to be chipped to allow for the 2 tension rods already in place which created an uneven attachment to the angle blocks. The angle blocks only had room for 2 tension rods and 4 compression braces. This was even more problematic in that the braces were expected to be held in place by friction rather than additional use of rivets or pins. According to Clarke, the bridge should have been apriori condemned, because it was in a constant state of unstable equilibrium.[66] As far as we know, no one at the 1877 meeting of the American Society of Civil Engineers outright named or condemned Stone, but they were left with the impression that whoever was in charge of erecting the bridge had

---

[65] Home (https://www.trb.org/) https://www.mytrb.org/Profile/AnnualMeeting/Registration

[66] MacDonald, 87.

made corrections, though while fixing one problem had created others, to the extent that the bridge was critically compromised.

A Reassessment of the Accident by Gasparini and Fields

In the 1990s, Dario Gasparini, an engineering professor at Case Western Reserve University, became the foremost authority on the construction elements and design of the Ashtabula Bridge. He and Melissa Fields, a working engineer, published a paper for the *Journal of Performance of Constructed Facilities* in May of 1993, titled "Collapse of the Ashtabula Bridge on December 29, 1876."[67] The authors reviewed the history of the truss bridge that by the time of the building of the Ashtabula Bridge in 1865, had become the most widely adopted methodology. However, in converting the bridge from wood to iron, little was known about tensile strength, fatigue factors, and ways to calculate the stress intensity of the various weight-bearing members of an iron bridge. The authors suggested at least two motives for Stone utilizing iron for his bridge: Stone's innovative, pioneering spirit that seemed to have never failed him up to this point, and the convenience of purchasing iron from a mill of which his brother A. B. Stone was President and part owner.

The authors reviewed Tomlinson's testimony, pointing out that the engineer (at odds with Stone) calculated the live load of the bridge as 2,500 lbs. per foot, and Stone testified that the dead load was a "ton to the foot run," but Gasparini and Fields were unclear as to how either of them arrived at their figures. Tomlinson believed the diagonal braces measuring six inches with four-inch flanges on each side to be insufficient. He recommended to Stone that the braces needed strengthening by "riveting plate" to them. More critically, Tomlinson complained to Stone that both the I-beams (for the chords and the bracing) were not full size. And, this may have angered Stone more than the disagreement as to what was required to make the bridge reliable. There seemed to be a contradiction in that Tomlinson pointed out the difference between his specifications and what was actually produced, as contrasted with his response to Albert Congdon, who informed the Legislative Committee that Tomlinson had designed and ap-

---

[67] Dario Gasparini and Melissa Fields. "Collapse of the Ashtabula Bridge on December 29, 1876," *Journal of Performance of Constructed Facilities*, (May 1993) 109-125.

proved the angle blocks. Tomlinson claimed that he "gave the pattern and the length of every part of it for the mechanics to work by."[68] He further stated, "I have only a kind of indistinct consciousness that the pattern makers were making some patterns which were not finished, but it was nothing essential." He finally testified that the bottom chords were "as good work as ever I saw made."[69]

Because Rogers knew nothing about camber and the chords had been prepared for a 6 inch camber, obviously the chords were too long. Gasparini and Fields are not exactly sure what was done, but believe that chord members were shortened and/or the lugs were planed down, "reducing the thickness of the vertical lugs," so that the middle of the bridge would drop. Evidently, the exact shortening and thinning were guess work and when the falsework was removed, there was hardly any camber in the bridge at all. Shims were added between the lugs and the I-beams, forcing the middle of the bridge up to Stone's 3.5 inches. When the falsework was removed the 2nd time, the braces buckled because they were put in horizontally, (flat) rather than vertically with the flanges on the top and bottom, rather than the side. The braces were turned, and Stone also added extra diagonal braces as suggested by Tomlinson. These changes required additional work as stated by Gasparini and Fields.

> To fit the additional I sections... two important modifications had to be made. One was to remove the angle block lugs intended to restrain the diagonal braces and another was to "chip" some ends of the braces in order to clear the vertical iron rods. This meant that the ends of the diagonal elements would, in some case, not have a "square" bearing and that there was no mechanical restraint – except friction from the axial forces – to prevent movement of the diagonals (this was largely true in wooden Howe trusses).[70]

When two locomotives were run over the bridge, it sagged 5/8 of an inch, but upon the passing of the locomotives, the bridge sprang back 3/8 of an inch, a differentiation of ¼ of an inch between load and non-load. (I ask, was this test ever subsequently run on the bridge? Did it retain this resilience throughout its 11-year existence?) The authors referenced the vague inspection record and recalled a conversation among Congdon,

---

[68] Ibid., 115.
[69] Ibid.
[70] Ibid., 116.

Stone, and Rogers, some two years after the bridge was put into service. Then Congdon stated:

> I examined it thoroughly, and a day or two afterwards, I told Mr. Stone, at his office, I told him I had examined the bridge and it looked very well; I didn't discover anything but one brace, and that was a but a trifle out, I could just jar it; it was probably loose enough – Oh! You couldn't slip a bank bill between the two wedges; it was an outside brace, and in about the third panel from the east end, as near as I can recollect. As I knew Mr. Rogers had the superintending of the erecting of the bridge, I told him of it. My idea was that he would go perhaps and turn that outside nut a little – it didn't require but a sixteenth turn, probably.[71]

Gasparini and Fields criticized Albert Howland for claiming that many of the compression members (diagonal braces bearing weight) had factors of safety less than one. The authors cited the MacDonald report to underscore the inability to calculate factors of safety for the compression members of an untried iron bridge in 1866. They noted MacDonald was the only engineer to point out a flaw in the 2nd angle block from the west end of the bridge. Gasparini and Fields observed,

> Indeed, the compressive braces and the top chords would have been stronger if the I- beams were continuously interconnected. Indeed, the safety of the top chords would have increased had they not been subject to bending from the floor beams. Indeed, the diagonals should have had better end bearing. Indeed, because of slight length differences, some I-beams were more stressed than others. Nonetheless, the bridge performed safely for over 10 years.[72]

In the end, Stone should have retained Tomlinson, but his impatience and temper got the best of him. Had Tomlinson overseen the work, the camber would have been retained and thus, the chords not shortened. Lugs would have not been removed or down-sized, so they could receive the diagonal braces. In short, adaptation between prefabrication and erection on almost any complex building project has to be made on the spot, and Tomlinson, a trained civil engineer, would have been more experienced in implementing these changes than the self-taught carpenter, A. L. Rogers.

In conclusion, Gasparini and Fields argued that MacDonald as well as the engineers Bowen, Johnson, and Graham, who wrote a report for

---

[71] Ibid., 117.
[72] Ibid., 121-122.

the Ohio Legislature, had no way "to show that the forces present in the compression members at the time of failure exceeded their "estimated capacities," and Gasparini and Fields questioned the strength estimates of Howland. They further agreed with MacDonald blaming the bridge failure on the stress factor, a defect which Gasparini refers to as "fatigue crack propagation at the void at the base of the lug and brittle fracture in the iron lug at the low temperature."[73] In the end, the engineers Gasparini and Fields exonerated Stone, by stating that "there is no doubt that the builders were technically competent, and wanted to achieve a 'first-class' innovative bridge, using the best workmanship and materials available."[74] As to all of the engineers who testified, none of them mentioned the effects of pre-stressing. "They disagreed markedly on how to estimate strength of slender compression elements and did not even allow the possibility of fatigue in iron."[75]

Gasparini and Fields correctly point out that nothing concrete or definitive in terms of ensuring bridge safety came out of the McDonald testimony and the Legislative Committee Reports. The Legislative Committee had recommended "a bridge design code, prescribing design loads, allowable stresses, and minimum strengths, and requiring expert review designs, construction supervision, and periodic inspections by engineers."[76] The recommendations failed to be enacted as law.[77] The authors suggest the inaction may have been due to Stone's influence and conclude that non-legislative action was the proper course "for a failure that was no one's fault." As I will argue in my final chapter, this last statement would have been more correct if it had read, "no one single person's fault."

A More Recent Observation

After careful study of the Ashtabula Bridge, Stephen Ressler in a lecture given at Lehigh University on February 18, 2022, and in a subsequent

---

[73] Email from Dario Gasparini to Darius Salter, January 31, 2022.
[74] Gasparini and Fields., 124.
[75] Ibid.
[76] Ibid.
[77] But not all was lost. According to Gasparini, the Legislature "affirmed the judgement of bridge engineers, that they should not use iron castings on railroad bridges. In fact, iron castings quickly fell in to disuse." Email from Dario Gasparini to Darius Salter, January 31, 2022.

zoom meeting with the author, outright condemned the bridge.[78] Ressler believed that instead of using chords made of one piece, Stone was persuaded to use multiple I-beams because his brother manufactured them. Also, using lugs rather than pins, bolts or rivets was a critical liability. A tubular chord with the same amount of material would have been "several times stronger." Also, rather than multiple I-beams for the compression braces, a singular tubular brace could have been used. In other words, as was pointed out many times in the hearings, there were simply too many independent parts not sufficiently connected to one another. In Ressler's words, the bridge was not an "integrated unit."[79]

The absence of evidence is not the evidence of absence. With that being said, I can find no one who has done a study of the Ashtabula Bridge, nor any other bridge engineer who has knowledge of a previous bridge similar to the Ashtabula Bridge using multiple I beams for the top chords. Thus, the assessment of Collins that the bridge was "experimental" is critical for any historical assessment. The design of using multiple I beams proved to be at least a liability, if not a fatal flaw in the bridge.

---

[78] Stephen Ressler, "The Ashtabula Bridge Disaster and the Advent of Civil Engineering Professionalism," The 2022 Khan Distinguished Lecture Series, Lehigh University, February 18, 2022.

[79] Email, Stephen Ressler to Darius Salter, April 6, 2022.

Chapter 11

Atonement

The Beginnings of Western Reserve College

At the turn of the century America was heating up spiritually, and would explode into what historians have labeled "The Second Great Awakening." A Presbyterian, Barton Stone envisioned a camp meeting unfettered by sectarianism at Cane Ridge, an open expanse of ground in Bourbon County Kentucky just outside of Paris, in August 1801. Seemingly, God has a sense of humor demonstrating irony in locating America's most significant religious event at that time in a county named for Kentucky's favorite whiskey, and a city best known for Voltaire's infidelism if not outright atheism. There Presbyterians, Methodists and Baptists met together for religious exercises, exercises in every sense of the word, the likes of which had never been witnessed in the young Republic, and possibly have not been duplicated since.

The Methodists and Baptists would take full advantage of the new soul harvesting technology, but not so much the Presbyterians. Camp meetings were the perfect social and religious gathering for Methodists, who adopted as their most defining sacrament the "altar call." Thousands were instantaneously converted to the extent that between 1800 and 1810 Methodist adherents grew seven times as fast as the American population. As a whole, Methodists were far too emotional for Presbyterians, who favored nurture of both the heart and head, and in the eyes of some could be somewhat snobbish about it. In 1835 a Presbyterian lawyer, Samuel Bigger, asserted that when Ohio University (Ohio Wesleyan at Delaware) needed a Methodist professor "they had to fetch one from Europe." Methodist historian William Warren Sweet claimed the Methodists believed the

"Presbyterians to be extremely arrogant, assuming themselves to be the only competent educators of the people."[1]

But there was nothing snobbish or arrogant about the Presbyterian David Hudson, who had recently converted out of agnosticism into a fervent piety. In 1799 he set out from his home in Connecticut to found a school in the Western Reserve, and the village where he landed would soon be named after him. In his elder years he recalled, "I asked the Lord for a home in the wilderness and He gave one to me. I asked Him for a church and He gave me that. I asked Him for a school and He gave me that, but the college, I never thought He would give me that - that is the child of my old age."[2]

When the cornerstone of the first building was laid in April 1826, a "stately procession" took place from "David Hudson's house to the village church, to the college site, and then back again to the church, with much prayer and sacred music and speaking along the way."[3] This kind of event would be replicated scores of times throughout the Midwest and beyond. The stated purpose of the college was the rule for early nineteenth century schools, not the exception: "To educate pious young men and pastors for destitute churches…"[4] At the heart of Western Reserve's educational philosophy was its theological commitment. In fact, they were one and the same.

> Sound learning & pure religion are mutual assistants & guards to each other, & both are essential to the wellbeing of Society, inasmuch as learning without correct religious knowledge would be dangerous to the best interests of Society and as religion without sound learning would be liable to degenerate into bigotry, enthusiasm & superstition, & thus endanger not

---

[1] William Warren Sweet. *Circuit Rider Days in Indiana* (Indianapolis: W. K. Stewart, 1916) 61.

[2] C.H. Cramer. *Case Western Reserve: History of the University, 1826-1976* (Boston: Little Brown and Company, 1976) 4.

[3] Ibid., 9.

[4] Richard E. Baznik. *Beyond the Fence: A Social History of Case Western Reserve University* (Cleveland: Case Western Reserve University, 2014) 19.

only the best interests of society, but also the interest of eternity. . .[5]

What became immediately apparent to Henry Hitchcock, the school's third president, was the school had a huge debt and was desperately short of funds. Of course, this desperation would only intensify during the 1873 financial panic. In 1870, Professor Carroll Cutler became President of the school, but the financial burden of keeping the doors open from day to day was so overwhelming that he resigned after two years. Not being able to find anyone else to take the job, the school trustees begged him to stay, which he did until 1876. At that point Hiram Haydn, who would later become President of Western Reserve University, was Amasa Stone's pastor.

On December 13, 1877, Richard Parsons (the same Richard Parsons we encountered in Chapter 2,) editor of the *Cleveland Herald,* wrote an article prophesying that if the school moved to Cleveland, it "would at a bound spring into full life and beneficent usefulness," and also suggested that, "One of our wealthy citizens should embrace the opportunity."[6] Not only would Amasa Stone embrace the opportunity to almost single-handedly save the school from extinction, but publicly identify himself as one of Cleveland's wealthiest citizens.

This plea came at a time of intense ongoing debate as to what was the most salubrious environment for a college. For Americans who were still mostly rural, urbanization threatened society with sin and dissipation. This was especially so for a "Christian college" where separation from the world was a must for the cultivation of morals. We of the twenty-first century tend to forget that during the post-bellum years there was little to no dissonance or distinction between the urban core (what we know as downtown today) and oil refineries, steel mills, and all kinds of manufacturing endeavors. To the average American mind, "the city" was dirty in almost every way, especially morally and industrially. Iron and steel mills as well as refineries were the worst culprits for spewing ashes and whatever impurities were extracted by furnaces using coal, leaving residue on the surrounding environment. Cleveland had not become Pittsburgh, but

---

[5] Ibid., 20. The original building which housed the chapel, classrooms, and administration still sits as the centerpiece of Western Reserve Academy in Hudson, Ohio. A plaque on the front reads: "The Chapel and architecture in tradition, recalling Old Yale College, the ideals of which inspired the establishment of higher learning in the Western Reserve was dedicated in 1836 to the service of Almighty God."

[6] Ibid., 67.

was headed in that direction. By 1890 the smokestacks of the Cleveland Rolling Mill, which employed about 8,000 people, darkened the sky. One Cleveland historian wrote, "The smoke of prosperity mingled with the odor of hemp and canvas, oil, and grease…. The air was filled with hoarse blasts from steamship whistles, the clang of ships' bells, and the hoot of tugs and locomotives. Industry was making men rich."[7]

The resistance to leaving Hudson was strong: "Leave it where it is, give it a good library, a museum, and more instructors, establish a girl's school as thorough as the college, with free access to lectures, library, and museum." Further, "Establish in Cleveland a 'polytechnic school of the highest grade with such men as Faraday, Tyndall, Loomis, Owen, Darwin, and Lyall as the chairs of instruction.'"[8] One has to be amused that the writer believed Michael Faraday and Charles Darwin would relocate to Cleveland. The contention as to what a college education would include, as we briefly highlighted in chapter 2, was still in play. Some suggested the college should be consolidated with Oberlin, but the memories of Oberlin abolitionism and Charles Finney's revivalism were far too hot to handle for staid Presbyterianism. They possibly remembered Finney's hyperbolic condemnation that there was a "Jubilee in Hell" every time the Presbyterians got together for their General Assembly.[9]

As the college relocation argument raged, one Clevelander broadcast that Hudson was a country village of "dull monotony" where the only recreation offered was one of dissipation. "Cleveland can afford a thousand means of culture that Hudson will never know, and college patrons will be doubled in a year."[10] A black janitor named Branch best summed up the contention, "No, sah; It don't go down like maple syrup no how, sah."[11] Staying with Mr. Branch's analogy, no one immediately stepped up to sweeten the deal.

---

[7] Carol Poh Miller and Robert A. Wheeler. *Cleveland: A Concise History, 1796-1996* (Cleveland: Case Western Reserve, 1997) 77.

[8] Hiram Collins Haydn. *Western Reserve University from Hudson to Cleveland: 1878-1890* (Cleveland: Western Reserve University, 1905) 40.

[9] Whitney R. Cross. *The Burned-Over District: The Social and Intellectual History of Enthusiastic Religion in Western New York, 1800-1850* (New York: Cornell University Press, 1981) 258.

[10] Haydn, 91.

[11] Ibid., 43.

Persuading Stone to fund a Cleveland College

Attempting to persuade Stone to become a prime financial mover in the transition from Hudson to Cleveland and giving him due credit, his pastor Hiram Haydn touted him as a "Man of wealth in Cleveland, of intelligence and large influence in practical affairs, educated as a civil engineer, who had already entered in the field of public charities in a generous fashion, a pioneer in the construction of railroads and telegraphs, began to turn over the project of founding a college in Cleveland, which would be a worthy memorial to an only and gifted son who was drowned, but a few years before while a student at Yale."[12] Haydn was partially accurate as to both motivation and identification, but either did not know or misrepresented Stone's history. As we have shown, Stone was not a "civil engineer," had little formal education, and had almost zero knowledge as to the purpose and curriculum of an institution of higher education. Haydn certainly did not advertise that the school was looking for someone who needed to lighten a huge load of guilt. Nonetheless, according to Case Western Reserve University historian Frederick Clayton Waite, "There is no doubt that Dr. Haydn was the moving spirit from June 1878 and probably before. He reported to the Board of Trustees early in March 1880 that a wealthy citizen of Cleveland, whom he did not identify, desired to know what amount would be required to effect removal."[13] The Board of Trustees on September 20, 1880, voted 14 to 2, to make the removal of Western Reserve College to Cleveland.[14]

We have no evidence that Amasa Stone ever visited Western Reserve College in Hudson, or had a formal meeting with its trustees. It does seem that the trustees put a $500,000 price tag on relocation, and this was in all likelihood communicated by Hadyn to Stone. The $500,000 gift came with a huge price tag, not strings, but dock-size lines attached. In a letter written by Stone to President Cutler, dated July 7, 1880 the would-be benefactor listed the following requirements for a half million-dollar gift:

> First - the college be removed to Cleveland to a site to be approved by me, to be furnished free of cost, and in close proximity and harmony with the Case School of Applied Science. Second - that the corporate name of the

---

[12] Ibid.

[13] Frederick Clayton Waite. *The First Forty Years of the Cleveland Era of Western Reserve University,1881-1921* (Cleveland: Western Reserve University Press, 1954) 5.

[14] Ibid., 6.

institution be changed to, "Adelbert College of Western Reserve University" Western Reserve University may or may not be a part of the corporate name. Third - the number of trustees to be reduced to eight (8) by resignation and their places be filled by persons to be proposed by me. Fourth - I, to donate the sum of $100,000, as the work progresses, for improvement of grounds, and for a college building after satisfactory designs, the sum of $50,000. For the improvement of grounds and for a dormitory building after satisfactory designs as the work progresses, to be rented to professors and students and securities to the par value of $350,000, so soon as the buildings are completed and occupied by the college Faculty.

This memorandum not to be made public.

Very respectfully yours,

A. Stone[15]

Amasa Stone and Leonard Case Jr.

Amasa Stone was a competitive person, and his competition was in the person of Leonard Case Jr., whose name the university would retain. According to historian C.H. Cramer, "Relations between Leonard Case Jr. and Amasa Stone were dyspeptic at best."[16] While they served as directors at the same bank, Case had accused Stone of financial impropriety, and Stone never forgot an accusation. It is probably more than just coincidental that Stone's gift for the removal of Western Reserve College to Cleveland was made just after Leonard Case's gift was made known in 1880. When his founding gift for "The Case School of Applied Science" was publicly announced, Stone would not be one-upped. He wanted the Hudson School upon its transfer to Cleveland to be named after him but did not make the request a stipulation for his gift. He was told that the albatross from the Ashtabula disaster would be too great for the new university to transcend, or words to that effect. But Stone did not initially envision a college named after him or his son by bankrolling the entirety of the transition. He began by offering $100,000 if four other sponsors could be found to give the same. There were no other takers.

---

[15] Ibid., 49.
[16] Cramer, 81.

Case had a much richer educational background than Stone. He graduated from Yale, but was disillusioned with the lack of practical knowledge offered in the Yale curriculum. Case was not a man of abstractions. He returned to his hometown of Cleveland after graduation, passed the bar exam in law but never practiced. In fact, he never practiced anything resembling regular employment. Though Case was not a playboy, such lack of responsibility brought scorn from Stone.

There were two other reasons that Stone would have disdained or been envious of Case. Leonard Jr. had inherited his money from his father, Leonard Case Sr. If any man ever pulled himself up by his bootstraps, that man was Amasa Stone. Another difference was the Case family, at least initially, was far more prominent in Cleveland than the Stone family. Almost immediately after Stone's arrival in Cleveland, William Case, Leonard Jr.'s brother, was elected mayor of Cleveland, 1850-1852. The Cases were social insiders in Cleveland, and far too elitist for Stone's taste. Both Case brothers were members of a literary society called "The Ark," so named after the boat built by Noah for having every specimen of animal on earth. The members of "The Ark," consisting of the "learned" of Cleveland, met monthly to discuss zoology, ornithology, archaeology, and many other areas of natural history, as well as play chess. The entire ethos of this endeavor was antithetical to Stone's life motif of constant work, which did not allow for playing much of anything.

The backgrounds of Leonard Case Jr. and Amasa Stone Jr. could hardly have been more unalike. Case was born into wealth; his father Leonard Case Sr. was an attorney, who being from Connecticut foresaw the potential of Cleveland, Ohio as a strategic port on Lake Erie, and invested every spare dollar he had in its real estate. His interest was far more than monetary, as he gifted trees that would line the dirt streets of an inchoate community. By the time he died in 1864, he was the "wealthiest man in Cleveland, and by far the largest land holder in Cuyahoga County."[17]

Leonard's brother William inherited his father's business acumen, enlarging the family fortune. While serving as Mayor, he led in developing the best city library west of the Alleghenys. He inherited his father's interest in horticulture and natural history, establishing a 40-acre nursery, and continued his father's tree planting project throughout urban Cleveland. He communicated with John James Audubon, providing the ornitholo-

---

[17] C. H. Cramer. *Case Institute of Technology: a Centennial History 1880-1980* (Cleveland: Case Western Reserve University, 1980) 2.

gist "with descriptions and drawings of Ohio birds and small mammals."[18] With William's interest in animals and plant life, the "Ark" became an informal forum for conversation, a unique entity for this frontier city and beyond the interest of Amasa Stone, who could barely keep himself afloat on the raging currents of his multiple business enterprises which were already emotionally driving him and demanding his every waking moment. Stone had no sympathy for such superficial endeavors, on which, in his perception, the Cases wasted time.

In Stone's view, Leonard Jr. was even stranger than his brother. Leonard had never done a day of hard labor in his life. He was enamored with mathematics, and was able to read scientific works in French, Italian, and German. Leonard Jr. was a recluse, and examining his actions from today's vantage point one would suspect that he was autistic. When invited to a friend's house he responded,

> Please believe me that I thank you for your kind offer to entertain me in your "cottage." Unfortunately for myself I am very unsociable. Visiting is very irksome to me.... I haven't tried it for a long time. Cottage or castle would mean little to me. I could be as contented in the one as the other, but ah me, I could be contented in neither. Are you sure you were sincere when you thanked God you were contented? I do not believe it is human nature to be contented. The nearest approach to happiness in this world is in hard and successful work and the higher the work the keener the enjoyment - but never contentment. [19]

What was Leonard going to do with the $1,500,000 he inherited along with the 400 acres in "downtown" Cleveland? In love with the sciences, and valuing technical knowledge, he donated both land and money for the establishment of The Case School of Applied Science. Upon his death at the age of 60 in 1880, the *Cleveland Penny Press* eulogized him.

> Since Case's death, instances of his great open-handed generosity are becoming talked of, which shows him to have belonged to a species of mankind almost extinct in this age, i.e., he was a man who, miserable in everything pertaining to his own life, lived with the determination of making just as many people happy as he could with the opportunities fortune and the Almighty had furnished him.[20]

---

[18] Ibid., 3.
[19] Cramer. *Case Western Reserve*, 200.
[20] Ibid., 202.

In spite of having endowed a school which would be named after him, Leonard Case Jr. died without a will, which led to all kinds of rediculous and fraudulent claims. A Myra M. Hastings claimed that she had been married to Leonard (who never married) and had borne him three children, which proved to be an outright lie. Another claimed in 1902 to have found a family Bible in a northern Ohio barn which led down various genealogical trails, giving birth to dozens of fraudulent claims by dishonest individuals who could find equally dishonest and greedy attorneys. Case became a very popular name to claim somewhere back in one's lineage.[21]

Another key difference between Case and Stone was that in appointing trustees they valued a different class of people. No one incarnated the difference more than did Rufus Ranney, the first chairman of the Board of Trustees of the School of Applied Science. With little formal education he "became a learned man in law, history, literature, and languages; he could read civil law in the Latin of Justian and the Code Napoleon with equal facility."[22] As a member of Ohio's first Constitutional Convention, he became known as a Hercules of the convention and for seven years served on the Ohio Supreme Court. When the new Cuyahoga County courthouse opened its doors for business on New Year's Day 1912, statues of Thomas Jefferson and Alexander Hamilton stood at the south door, and statues of John Marshall and Rufus Ranney stood at the north door.[23] Stone valued his friends and Case his heroes.

A School Named for Adelbert Stone

Stone had at least one very understandable motive for founding Adelbert College of Western Reserve University. Adelbert, the oldest child and only son of Amasa and Julia, called Dell, desired to enlist in the Union Army. The father objected and instead shipped him off to an academy in Middletown, Connecticut to be groomed for entrance into Yale. To the father's credit he helped raise $60,000 which would have been sufficient for recruiting 200 young men in the cause of the Union. According to John Hay:

> During the war for the Union, Mr. Stone was an active and ardent supporter of the administration of Mr. Lincoln, of whom he was a trusted

---

[21] Cramer, *Case Institute of Technology*, 7-8.

[22] Ibid., 10.

[23] Ibid., 11.

friend and counselor. The President frequently sent for him to come to Washington to advise him in the most important problems of supply and transportation of the army. He tendered him an appointment as brigadier general for the purpose of superintending the construction of a military railway from Kentucky to Knoxville, Tenn., a project which was, on Mr. Stone's advice, afterward relinquished by the government.[24]

Dell, a handsome young man, was something of a prima donna, at least to the extent that he was not hazed as a Freshman upon entering Yale in the fall of 1863. He wrote his 11-year-old sister Flora, addressing her in French, "Ma chere soeur" in which he informed her, "The other evening some Soph's took a Fresh by force about two miles out of town and having stripped him, they blacked him all over with burnt blacking and then left him to find his way home. How would you like to have them do that to me? But there is no fear of that, for they will not touch me."[25]

The Stones didn't have long to celebrate the end of the Civil War on April 9, 1865 and the return of Ohio's gallant soldiers. The son never returned from Yale, at least alive. As a Junior, his third year at Yale, Dell wrote his father June 25, 1865, "Tomorrow prof's Dana and Brush are going to take about 20 of us up to Middletown and vicinity on a mineralogical expedition. We will have a pleasant time."[26] Two days later Amasa Stone was notified by telegram that his son had drowned. The next day, June 28, more details came, "Your son was swimming alone at some dis-

---

[24] John Hay, "Amasa Stone,*" Magazine of Western History* Vol. III (November, 1885-April, 1886) 110. I do not believe that Hay would have fabricated this narrative, but because there are no letters between Lincoln and Stone regarding any of these claims in the National Archives, it must be accepted with caution. Stone wrote three patronage letters to Lincoln: March 8, 1861, requesting that Fayette Brown be appointed as Collector of the Port of Cleveland; August 21, 1862, requesting that A. Everett be appointed as a Tax Collector for Cleveland, and May 4, 1865, requesting Capt. John H. Young be appointed as a Quartermaster. Letters, courtesy of the Lincoln Presidential Library, Springfield, Illinois. There is no doubt that Stone knew Lincoln or, at least, met him. On Lincoln's inaugural trip to Washington in 1861, he stopped in Cleveland. Stone rode the Presidential train from Cleveland to Erie. Letter from Amasa Stone to George W. Patterson, February 7, 1861. Courtesy of Melissa Mead, University of Rochester Archives.

[25] Gladys Haddad. *Flora Stone Mather: Daughter of Cleveland's Euclid Avenue & Ohio's Western Reserve* (Kent, Ohio: The Kent State University Press, 2007) 14-15.

[26] Ibid., 16.

tance from his comrades and was probably taken with cramps. As he threw up his hands and uttered a cry for help. He sank several minutes before any swimmer could reach the spot. And did not once rise."[27]

Adelbert would have celebrated his 21st birthday on July 8, but instead was buried at Lake Side Cemetery on July 2, as his sisters, Clara, 16, and Flora, 13, looked on with disbelief. Flora's biographer Gladys Haddad is no doubt correct that Amasa never recovered from the loss. Adelbert College would fade from memory as his son's name was relegated to a building on the campus of Case Western Reserve University.

Appraisal of Stone's Gift

Whatever Stone's motivation for enabling Western Reserve University's move to Cleveland, the enduring legacy has been two-fold: propitiation for the Ashtabula Disaster and jealousy of Leonard Case Jr., who also attended First Presbyterian Church, also known as Old Stone Church where Amasa Stone was a leading layman. It is doubtful the two of them sat together. Case Western University Sesquicentennial historian C.H. Cramer wrote in 1976 that after the Ashtabula Disaster, Stone's name "became anathema throughout the United States; he was regarded by many as a murderer and endured agony during the seven years left in his life. This had two effects: One was on the fortune of Western Reserve University; the other was a transformation of Stone's personality that brought him to an unnatural death."[28]

On February 4, 1996 an article appeared in *The Cleveland Plain Dealer* titled, "Bank Rolling Higher Learning Philanthropists' Feud Led to Founding of Two Schools." The author Bob Rich referred to Stone as someone who was as "flinty as his name," and stated, "Stone's arrogance had made him unpopular, even in his own top-drawer society circle. But it also drove him as a philanthropist to do things for which only he would get the credit, and that included competing with the Cases by bankrolling a successful university in Cleveland."[29] The bottom line was that Leonard Case Jr. may have been dead, but Amasa Stone was still competing with the family name. A more realistic and objective description of the two men was given by Hiram Haydn:

---

[27] Ibid., 16-17.

[28] Cramer. *Case Western Reserve: A History*, 84-85.

[29] Bob Rich. "Bank Rolling Higher Learning Philanthropists' Feud Led to Founding of Two Schools." *Cleveland Plain Dealer* (Feb 4, 1996).

Mr. Case was a recluse, a student, of frail health, and content with a small circle of friends who were wont to meet often, if not daily, in what was known as "the Ark," a one-story building erected on the old homestead of the Case family. The people at large knew little of him. They had heard of a great landed estate that went by this name, and they saw the record of a generous gift to amply endow a great school, but they had not as yet got its bearings as a public benefaction which might claim their sympathy and co-operation.

Mr. Stone was a man of affairs, known and appreciated by the men of his generation that knew him, both for his sterling integrity and his executive ability. He was self-reliant, of immense energy and perseverance along lines once laid down; a man who brought things to pass and wrested success out of difficult undertakings. But he was a man of reserved manner, reticent, and yet not as difficult of approach as he would seem. He had a warm heart, though "not worn on his sleeve," not caring to be at home to all sorts and conditions of men, yet making the worthy poor, and young men in whom he believed and who needed a friend, so much at home with himself that, after many years, recalling his helpful consideration in their day of trouble, they mention his name with gratitude.[30]

The latest chronicler of Case Western Reserve University, Richard Baznik, interprets that Stone would play the last hand, trumping Case after his death and "The Case school provided him with a measure of control over the implementation of Case's trust by requiring the Case school to locate next to Western Reserve on land other than the property Case had provided. He could super-impose his will on Case."[31] Thus, financially capable Clevelanders including one John D. Rockefeller, contributed to the purchase of 43 acres just east of Center City that would accommodate both schools. The story is told, probably apocryphal, that the School of Applied Science desired the western half of the property closest to downtown. "Knowing Stone's need to dominate every transaction however, they proposed Case be given the eastern half of the property to which Stone predictably objected, demanding that they accept the western half."[32] Thus, the School of Applied Science received the plot of land which the Trustees really wanted.

---

[30] Haydn, 44-45.
[31] Baznik, 69.
[32] Ibid., 71.

For decades the relationship between the two schools was less than amicable, and continued in a spirit of animosity incarnated by the two founders, Case and Stone. When the two schools settled into their new locations, a six-foot-high fence separated the two schools. Presidents of the respective schools, Winfred Leutner and T. Keith Glennan, reached a mutual agreement to dismantle the fence in 1941 after a six-decade grudge match.[33]

The Question of Motivation

The eighteenth century Congregationalist pastor Samuel Hopkins championed a theological perspective which came to be known as "disinterested benevolence." Hopkins believed that one could be enabled in this life by God's grace to act in such a way that the actor would have no concern for the return of that act upon himself. Or to put it another way, acts of altruism would be purely motivated with no consideration for how kindness and charity would advantage the self. We do not have room here to discuss the soundness of Hopkins' optimism, but only to make a couple of observations. The first problem that disinterested benevolence runs into is epistemological, the philosophy which attempts to answer the question, "How do we know something?" How can one have perfect knowledge of the inner self? How can one know that he or she is perfectly free from self-advantaging motives?

Second, not being able to know our own motives, how can we sit in judgement of others? Stone's personality, which was anything but gregarious, made him seem self-absorbed, disregarding the presence and desires of others. All individualists seem self-absorbed, absorbed by their own goals, values, gifts, and definitions of achievement. Ford, Edison, and many other inventors and industrialists of the Gilded Age would have been assessed in the same way. Haydn admitted concerning Stone: "He was a very difficult man of whom to make game, and any attempt to overreach him, or to draw him into an alliance or an enterprise that he did not approve aroused the antagonism of a very strong nature; and it is not to be gainsaid that enemies were made."[34]

After Leonard Jr's death, on January 10, 1880, *The New York Times* captioned his obituary with the following headline, "Generous gift to sci-

---

[33] Ibid., 194.
[34] Haydn, 85-86.

ence - Leonard Case leaves property valued at $1,500,000 for a 'scientific school'" and went on to state, "In the institution is to be taught mathematics and civil engineering, chemistry, mining, modern languages and similar branches."[35] After it was all said and done, Case had trumped Stone. The University as a whole would be named Case Western Reserve University and gain the reputation of perhaps, other than the University of Chicago, being the most reputable academic institution in the Midwest. The name of Adelbert College would be reduced to a building on campus named after Amasa Stone's son, and few people other than historians know when and how the name of Adelbert Hall, still on campus, originated.

To completely attribute Stone's initiating gift for the founding of an academic institution in Cleveland as expiation for his guilt incurred by the Ashtabula Bridge disaster misses the mark, if not being defined as recklessly judgmental. Of course, Stone wanted to pay tribute to his son as well as create his own legacy. It was at least partially "noblesse oblige," the *sine-qua-non* of having arrived, of having treaded on and conquered every obstacle in the way of success. As Haddad states, "noblesse oblige" implied

> that with wealth, power, and prestige came social responsibilities: if one was capable of doing a task to benefit another, one was obliged to do so. In the Old-World noble birth carried these responsibilities. Transported to American shores and lacking a hereditary aristocracy, a new concept emerged: civic stewardship - the notion that successful citizens owe a dual obligation of time and money to the community in which they have prospered is a uniquely urban interpretation of the ancient ideal.[36]

Stone fulfilled his obligation to Cleveland, his adopted hometown. He was on his way to wealth before he ever arrived in Cleveland. Contrast funding a major academic institution not far from his house with Rockefeller founding an academic institution in Chicago. Rockefeller did get his start in Cleveland, and all of the major ideas for becoming the wealthiest individual who ever lived were conceived in the city where he got his first job. It wasn't that Rockefeller totally forgot Cleveland, but his name would primarily be attached to the University of Chicago as well as such iconic venues as Rockefeller Plaza in New York City.

---

[35] Baznik, 66.
[36] Haddad, 114.

In a well-documented and intriguing presentation, Kenneth W. Rose explained that Rockefeller gave to sectarian enterprises, i.e., the name Baptist had to be attached to the institution or endeavor. He was especially influenced by Henry L. Morehouse and Frederick T. Gates, which led to gifts for Atlanta Baptist Female Seminary (later Spelman Seminary, his wife's maiden name) and Denison College in Granville, Ohio, a Baptist School. Gates persuaded Rockefeller to bankroll the University of Chicago by emphasizing the burgeoning West was without a thoroughly religious university. (He failed to mention Northwestern University in Evanston, Illinois, just outside of Chicago, but of course that was Methodist.) Gates argued the vacuum of sound religious training meant that youth were attending "state higher schools of irreligion."

Gates prophesied to Rockefeller that a university in Chicago would be the "fountain of western life" and would "lift so far aloft a Baptist College as an intellectual and religious luminary, that its light would illumine every state and penetrate every home from Lake Erie to the Rocky Mountains."[37] This was sheer myopia, or ignorance of the fact that almost all major educational institutions raised up for sacred purpose ultimately profane themselves. This pristine picture of Chicago would be lost by the time of Rockefeller's death. Al Capone's liquor traffic would conquer prohibition, and the city would host the most notorious house of prostitution in America, "The Everleigh Club."[38]

But even more curious was a secondary argument offered by Rev. C. O. King that the "atmosphere of city social life (meaning Cleveland) is unfavorable to correct habits of study, to those who intake it freely."[39] Gates compared Denison students to those at Cleveland High School, the latter being characterized by "worthless farce full of love and slang," and expressed fear that "our high school (Cleveland) is taking the moral stamina out of the youth of this city."[40]

---

[37] Kenneth W. Rose. "Why a University for Chicago and Not Cleveland? Religion and John D. Rockefeller's Early Philanthropy, 1855-1900," *From All Sides: Philanthropy in the Western Reserve* (Cleveland: Case Western Reserve University, 1995) 10.

[38] See Karen Abbott. *Sin in the Second City: Madams, Ministers, Playboys, and the Battle for America's Soul* (New York: Random House Trade Paperback Edition, 2007).

[39] Rose, 12.

[40] Ibid., 13

Rockefeller and Stone gave about the same amount of money to begin their respective schools, $600,000, given that an additional $100,000 was bequeathed to Adelbert College in 1883 when Stone died. Not that Rockefeller forgot Cleveland; he gave $1,625,522.57 to the Rockefeller Park, Women's Temperance Union, YMCA and Lake Side Cemetery, as well as many other entities.[41] Stone probably gave a similar amount to the charities to which we refer below. But Stone's worth at his death of about $10,000,000 was mere chump change compared to the almost $700,000,000 that Rockefeller had given to charity or left to his son, John. This enormous wealth was enabled by one single development unforeseen when the main product of Standard Oil was kerosene. Henry Ford, William Durant, and the Dodge Brothers as well as dozens of others, would produce millions of cars, all of them running on the world's most valuable commodity on which Standard Oil had a monopoly. Good thing. Edison did away with the need for kerosene. Maybe God really was on Rockefeller's side.

But Stone's philanthropy had not begun with a gift for a major university in Cleveland. As we have seen, he was one of the leading givers and oversaw the construction of the Cleveland Academy and generously gave for the rebuilding of Old Stone Church in 1857. In 1867, he was the primary financial source for the founding of a YWCA in Cleveland and Saint Vincent de Paul Charity Hospital. In 1869, he was one of the leading givers for an "unwed mother's retreat." He built an industrial school for the Children's Aid Society. In 1876, Stone initiated and financed a home for "aged, Protestant gentle women" (a nursing home for widows) which was eventually named the Amasa Stone House, but today is part of the A. M. McGregor Nursing Homes system.[42]

Building Adelbert Hall, the first building on Case Western Reserve's campus, was a tonic, a rejuvenation for Stone. Stone oversaw the project, the construction overrunning the cost estimate, and he advanced the additional $15,000. He made many onsite adjustments to the original plan, and in spite of a shortage of brick, the building was finished October 26, 1882. On that date an elaborate dedication ceremony took place, including Daniel Gilman, President of Johns Hopkins University, delivering the main address. Then a one-mile procession took place from Euclid Avenue

---

[41] Ibid., 20-23.
[42] Haddad, 71, 75, 80, 98.

to Adelbert College. There Amasa Stone presented the title deed to President Cutler and the Board of Trustees, and stated:

> GENTLEMEN: I take this occasion to deliver to you the title deed to these premises, and it is proper to state that the land on which these building are erected is the gift of fifty-three generous citizens of the City of Cleveland, who, with myself, appreciated the high and honorable status of the Western Reserve College, took measures to aid in bring it to this spot, where we hope its sphere of usefulness would be much enlarged and carried as an educational institution to as high a standard as any in this broad land.
>
> Our responsibility will cease with the completion of three structures, and its prosperity then rests with you. Yours is a grave responsibility, which we shall watch with much solicitude. It would be a great pleasure could I present to you these structures at this time entirely completed, but delay in getting the work started and the limited supply of brick last autumn, prevented our getting them under roof before winter; also a delay of more than six weeks by the impossibility of burning new brick last spring by the inclemency of the weather, delayed our progress this season, but through a kind Providence in having better health than I anticipated, the work has been pressed forward as fast as was consistent and secure good results in its execution.
>
> I could say much more as to my anxiety to have this institution brought to and maintained at the highest standard, but the high character of the board of trustees may put to rest any forebodings on this point.[43]

Perhaps a measure of peace entered into Stone's soul, and he listened to Cutler's prayer of dedication, "In memory of one who was suddenly called away from life, with all the high hopes which centered in him unaccomplished."[44] Former U. S. President and friend of Stone, Rutherford Hayes, responded to the benefactor by exalting "the noble gift of Mr. Stone and indicated the Board's belief that the beneficial influence of the college would be widely extended and would reach far into the future."[45] But his peace may have been disturbed during the lunch at Doan Armory when Judge Ranney made a toast to Leonard Case Jr.:

---

[43] *The Dedication of The New Buildings and Inaugural of Adelbert College of Western Reserve University, Cleveland, O., October 26, 1882, (Lately Western Reserve College, Hudson, Ohio)* (Cleveland: A. W. Fairbanks, 1883) 35.

[44] Ibid., 92.

[45] Ibid., 93.

Mr. Case, upon whom a vast estate had fallen, died a little over two years ago. Some two years before his death he made deeds for the purpose of founding this institution of applied sciences. He was distinguished as one of the best educated men of the country. His great hope was to found an institution that should teach science to be applied to the actual industries and uses of life. He contemplated that students should come prepared from other institutions as well as from the high schools of this city. The Trustees have undertaken, in a small way, to carry out his wishes. He proposed to found an institution that should be a benefit to the city in all future time.[46]

In spite of the homage paid to Case, the founding of Adelbert College of Western Reserve University was the high point of Amasa Stone's life. There were no more rivers to cross with bridges, no more buildings to erect, no more edifices to build and engineering challenges to solve. He committed suicide little more than one-half year later.

---

[46] Ibid., 58.

Chapter 12

Her Father's Daughter

The Iron Ore Business

Samuel Mather was a descendant of the prototypical Puritan family in seventeenth century America, Richard, Increase and Cotton Mather. However, Samuel was a descendant of Timothy Mather rather than his brother Increase, the latter being the father of Cotton. Increase and Cotton would go down in history as advocates of the Salem Witchcraft Trials, when 19 "witches" were hanged in 1693. Historians regard the Mathers as the most influential family in New England Puritanism. The ancestry of Flora's husband included five Samuels continuing through Samuel Livingston Mather, father of Samuel, Amasa Stone's son-in-law. Samuel Mather's grandfather had participated in the Connecticut Land Holding Company which bought 3,000,000 acres of land in the Western Reserve. By investing $18.46 in the company, Samuel Jr. became a "Director." In 1843, the father sent his son, Samuel Livingston, to investigate the land holdings, and he stayed.[1]

Fascinated with Cleveland's shipping possibilities and access to iron ore, Samuel Livingston Mather founded the Cleveland Iron Mining Company in 1853. This fortuitous initiative would place the Mather family in position to provide iron for the North during the Civil War and the ever-expanding needs for iron in post-bellum America. Iron ore in vast quantities was shipped to Cleveland from Minnesota, Wisconsin, Michigan, Ohio, and Pennsylvania. By the time Amasa Stone moved to Cleveland in 1850, Samuel Livingston Mather had already become wealthy and built a mansion on Euclid Avenue.

---

[1] Kathryn L. Mackley. *Samuel Mather: First Citizen of Cleveland* (Cleveland: Tasora Books, 2013) 2.

In Cleveland, Samuel Livingston and Georgiana Pomeroy Woolson raised two children, Samuel, born July 13, 1851, and Katherine Livingston born on September 3, 1863. When Samuel was two years old, his mother Georgiana died. Samuel Livingston remarried Lucy Gwinn who gave birth to one child, William Gwinn. William lived to be 95 years old, and died a wealthy man, having earned much of his money through the Cleveland Cliffs Iron Company. William Gwinn Mather became a civic leader, patron of the arts and a heavy giver to and trustee of Case Western Reserve University, Kenyon College, Trinity Cathedral, Lake Side Hospital, and other endeavors.[2]

Samuel, Flora Stone's future husband, was not a robust child, being small of stature with poor eye sight. His physical limitations became even more pronounced when he was almost killed in a mine explosion in Ishpeming, Michigan, at the age of 18. The accident fractured his skull, both arms and his spine. But the unfortunate event did not deter him from joining his father's iron ore business at the age of 22, and with that and other business interests, he would become the wealthiest man in Ohio. And other than John D. Rockefeller and J. P. Morgan, he was perhaps the richest man in America by the time he died in 1930, worth approximately $100,000,000.

The Romance of Cleveland's First Couple

Living only a few houses apart, with Flora being one year younger than Samuel, it was not difficult for Samuel and Flora to meet and eventually marry on October 19, 1881. Samuel was 30 and Flora 29 years old, a late marriage age for that period. They were inveterate letter writers, exchanging thousands of letters, each of them disappointed if they did not receive at least one letter per day when apart. The pre-marriage letters are bathed with pathos and often verge on the comedic with a hint of desperation. On January 14, 1880, Samuel abandoned all façade and threw his dignity with complete vulnerability at the feet of the woman he coveted:

> Flora, I am longing to have you love me, and to have you show it, and to have you think more of me than of anyone else. But unless you can do this, and can let the past be an oblivion without allusions thereafter, it will be better I fear for both of us to have this settled at once. If you can love me and I long to hear and know that you love me as I love you and as I want

---

[2] Ibid., 6.

you to love me, it will be as a tonic to my whole man and make me a new man – for I have been sick at heart, discouraged and dispirited and constantly fighting against the blues and not in the best of health.

I have broken all the barriers that my pride had set up for me – for it is far easier to entreat and plead when one is "up" than when "down." It always has been for me. I will not read this over or hesitate, but just send it and pray that my heavenly Father may bless to both our happiness.[3]

If making himself completely vulnerable would not work, perhaps making Flora jealous would. In the following August, Sam found himself in a compromising situation and was quite honest about it. Whether he exaggerated his wandering eye or not, is left to conjecture.

Last evening I strolled over to neighbor Chamberlain's to inspect the "renovation" under progress there, and while poking about was caught by Mrs. C. – who took me in hand and conducted me all over the place – inside and out. The time, the hour, the place, all but a mysterious halo of romance. I saw and cried "spare-oh-spare-oh-bird! (slang for "girl" or "young woman") but they heeded not my mellow wit. Flora, I had an old and ancient habit, when nought had happened and still to write – I had an inclination, to make "summat (something) out of nothing." This will, at all events, serve to turn your thoughts, for the time necessary to read this, upon me, and if you laugh at my foolishness, yes you will believe my love, will you not![4]

Flora in personality was loquacious, and apologetic about her talkativeness. She attributed this tendency to the solemnity and silence in the home in which she was raised. "We never were a talkative family, until lately, among each other. My Father, when we were little, was too busy & preoccupied to take any notice of us, & you know that my mother is usually very quiet. Since we have grown up my father has tried to talk more, and when he and I are alone together we get on beautifully. Clara & I often walk downtown together & never say a word & yet I enjoy being with her."[5]

After Sam accused her of vacillation, Flora responded:

I would be more than human, I think, if I did not feel a little vexed at the

---

[3] Letter, Samuel Mather to Flora Stone, January 14, 1880, Samuel Mather Family Papers, Container 2, Western Reserve Historical Society.

[4] Letter, Samuel Mather to Flora Stone, August 17, 1880, Samuel Mather Family Papers, Container 2, Western Reserve Historical Society.

[5] Haddad, 53.

moment of correction, so if you will tell me *now*, before I offend again, it will save that little bit of temper, for if you tell me *in cold blood*, I promise you I will not be in the least vexed. I *know* you can think of a half a dozen little habits which, perhaps, could be easily corrected. Give me a chance, my dear, & if then I annoy you again, give me a scolding. I shall deserve it, & if you *don't* tell me, you must bear half the blame of your own ill-temper, sir.

But now may I find a little fault with you, Sam? Why did you accept & then change your mind? That is quite a habit of yours, do you know? You said you had no good reason for declining. A good reason is not necessary for a large party, I should think. It seems to me that you ought to make up your mind that you will or will not do a thing, & then unless you have a very good reason for changing, *stick to it*. I know you often put the family out because they don't know what you're going to do, you will not say.[6]

Few people "stood up" Amasa Stone. Being a "no show" demonstrated either Samuel's clout or revealed that he was not all that comfortable in social gatherings. But neither of these traits prevented his entrance into the Stone family. Perhaps Samuel became more socially sensitive and people-oriented under Flora's tutelage.

The social gossip freely flowed in Washington D. C. papers and made its way to Cleveland. The reserved Sam felt his privacy had been invaded. Did he know this was the kind of news the public craved and sold more newspapers than either business or politics? Tabloids are the price of fame. Sam wrote Flora:

The morning "Leader" states than "Misses Stone & Mather are mentioned in all the accounts of Washington Social gayeties." And is this the outcome of the Mather family attempts at social distinction? I shall take my "walking cane" tomorrow morning and have an "interview" with the Washington correspondent, through the local editor. I do not intend to kill him outright.[7]

On October 19, 1881, Flora and Samuel were married at the Stone mansion in what *The Cleveland Herald* called, "A Quiet but Elegant Wedding of The Avenue Last Evening."[8] It was truly a Euclid Avenue wedding,

---

[6] Ibid., 56.

[7] Letter, Samuel to Flora, August 17, 1881, Mather Family Papers, Container 2, Western Reserve Historical Society.

[8] Haddad, 60.

as indicated by gifts from friends and neighbors. Haddad states, "Sam and Flora received paintings, etching, engravings, photogravures, books, an encyclopedia set, a full set of Shakespeare's works, and *The Encyclopedia of British and American Authors*. One such gift in particular delighted Flora, who wrote to Sam, 'Could anything be more delightful & elegant than a complete set of Thackeray in half calf?'"[9]

Flora indeed had married into American aristocracy, not only on the Mather side, but Samuel's grandfather had married into the Livingston lineage. Stephen Birmingham argues that the Livingstons are the most aristocratic family in American history.[10] Robert R. Livingston along with John Adams, Thomas Jefferson, and Benjamin Franklin, was on the committee to draft the Declaration of Independence. Flora's marriage was the most significant event in Amasa Stone's life to demonstrate that he had "arrived." But the arrival would be a blow to his ego, as he attempted to place Flora and Samuel within the Stone fiefdom. Feudal estates had fallen out of favor a long time ago, though they still lingered in the Anglophile mind. But the Mathers had a mind of their own.

Samuel did not need Amasa Stone as did John Hay. Samuel was under the tutelage of his father until Samuel Livingston's death in 1890. The son had to explain to his father why he was lending so much money to his wife during their honeymoon. The newly married couple must have kept their finances separate, perhaps the entirety of their marriage. Not that Samuel did not attempt to get on the good side of the Stone family. On January 14, 1881 he wrote Flora, "This morning I called upon your uncle, Mr. A. B. Stone, and sold him 40,000 tons of iron ore. So, you see, I am opening

---

[9] Ibid, 58-60.

[10] Stephen Birmingham. *America's Secret Aristocracy* (Boston: Little, Brown & Company, 1987). Here our story circles back to Cornelius Vanderbilt, who in working as a simple ferry man for Thomas Gibbons, found himself in a fight with the Livingston clan over rights to the New York waterways. Thomas Gibbons sued the Livingston family and won the first anti-monopoly case in American history, handed down by the United States Supreme Court on March 22, 1824. The outcome of Gibbons vs. Ogden (the Livingstons had leased what they thought to be their exclusive navigational rights to Aaron Ogden), was a slam dunk in today's parlance, since Gibbons was represented by Daniel Webster and William Wirt, Attorney General of the United States. See T. J. Stiles. *The First Tycoon: The Epic Life of Cornelius Vanderbilt* (New York: Vintage Books, 2010) 75.

up family relations with your family, young woman. He was pleasant and very well."[11]

Amasa Stone's Attempted Control Over His Daughter and New Son-in-Law

After the wedding, the Mathers immediately left for Europe but did not escape the contention about where they would live upon returning home. Flora's controlling father wanted Samuel and Flora to live with him and Julia. Thus, three days following the wedding, he wrote a letter, intending to put Flora on a guilt trip:

> I fully appreciate your kind and affectionate words and believe them to be sincere and I need not assure you they gave me great comfort.
>
> It may at times have seemed to you austere and harsh in expression, but I am unconscious of having uttered or done anything that I did not think was for your welfare & happiness, and you cannot in any manner so practically reciprocate what your mother and I have done for you and so place your future as to be as near us while we sojourn here below as practicable which cannot in the common course of things be for a great while. I expected that extra efforts of your mother and her position, would make her sick, but she has stood up nobly and seems very cheerful and well.
>
> We shall hope to hear from you very often. Please say to Mr. Mather that my leaving the station without bidding him good by was absence of mind on my part, and unintentional, but I got to the top of the hill before appreciating my neglect.[12]

Amasa and Julia Stone were rolling around in an 8,500 square foot house, mourning the loss of their last daughter. They had exerted sufficient pressure on John and Clara Hay to live next door. Now, Amasa would attempt patriarchal influence on his new son-in-law not to live next to them, but to live with them. He continued to communicate this domestic proposal to Flora. While in Italy, Samuel was so perplexed that he wrote his own father on January 12, 1882:

> I had a letter from Mr. Stone a few days ago in which he said "please let me

---

[11] Letter, Samuel Mather to Flora Stone, January 14, 1881, Samuel Mather Family Papers, Container 2. Western Reserve Historical Society.

[12] Letter from Amasa Stone to daughter, Flora, October 24, 1881, Samuel Mather Family Papers, Container 2, Western Reserve Historical Society.

know whether on your return you will take up your quarters, with *which we will be very glad to have you do.*" The understanding is his. I replied today, thanking him very sincerely, but saying that "it has always been my desire to have a little home of my own - in which I find Flora sympathizes with me and shares in." Therefore "that I can only say at present, in reply to your very kind invitation that we will be pleased to spend next summer with you in which time I hope to be able to find a home, suitable to my means, in such proximity to both our homes as will enable us to be in & out, daily." I answered thus because Flora said her Father's words mean a desire for us to live with them indefinitely as long as they lived.[13]

The ultimate decision would be to live in the Hay's house until the owners returned from Europe. Sam was torn between desiring to please Flora and maintain his own independence. He needed both psychological and physical distance between him and his new in-laws. He may have been initially intimidated by Stone. Samuel's 5 feet 6-inch stature, slight build, mustached, spectacled face was a study in contrast to the masculine features of Amasa Stone. He chaffed under the pressure put on Flora by her father. Again, Samuel wrote his father:

We think it is best to come home. Col. & Mrs. Hay have written Flora that they intend going abroad in June, if nothing prevents, & would be glad to have us take their house during their absence.

Flora had written her mother, as I had to Mr. Stone, that we would come there for the summer (if we came home, which, at that time, we had not determined upon) but as Mrs. Stone seems to think it would be advisable, it is probable that it will result in our going into the Hay House for the summer.

I would suggest not speaking about it, as Flora wrote Mrs. Hay at first telling her that we had already promised to come to Mr. Stone's and it stands just that way now.[14]

A Loving Marriage

For most of their marriage, Samuel Mather was on the road half of the time, making rounds to his mines, mills, and other business appointments: Detroit, Marquette, Escanaba, Chicago, Youngstown, New York, Grand Rapids, Tarrytown, Catskills, Pittsburgh, Washington, St. Louis, and Hot

---

[13] Haddad, 63.
[14] Ibid., 65.

Springs, Arkansas for his health. Thus, there are thousands of extant letters between Flora and Samuel, most of them located in the Western Reserve Historical Society. The couple was truly in love with one another.

One July 27, 1883, Sam wrote:

> Dearest Wife:
>
> It was a real grief to me not to get a letter from you while off on my two day trip among the mines – and not to have opportunity also to write you and tell you how I am constantly thinking of you and loving you....
>
> Upon my return I found your two letters of Sunday and Monday, and was much refreshed thereafter....
>
> Your dear letter of Sunday went right to my heart, and I wanted to kiss you and tell you that I needed charity – more than you....[15]

Even before marriage, Sam and Flora were spiritually introspective, if not theologically profound; Sam wrote to Flora:

> *Are* we to be held strictly to account for idle hours, idle words, and all mis-spent opportunities? These, if heaped on top of actual transgressions, will make a burden heavy indeed to bear. And how about the sins we commit in thought, even though not in deed?
>
> We are all in the same category, poor human mortals all of us. And at least can have full sympathy and mutual love and forbearance one with another. Knowing our own shortcomings so well how is it that we are otherwise than kindly and gentle towards others?[16]

Flora responded on September 13, 1881:

> Sam, my own dear love, no, we have *not* to give account for all our mis-spent moments & opportunities in the way you fear. I suppose we must give an *account*, for the Bible says it will be for "*every idle word.*" Poor me! But that is to God & your idea is that because of the ill-spent hours we have less chance of spending the future ones properly. That is true no doubt, the groove is worn & the water naturally flows there, but it would be very weak character, you are not that would make such an excuse. Do you

---

[15] Letter, Samuel to Flora, July 27, 1883, Samuel Mather Family Papers, Container 2, Western Reserve Historical Society.

[16] Letter, Samuel to Flora, September 28, 1881, Samuel Mather Family Papers, Container 2, Western Reserve Historical Society.

know, I believe God pitied the depression & discouragement that comes from physical exhaustion just as a mother loves & pities her little feeble child. I don't believe He counts as sins of omission or commission all our foolish & rebellious acts & thoughts. Perhaps I am getting into deep water. *I* know what I mean, but I fear I'm not making myself clear. Besides, you don't want it now – maybe - if the weather is cooler & you have slept well you are properly working away with energy at all sorts of disagreeable tasks, just to prove to yourself that you *can* overcome them.[17]

Devoted to Her Father

Samuel would never have the same relationship to Amasa Stone as did John Hay. And, Samuel Mather would have been a multi-millionaire outside of his marriage to Flora. Not so of Hay. Flora was a highly sensitive, introspective creature, always processing her motives. She reconciled the guilt of her wealth by giving and depreciating the gift. Flora would also carry out the memory of her father, more than did Clara. She wrote her former teacher Linda Gilford on July 4, 1891:

> Giving is such a cheap, easy virtue, a sort of penance, to quiet one's conscience. Character is the great thing, isn't it? But in this matter of possession, Miss Guilford, you know I have a double duty I want to do, as Father would wish with what he left me....
>
> [Mother] was a student of Biblical literature, always searching the best commentators. Both parents read aloud often, and well, - and how we used to enjoy Biblical Twenty Questions games on Sunday night! All good literature interested them, including particularly history, biography, art, essays, travel and current events. Mother's days were all too short for the reading she wanted to do.[18]

Shortly after their father's death, Clara and Flora commissioned John LaFarge to create a stained-glass window, gracing the back wall of Old Stone Church. "The window's over-all design is composed of three individual but related panels, approximately eighteen feet high, a taller and wider center section, flanked by a narrower lancet window to each side."[19]

---

[17] Haddad, 58.

[18] Ibid., 85.

[19] Jeannine Love. "John LaFarge and Cleveland's 1885 Amasa Stone Memorial Window: A Case for Re-evaluation?" 7. Courtesy of Don Guenther, archivist at Old Stone Church.

At the top of the center window is the word "benevolence," and at the bottom, "I delivered the poor and fatherless, and I caused the widow's heart to sing with joy."[20] The window depicts two women, one standing – the other kneeling. The standing woman is of more regal appearance, perhaps Mary, the mother of Jesus, and the other evidencing poverty and receiving Mary's blessing. The window is interpretated as Stone's founding of the Home for Aged Women, as well as other philanthropic endeavors.

Other interpretations claim the woman is St. Catherine, and the kneeling person receiving her blessing is Amasa Stone himself. If this is true, LaFarge scholar Jeannine Love believes that the family decided the depiction of Stone shouldn't be overtly blatant given his reputation for the Ashtabula Bridge and subsequent suicide. It was easy to transform Stone into an impoverished woman, by placing a cap on his head covering his bald dome.

In the side panels are chariot-like figures, possibly representing Flora and Clara, but Haddad believes the daughters to have been too modest for such ostentation. Less given to speculation is the inclusion of tools of the building trade depicted in the center window with which Stone would have been familiar – a trowel, a compass, and a hammer.

While we have a magnificent window, we also have a piece of art shrouded in mystery and controversy. Stone's life defied simple interpretation. Love argues that LaFarge's use of small pieces of multi-colored glass, textured sheet glass and layered glass to intensify the richness of color was for the purpose for outdoing anything done by Tiffany or other glass artists. This competitive motivation may be the facet of this creative enterprise which best represented Stone. He never intended to be second in anything.

Flora's last act of devotion to her father was building a chapel on the campus of Western Reserve University named for her father. It was designed in English Gothic style, by the famed architect Henry Vaughn. Charles Thwing, who was University President, envisioned a chapel as a "revival of spirituality, which must be the fundamental element in each generation."[21] Flora believed the chapel to be an appropriate memorial be-

---

[20] Ibid.

[21] Walter Leedy. "Henry Vaughan's Cleveland Commission: A Study of Patronage, Context, and Civic Responsibility." http: //architronic.saed.kent.edu/v3n3/v3n3.04.html. Flora initiated the project although Clara helped with the finances.

cause, "Our father had a strong conviction of the value of Biblical instruction and of church services. He believed…religion to be the only foundation of the individual or for the community life."[22] Historian Walter Leedy
may have been more correct when he stated, "WASP elitism reinforced the
prejudice for English Gothic."[23]

Nonetheless, the Chapel completed in 1911 served as the gathering
place for "compulsory chapel." Not a seat of the 500-capacity sanctuary
was vacant. With an increasingly pluralistic setting for a worldwide university, requiring students to attend chapel was discontinued by the late
1920s. A survey done by the students in 1924 indicated that "they perceived the importance of chapel "Nil" and "a disagreeable compulsory exercise breaking into the morning's work."[24] At the time of the Stone chapel
completion, Western Reserve University and the Case School of Applied
Science were still separate schools, and students quipped that a gargoyle
sitting on the west top of the tower was "spitting on their neighbors and
rivals at the Case School."[25]

Sitting in the shadows of the foyer is a marble bust of Amasa Stone.
Dressed in the style of Greek antiquity, he has on a Roman toga revealing
a muscular neck and upper chest. There is a disconnect between a draped
robe of the first century and the sideburns and mustache of the nineteenth
century. The gleaming white marble accents Stone's expansive forehead,
straight nose, and turned-down mouth which rarely smiled. Nineteenth
century likenesses forbade smiling with the exception of Abraham Lincoln
immediately after the Civil War. The contrast between the two is that Lincoln exhibits deeply etched lines in his face while Stone's likeness is void
of the anxiety and pain that constantly besieged him. Marble did away
with the worries and guilt which finally killed him. This was not possible
in photography, which may be the reason that the most oft seen picture of
Stone is a side head-shot.

Samuel Mather served as the keynote speaker for the dedication service. After listing his father-in-law's many accomplishments, he gave the
following tribute:

> It would have seemed superfluous to have tried to say anything to the citi-
> zens of Cleveland on that date in regard to Mr. Stone's personal traits of

---

[22] Ibid., 3.
[23] Ibid., 5.
[24] Ibid., 9.
[25] Ibid.

character, but I may perhaps be allowed to say at this later date that the impression which he made upon everyone that came in contact with him was of absolute personal integrity. It seemed to be impossible for him to deviate in the slightest degree from the strictest veracity, so much so that his absolute adherence to truth and accuracy was almost irritating to those who were not quite so rigid in such matters. In his personal intercourse he was one of the kindest, simplest and most unassuming of men. A brakeman of his railroad was always as sure of a courteous and considerate hearing from him as a senator or a millionaire. There was no man in the country so great as to daunt him, and none so simple as to receive from him the treatment of an inferior. He was a man extraordinarily clean in heart, in hands, in lips. His closest intimates never heard a word from him that might not have been spoken in a drawing room. Of his domestic life, I will simply say, a better husband, father and friend never lived![26]

The Mather Mansion

Samuel and Flora Mather had four children, the first named Samuel Livingston (1882 after Sam's father) and the second, Amasa Stone (1884 after Flora's father), a daughter Constance born in 1889, and another son, Philip Richard born 1894. The family lived with Julia after Amasa died, and split their time between Euclid Avenue and a retreat-like setting named Shoreby located at Bratenahl, Ohio, on Lake Erie.

In 1906, Samuel and Flora decided to build their own house at 2605 Euclid Avenue. They contracted architect Charles Schweinfurth to design the 43-room mansion at a cost of over $1,000,000. The Tudor-gothic Revival house sat on 2.4 acres of land which included formal gardens, a squash court and an 8-car garage.[27] The mammoth structure stood 180 feet deep, 90 feet wide, and 52 feet tall. The squash court was for men only, and one wonders how much squash the diminutive, near-sighted, partially crippled Samuel Mather played.

The entrance hall was 40 x 24 feet, and Samuel's library was 38 x 23 feet, which he filled with books from around the world. In the fireplace room stood a marble figure sculptured by Antonio Rossi in 1870, and purchased by Amasa Stone in 1877. It remained in the Stone mansion until the Mathers built their house. A ballroom occupied the third floor, measuring

---

[26] Samuel Mather. "The Amasa Stone Memorial Chapel," *Western Reserve University Bulletin* Vol. XIV, No. 6 (November 1911) 100.

[27] Mackley, 63.

27 by 65 feet. In it hung an eight by five-foot painting, "Emigrants Crossing the Plains," painted by Albert Bierstadt, who loved the American west, yet painted the scene while in Paris in 1860. On his first trip to Europe in 1870, Amasa Stone purchased the painting for $15,000 and it now hangs in the Cowboy Hall of Fame in Oklahoma City.[28]

Flora's bedroom was designed as a suite of rooms: a dressing room, a bedroom, a marble bathroom, and a sitting room, 25 feet by 21 feet with fireplace. Samuel had his bedroom suite adjacent to Flora's. How well did Samuel sleep in a house completed in 1911 after Flora's death, never joining her husband in this ultimate symbol of the Gilded Age? Such was the opulence and grandeur of Gilded Age materialism as contrasted to the Mathers' philanthropy. The house still stands on the campus of Cleveland State University for the occasional class lecture or social gathering. The visitor may be awed by this symbol of mammon, but will probably also sense a cold and vacuous gloom left by the unfulfilled aspirations of wealth trying in vain to define itself.

In the decades between Flora and Samuel's death, the family, including children and grandchildren, gathered at the mansion for Christmas. All of the family members with candle in hand, marched through the rooms of the first-floor singing, "It Came Upon a Midnight Clear," the procession ending around the table spread with a bounty of a Christmas Eve dinner, provided by servants.[29]

The First Couple of Philanthropy

Samuel and Flora Stone Mather became the first couple of philanthropy in Cleveland, as well as the most connected to whatever was happening in Cleveland. When Flora gave to something she became involved in it. Her 1902, $75,000 gift for a lady's dormitory on the Western Reserve campus lingered as a remembrance by an alumna:

> Mrs. Mather was a small woman with black hair whose attractiveness lay largely in her very expressive and kindly eyes. She was frail but seemed to have great nervous energy. Even in our thoughtless youth we used to wonder that she could come to us so often. She was a busy woman, mistress of a large house with a family of young children and she was actively engaged in church and civic affairs. Then too, she lived at a considerable distance and the drive out and back with horse absorbed much time but we

---

[28] Ibid., 76.
[29] Ibid., 77-79.

were never forgotten.[30]

Charitable causes had been instilled in Flora Mather by her parents. She was an organizer of the Young Ladies Temperance League, financially supported the Children's Aid Society, Home for Aged Women, The YWCA, The Bethlehem Day Nursery, and gave $50,000 to endow a history chair named for her pastor, Hiram Haydn. When praised for endowing the College for Women, she responded, "The praise belongs not to me; I am but the dispensing hand of a Father's bounty."[31] By "Father" she meant Amasa Stone. She gave heavily for the building of "The Goodrich House," named for her girlhood pastor on the campus of "Goodrich Social Settlement," in the vicinity of Old Stone Church. The model of Jane Adams' Hull House in Chicago had made its way to Cleveland.[32]

"Patrician"

Samuel Mather's contributions to the city of Cleveland were dizzying. His giving was eclectic: the Cleveland Museum of Art, Western Reserve University, Trinity Cathedral, Lake Side Hospital, Western Reserve Medical School, Cleveland Unit of the American Red Cross, the Municipal Association, the Civic League Community Fund (predecessor to the United Way,) and Kenyon College in Gambier, Ohio. According to his biographer, "During World War I the Cleveland Red Cross War Council set $2,500,000 dollars as their goal for a victory chest, but under Mather's leadership, the Red Cross raised $11,000,000."[33]

One story has it that Charles Thwing, President of Western Reserve University, visited Mather's office to plead for a $500,000 gift. He quietly responded that "He guess he could." He called his secretary and ordered

---

[30] Haddad, 74.

[31] Ibid., 73.

[32] According to Judith Freeman Clark, this idea did not originate with Adams. "In 1886, the first settlement in the United States, the Neighborhood Guild, was founded in New York City by Stanton Coit. This simple effort was patterned after similar settlements in England — namely London's Toynbee Hall. It was among the first of many such attempts by a growing number of college-educated individuals to help the urban poor. Vida Scudder, a Smith College alumna, founded the College Settlement Association in 1887 along with several other young women who had also visited London to view Toynbee Hall." Judith Freeman Clark. *America's Gilded Age: An Eyewitness History* (New York: Facts on File, 1992) 97.

[33] Mackley, 25.

her to draft a $500,000 check. After returning to her office, she opened the door and asked, "On which bank?"[34]

One brief biography stated that Mather, at different times, "was an officer or director in more than 25 corporations, and was intimately connected with the commercial industrial growth of Cleveland."[35] He was awarded the "Cross of Legion of Honor" by France in 1922, and the "Serbian Cross of Mercy" in 1920. His interests were varied, indicated by memberships in the "Archeological Institute of America," the "American Natural History Museum," the "New England Genealogical Society," and the "Pilgrims of the United States."[36]

William McKinley's presidential campaign manager, William Hayward, said of Mather following his death, "An aristocrat by nature and by inheritance, the hail-fellow methods of the ebullient and western democracy found him unresponsive. He was . . .the antithesis of the profane, red-blooded, swash-buckling of Mark Hanna, who gave up business for politics."[37] The *Cleveland News* said of him, "But for all of his wealth, his power in the steel and mining venues, and his gifts to charity, and the arts, Samuel Mather was a modest man. He disliked publicity, rarely was interviewed, and seldom made speeches."[38]

> Born a patrician, he remained a patrician through life. He respected himself, each task that came to him and his ability to perform that task well. Of deep religious faith, recognized his church as his first responsibility and became its foremost lay figure. A stranger to sham, he carried with him always the quiet, sincere dignity that was a part of his nature. [39]

When Mather died, he was worth between $80,000,000 and $100,000,000. He employed thousands of workers in his mines, on his ships, and on his docks. Even with all of his philanthropy, there is little, if anything, to indicate that he contributed to the betterment of his laborers. Such were the contradictions of the Gilded Age.

---

[34] Ibid.

[35] "Obituary, Samuel Mather: 1851-1931," Samuel Mather Family Papers, Container 2, Western Reserve Historical Society.

[36] Ibid.

[37] Quoted in Mackley, 26.

[38] "Samuel Mather Was Leader in Industry, Charity, Church," *The Cleveland News* (October 19, 1931) 2.

[39] Ibid.

Flora's Death and Funeral

Flora aged before her time. At age 54, she looked to be a very old woman. Her face was etched with deep lines, her eyes were sunken and surrounded with dark circles, cheeks drawn in towards her small mouth, which was closed to cover her ever-protruding teeth, of which she was always self-conscious. Her visage was glum and joyless. Life was ebbing away. In August of 1908, Flora was diagnosed with breast cancer. Sam was disappointed that it had not been discovered earlier. Sam wrote Clara that it was "unaccountable" that the cancer had not been previously found. He explained the lack of detection.

> Dr. Crile operated on her last January to cut off a mole on the breast and saw nothing else. Dr. James treated her in February in N.Y. & apparently saw nothing. Dr. Cushing looked superficially at her breasts about June 1st when she complained to him of a little pain; he says he understood her to say that the "scar" pained her - but seemed to see nothing to cause alarm then. Yet now he says it is either a *very* rapidly spreading tumor else it must have been there at times of previous examinations & yet was undiscovered. I cannot understand it and am distracted.[40]

Flora explained her situation to Clara:

> The breast is quite hard (but movable) & I've had considerable stabbing pain. They agreed that something should be done at once, & as Sam wrote you, named the best person to consult. We told the doctor we were both opposed to operate on people of our age. Dr. Cushing & Dr. Crile said he hoped an operation would be a cure:- that X-rays were used after to kill any possible traces, but that X-rays used externally had not been found to effect cures. He said an operation would be a cure.[41]

But there was no cure. Within six months Flora Mather was dead. She died on January 19, 1909, at the age of 56. On September 22, 1895, she had expressed her desires for her funeral. As always, the expectations of her father, Amasa Stone, had been her major responsibility in life, other than being a devoted wife and doting mother to her four children. "I wish no remarks made at my funeral, and that the prayers be entirely impersonal. I prefer that the prayers of the Episcopal burial service be used....I have tried to be a faithful steward of the means left me by my father. There

---

[40] Haddad, 99.
[41] Ibid., 101.

are higher virtues and humbler lives than mine; far more admirable, far more lovable."[42]

Her expressed desires were respected. The funeral service was held at Old Stone Church, the funeral ritual from the Book of Common Prayer was read, and there was no eulogy. The one half-hour service included selections sung by the church choir, Tennyson's "Crossing the Bar," "Rock of Ages," and "Jesus, Lover of My Soul." The burial service took place at Lake View Cemetery, and her remains were interred in the Stone plot, with her father, her mother, and her brother, Adelbert. Samuel would be laid beside her 21 years later. He never remarried. Flora had truly been the love of his life. Though nothing had been said about her at the funeral, there were many tributes expressed afterwards. Typical was the testimony paid by Western Reserve University President emeritus, Charles Thwing:

> Mrs. Mather was one of the most self-less women ever known to me. She had in a sense thrown herself outside of her selfhood. Yet she was a full, noble, and complete self. The mystery is explained somewhat by the fact that her purposes, her work, her interests, were found not in herself but in a life beyond her own. The intellect was keen, broad, deep, and high in its thinkings and aspirations. Her intellect was devoted to the understanding of great causes. Her intellect also was just and discriminating. Her heart was a worthy companion of the intellect. It overflowed with love to all and to each, to great causes and to all personalities.[43]

A Life of Sublimation

In vain does one search for any reference to Amasa Stone's death in the letters of Flora and Samuel Mather. The subject was simply taboo. The family was initially in a state of shock and continued in denial. Thus, it deserves to be asked, was all the giving and doing compensation for her father's suicide? Did she take it upon herself to prove to the whole world that her father was a person who could be trusted, rather than a person who through carelessness, had killed 98 people? Did Flora and Samuel ever visit Ashtabula? There is no evidence that they were present at the unveiling of the monument at Chestnut Grove Cemetery in 1895. Flora worked

---

[42] Unaddressed and untitled letter, Samuel Mather Family Papers, Container 10, Western Reserve Historical Society.

[43] Haddad, 110.

herself into an early grave just as did her father. Samuel was not oblivious to his wife attempting to become the ideal self and always falling short.

> In one of your letters from New York you spoke of your complaining about being tired & etc. I do not think you are at all of a complaining disposition – the contrary you have always been very brave & spirited & have never spoken half so much of being tired as most people do. Myself for instance – even though you must often have been weak & exhausted. I do occasionally feel like complaining that your energetic and unselfish nature impels you to do more than you are really equal to – I never think you do less than you are able to or that you make any ado about it.[44]

Denial of an event does not make it go away, but leads to repression, events embedded in our unconscious, beyond our awareness. The defense mechanism of sublimation becomes an unruly master of motives and actions. Its subterranean interiority harkens back to a time and place which we have tried to block out but becomes a driving force for much of what we do. Freudian psychoanalysis in the form of depth psychology had yet to invade the United States. As an introspective person, Flora was all that much more troubled, and there was no one to provide the feedback which she desperately needed.

Amasa Stone would have been proud of Flora. No daughter could have been more faithful to a father whom she deeply loved and respected. But she was not totally unaware that she was doing penance for the sins of the Father. One obituary stated that Stone "had had no extravagant habits, and being a man of no great depth of culture as regarded books, he had little to call his mind from business. He was not a person to set much store by social recreation. His daughter, Mrs. Mather is like her father and mother – plain and unassuming".[45] Flora was her father's daughter.

---

[44] Letter, Samuel to Flora, July 27, 1883, Samuel Mather Family Papers, Container 2, Western Reserve Historical Society.

[45] "More About Stone, Something of His Financial Dealings and Relations," unidentified newspaper article. Courtesy of Case Western Reserve University Archives.

Chapter 13

Two Tragic Deaths

Charles Collins' Death

Charles Collins was born in Richmond, New York, in 1826. He evidently attended college elsewhere before graduating from Rensselaer Institute as a civil engineer. He then worked throughout New England and in particular on the Boston-Albany Railroad. He came to Ohio in 1849 to supervise the construction of the Cleveland, Columbus, Cincinnati, and Indianapolis Railroad.[1] Collins had been Chief Engineer of the Lake Shore since its inception in 1858. George B. Ely, who was a one-time director for the Lake Shore and close friend of Collins, told a reporter that Collins was

> an extremely proud man and thought more of his honor than of his life. He was of a very nervous temperament, and the worry and anxiety connected with the Ashtabula accident has worried him terribly. So much has it pressed upon his mind that he could not sleep.

> He told me the other day that he had not slept since the trouble occurred and the comments which have been made by the public, many of them throwing the entire blame upon him as the engineer of the road, have driven him crazy without a doubt. A few days ago, in conversation with Mr. Paine, our general superintendent, he made the remark, "Here I have been working thirty years for the protection of the public and now they turn right around and kick me for something which I have had nothing whatever to do with."[2]

Collins' death was anything but a straightforward event. It remains shrouded in mystery to this day. Further clouding the picture is that no

---

[1] "The Death of Mr. Charles Collins," *Railway World* (February 3, 1887) 99.
[2] Ibid.

inquest was done at the time of Collins' death and whatever police reports were filed have since been destroyed by a flood which ruined the records of the Cleveland Police Department. Thus, all we can learn comes from newspapers and subsequent autopsies performed by two doctors. Collins had last been seen on Wednesday, January 17, but was not found until Saturday morning, January 20. The coroner concluded that Collins had died some 36 – 48 hours before being discovered. "The body lay in the bed on the front side, the pillow and sheets being drenched in blood. At left of the corpse were a large revolver, and an unopened razor. At the right was a double-barreled pistol. Three chambers of the revolver were found to be empty, but only one wound was discovered."[3] According to the coroner, Collins "thrust" the muzzle of the pistol into his mouth, with the ball going through the roof of his mouth, and coming out the head near the back.[4] The scene was truly macabre.

> The corpse presented a most ghastly sight and the body had already begun to decompose, filling the room with a nauseous odor. The blood flowing from the mouth and ears not only soaked the bed but left streaks across the face, presenting a repulsive spectacle. The features would identify the body but the livid face, distorted, discolored and stained, was terrible for a friend of Charles Collins to look upon.[5]

Collins did not leave a suicide note, but left an envelope seeming to have been a memory device for himself. It simply read: "No. 10 will leave at 11:15, No. 8 at 9:45." This would possibly indicate that Collins planned to travel the next day, and did not intend to commit suicide but was murdered, later argued by two physicians.

Two funeral services were held for Collins. The first took place on Tuesday, January 22, 1877 at the Collins' residence in Cleveland. Hiram Haydn presided over the Cleveland service, and said to a packed house, "Here is a gentleman who could not stand to bear suspicion that he was responsible for a matter that it had been clearly proven no responsibility could rest upon him."[6] The next morning a special train left Union Station at 8:30 a.m., and a service was held that afternoon at the house of Albert

---

[3] "SUICIDE! Charles Collins, Engineer of the Lake Shore Railway Dead!" *The Plain Dealer* (January 20, 1877).

[4] Ibid.

[5] Ibid.

[6] "The Funeral of the Late Charles Collins," *The Plain Dealer* (January 22, 1877).

Harmon, brother of Mary Collins, Charles Collins' wife. (Mary Harmon Collins was raised in this house, and it still stands today in Ashtabula.) Representing the Lake Shore was J. H. Devereux, who "made a few remarks on behalf of the railroad officials."[7] The casket was then taken to the family crypt at Chestnut Grove Cemetery, where it still rests today, only approximately one hundred feet from the monument dedicated to the Ashtabula Train wreck victims. What is strangest about all of this is that Amasa Stone never showed his face. He did not attend either service. He was not a pallbearer. There is no historical record of any expression of sympathy or any statement on the service of Charles Collins to the Lake Shore and Michigan Southern Railroad.

The Suicide Investigation

Embarrassed by the diagnosis of suicide, Collins' family hired two of America's most prominent physicians to conduct an autopsy on April 26, 1878. Stephen Smith was a surgeon to Bel Air and St. Vincent Hospitals, New York, and Professor of Surgical Jurisprudence at University Medical College, New York. When Smith died in 1922, the *New York Times* headlined his obituary with "Dr. Stephen Smith Dies in 100th Year; Famous Physician Was a Pioneer in Sanitary Reforms in New York City. HIS RULES FOR LONG LIFE Organizer of Local Board of Health and American Public Health Association Was Long at Bellevue."[8] The article went on to say that Smith was a professor of anatomy at Bellevue Medical College and had published a "Handbook of Surgical Operations" which was issued to every surgeon in the Federal Army. Smith was well qualified to do an autopsy.

At the Centennial celebration in Philadelphia, a British surgeon, Joseph Lister from England, lectured on the uses of carbolic acid as an antiseptic. Among the many physicians at the "Medical Congress," who strongly objected to Lister's lecture on germ theory, was Dr. Frank Hamilton of Bel Air Hospital in New York. Hamilton insisted that his method of leaving "the wound open in the air," would make the infection go into its "proper receptacle," or to the surface where it could be easily drained. In the summer of 1881, Hamilton along with the physicians William Bliss and David Agnew, attended the wounded James Garfield. They did use Lister's

---

[7] "The Last Tribute of Respect: Funeral Services of the Late Charles Collins at Ashtabula," *The Plain Dealer* (January 25, 1877).

[8] "Dr. Stephen Smith Dies in 100th Year," *New York Times* (August 27, 1922).

recommended antiseptic for spraying the wound and dressing it. Unfortunately, they were not sufficiently convinced of unseen microbes to sterilize their own hands. They repeatedly searched for the bullet with unsterilized fingers spreading infection throughout Garfield's body. The President died of sepsis on September 19, 1881. The three doctors had killed him. To add insult to injury, Hamilton and Bliss requested the U. S. government to pay them $25,000 each. Congress eventually paid them about one half of the requested amount, while the *Chicago Medical Review* skewered them for their incompetence.[9]

Hamilton, 56 years old when he examined Collins' skull, was born in Wilmington, Vermont, graduated from Union College in New York, and the Medical College of the University of Pennsylvania in 1835. Like Stephen Smith, Hamilton was employed by Bellevue Hospital in New York City. He was a surgeon for the North during the Civil War. (Remember that thousands died, not by amputation, but by the infection caused by surgery, as did the Confederate General, Stonewall Jackson.) Hamilton at some point served as President of the New York Society of Jurisprudence. By the time that he and Smith did the autopsy on Collins, Hamilton had written three widely accepted medical texts including *The Principles of Practice of Surgery*. Upon her husband being shot, Lucretia Garfield immediately called for Frank Hamilton, and a special train delivered him to Washington, D. C.[10]

Smith and Hamilton were the best that money could buy. In spite of their impeccable credentials, one wonders first as to their training in forensic science, and second, how unbiased would a physician be if he was paid to disprove that a person committed suicide. Nonetheless, the examination and subsequent reports were thorough and convincing that Collins was murdered, rather than taking his own life.

Smith and Hamilton denied that Collins shot himself in the mouth, leaving a hole in the back of his head. By the time the two physicians examined Collins' skull, it had been horizontally cut in two, i.e., the dome had been removed and re-attached. Smith found a hole about four inches behind the left ear. The hole was cleanly cut, about ½ inch in diameter. On the opposite side of the skull was a hole more irregular and jagged,

---

[9] See Candice Miller. *Destiny of the Republic: A Tale of Madness, Medicine, and the Murder of a President* (New York: Doubleday, 2011) 16-17. 159, 195-196, 210, 223, 227-228, 231, 257, and Ackerman, 439-440.

[10] Courtesy of Leonard Brown.

"Two and one-half inches in length, anteroposteriorly and one and one-half inches in breadth, vertically;" indicating that this was the exit hole for the bullet. "On examining the margins of these bones, it is apparent that they were driven outward." Smith concluded that the "left opening was made by the entrance of the ball" and "the right opening was made by the ball by its exit or its attempted exit." Because Smith was not at the death scene, he could not verify the hole that had been made by a bullet in the wall. (It seems that Smith would have gone to the Collins' house, but we have no record of that.) Smith even experimented with an 11-inch Navy revolver. How he did this is difficult to fathom, other than mutilating the skull of an otherwise good cadaver.

> In the several experiments with a navy revolver eleven inches in length, seven shooter, it was uniformly found that when a ball entered the skull at the left opening & in the direction marked by the groove, struck the inside of the skull at a point corresponding with the centre of the right opening, and either fractured the bones outwardly, and escaped, or fell back into the cavity of the skull. In every case a broad central fractured [sic] was found extending from the left to the right opening, in a zigzag direction, with fissures branching at various points, and a continuation of the central fracture beyond the right opening through the right half of the frontal bone.[11]

Smith concluded the ball did escape, and for it to have gained sufficient momentum, the revolver would have been held more than four inches from the skull, which is inconsistent with a person committing suicide, who would have placed the barrel against his skull (Seemingly Collins would have put the barrel of the gun at his temple rather than four inches behind the left ear.) But why the arbitrary four inches, Smith did not explain. The trajectory of the bullet led Smith to believe that the wound was not immediately fatal. Thus, after being shot by himself, Collins's body would have demonstrated thrashing, convulsion, and other reactions, but Smith did admit the body could have been quiet from paralysis, rather than "restless, and excited from the injured brain." But whichever, there certainly would have been a convulsion of both hand and gun when the weapon was fired. This raised one of the two most critical issues in deciding whether Collins' death was murder or suicide. How could a large revolver have been neatly

---

[11] I have had both of the original reports of Smith and Hamilton in my hands. They are located in the Ashtabula Historical Society in the Jenny Munger Museum at Geneva-on-the-Lake, Ohio. I have used the transcriptions found in appendices B and C, pages 125-136 in Thomas Corts' *Bliss and Tragedy*, cited throughout this book.

in his hand in perfect composure, lying inside the flank of his left leg, after committing suicide?

The other critical issue was that Collins was right-handed, and it would have been highly irregular for a right-handed man to shoot himself with his left hand in the side of the head. "The body was found in the bed where he was accustomed to sleep, several days after he disappeared, in a very natural position, the clothes smoothly drawn up to the axillae (armpits.)" Thus, Smith further concluded, "The place of entrance of the ball, the position of the left hand partially enclosing the handle of the pistol, the undisturbed state of the body and of the bed clothes, are entirely inconsistent with the theory of suicide." Smith further concluded that by either complete paralysis or contrary convulsions caused by a self-inflicted wound, "the hand would have fallen from the side of the bed in a state of extreme abduction of the arm at the moment." Smith unequivocally stated an "opinion" that "Mr. Collins came to his death by a shot wound inflicted by other hands than his own."

It is unclear whether Dr. Frank Hamilton accompanied Dr. Stephen Smith or did an independent investigation. His report was not nearly as thorough or technical as Smith's. The only wrinkle which Hamilton added was that if Collins did not immediately kill himself and remained conscious, he would have resorted to the other weapons, the derringer and knife nearby. But the reader is possibly asking why would he have not fired another round as there was a bullet left in the chamber? "If he recovered partially, the general condition of repose is not explained. If he recovered consciousness completely, everything remained unexplained." Hamilton was even more emphatic than Smith that the profuse hemorrhage and other facts in the case "are utterly irreconcilable with the theory of suicide while they do not present the shadow of an objection to the theory that his life was taken by another person while he was lying asleep in his bed."

Thus, this episode can only be left with questions; if not suicide, who did kill Collins? Who would have had a revenge motive? Would Collins have revealed further details that would have implicated Stone or someone else? The answers to these questions are buried with Charles Collins and possibly a murderer whom he may or may not have known. All of this took place before fingerprinting, DNA tracing, and a computer bank profiling sinister persons.[12]

Amasa Stone's Will

Whether Amasa Stone had planned suicide or at least had a premonition that he was soon going to die, when he made out a new will on November 30, 1882, we will never know.[13] What we do know is that he was a thorough business man, and would cover what needed to be taken care of in death even as he had in life. His wife, Julia, was to be given $25,000 per year for the rest of her life, divided into monthly installments. She would also receive the residential mansion, and after her death it would go to "Flora and her heirs." The rationale behind this bequest is that Amasa had already built John and Clara a palatial residence. Also, to Julia went all the furnishings of the house as well as the marble bust of himself. This means that the bust was carved during Stone's lifetime.

To Clara and Flora, the father bequeathed $600,000 each and to his sons-in-law, $100,000 each. Emma Stone, daughter of Daniel, who had come to live with Amasa following her mother's death, had married Stone's secretary, Samuel A. Raymond, and each received $25,000. To his brothers, Liberty and Andros, he left $7,000 each and the same to his sister and his wife's sister. Eighteen nephews and nieces were left $5,000 each. He left to Rebecca and Albert Blood $6,000. (Albert Blood had married Rebecca. the daughter of Amasa's sister, Lavinia.) Stone did not forget his gardener and coachman, to whom he left for "faithful service" the houses they now lived in. His favorite charities were Adelbert College, $100,000; the Home for Aged Women, $10,000; the Childrens Aid Society, $10,000. Stone stipulated that the principal given to these institutions was not to be spent, only the interest. He was naïve at this point, as there is no way to guarantee that said gifts would remain in perpetuity. The will voided all previous wills, especially the one made on June 21, 1879. He appointed

---

[12] A similar incident took place during the scandal plagued administration of Warren G. Harding. A Harding operative, Jess Smith, seemingly took his life on May 30, 1923. He too shot himself in the left temple, but was right-handed. The gun disappeared and no autopsy was performed. What is known about the event points to suicide rather than murder. See Jared Cohen. *Accidental Presidents: Eight Men Who Changed America (New York: Simon & Schuster Paperbacks, 2019)* 223 – 229.

[13] There is no title on the will. It begins with the words, "In the name of the benevolent Father of all." Amasa Stone papers, 1874-1881. Container 1. Western Reserve Historical Society.

John Hay and Samuel Mather as the executors and "empowered them to release and discharge all debts or claims due or owing by or to me." Always private, he desired "that no inventory, appraisal or public sale of any of my personal estate be made."

Other than the $10,000 to Augustine G. Stone, the will revealed no surprises. Born in 1840, Augustine was the son of Amasa's brother, Daniel, and was 14 years old when his father died. Augustine, who would have been 42 at the time of Amasa Stone's will, was a favorite of his uncle. Amasa named him as the executor of his estate, in case Mather or Hay should die before the will was probated. He worked for his uncle Andros, Amasa's brother, who was President of the Saginaw Mining Company. Because of his industriousness and business savvy, Augustine became second in command. His diligence impressed his uncle Amasa. Augustine, never marrying, had become financially secure in the "iron, steel, and iron ore business" by the time he died at the age of 70.[14]

The New York Central

As well as being a business partner, Amasa Stone's relationship to Cornelius Vanderbilt was a friendship, with mutual respect for each other. When Stone testified to the integrity and lucidity of Vanderbilt at the probate of the latter's will in November 1878, he stated before the court that he "received his instructions from the Commodore and always obeyed them without regard to the other officers of the road." He further stated that he always considered the railroad management of the Commodore to be "eminently wise."[15] Stone also told the judge that the Commodore had intended to leave the management of the Railroad in the hands of his son, William Vanderbilt.

But the relationship Stone had with Cornelius would not continue with William, who had little regard for the Lake Shore and its one-time President. The final fallout came with William's decision to buy the Nickel Plate, the parallel line to the Lake Shore. Out of protection for the most prized possession of his long and frenzied business career, and claiming the purchase was not profitable or legal, Stone resigned as the Director of the New York Central on November 29, 1882.[16] He must have hand-

---

[14] Bartlett, *Simon Stone Genealogy*. 475.

[15] "The Commodore's Roads," *New York Tribune* (November 23, 1878).

[16] Letter, Amasa Stone to William Vanderbilt, November 29, 1882. Courtesy of John Hay Library Archives, Brown University.

delivered the letter, because on the next day, out of political correctness if not regard for Amasa Stone, Vanderbilt responded:

New York, Nov. 30, 1882

Amasa Stone Esq:

Dear Sir:

I received your letter and resignation. Your connection with the company has been long and your services in its behalf so important that I could not contemplate without an interview your final action upon so serious a step. We have always acted in entire harmony and if in the present instance your vote cannot be given in accord with the majority, you can attend at the next meeting and there have your vote recorded as you think right, after hearing the report of the committee of lawyers to whom the question was referred and the advisory council of Mr. Bullitt, a distinguished lawyer of Pennsylvania, who was called in.

Whatever action you may then take will not and ought not to disturb our personal or official relations, and I assume to withhold the presentation of your resignation believing it was activated solely by a delicacy which prompted retirement rather than differ by your vote with those with whom you have always been in harmony on all questions heretofor submitted. I therefore can justly request that you attend the next meeting of the Board and there act as you think proper.

Yours very truly,

Wm Vanderbilt, Pres.[17]

The purchase of the Nickel Plate was more than a "delicacy," and there would be no next meeting. Since his election as a director in 1869, Stone had attended 18 Board of Director meetings, most of them in New York City.[18]

---

[17] Letter, William Vanderbilt to Amasa Stone, November 30, 1882. Courtesy of John Hay Library Archives, Brown University.

[18] I compiled the number by going through the Penn Central Transportation Company Records, 1835-1960, Oversize Vol. 24. Courtesy of Bentley Library, Ann Arbor, Michigan.

Deterioration of Health

On October 7, 1867, Amasa Stone was driving his carriage through the center square of Cleveland, the same square that exists today. Something startled the horse and Stone and as his passenger, J. C. Buell, were ejected from the back of the carriage, hitting their heads on the cobble-stoned street. Both of them were knocked "senseless" and were unconscious or disoriented for perhaps a minute or so.[19] Stone was hurt worse than Buell. The injury would probably be diagnosed as a concussion today. This was another incident in Stone's life when "progress" was detrimental. For several months, the "Cleveland Square" had been closed off for removing the plank streets and paving them with cobblestone. Only a few days before the accident, Ontario and Superior Streets opened.[20]

Stone was only 49 years old at the time, and to say that he never fully recovered from this accident or that it brought some type of personality change is conjecture. We do know that he walked with a cane in the later years of his life. Buell was only 32 years old at the time of the accident, and just a little over two years later committed suicide by shooting himself through the heart, November 15, 1869.[21] The Board of Directors of the Second National Bank, of which Stone was President and for which Buell was a cashier, suspected that Buell was guilty of embezzlement. The Board of Directors had planned a meeting that Monday evening to investigate the matter. Rather than face an inquiry, Buell took his life that morning. Over the weekend, Stone had refused to release any information to the press.[22]

As early as January 4, 1882, Stone began seriously to complain about physical problems which were later intensified by Hay's absence on a European vacation. He complained of lack of bowel movements and lack of sleep. He conferred responsibility for his enterprises upon his son-in-law. "…should I be taken away, you are the one to take the helm, as you and Raymond only are posted in detail as to my affairs."[23] Towards the end of 1882, Stone became more alarmed about his health and constantly urged

---

[19] "Painful Accident" *The Cleveland Leader* (October 17, 1867).

[20] W. Scott Robison. *History of the City of Cleveland: Its Settlement, Rise, and Progress*. (Cleveland: Robison and Cockett, 1887) 144.

[21] "A Shocking Tragedy," *Cleveland Plain Dealer* (November 15, 1869).

[22] "The Buell Deficits," *The Cleveland Herald* (November 16, 1869).

[23] Letter, Amasa Stone to John Hay, January 4, 1882. Courtesy of John Hay Library Archives, Brown University.

his son-in-law to return from his European vacation. On December 10, 1882, he wrote: "Had you got through your visit abroad, it would be gratifying were you here, as there is no one I have confidence in with whom to submit my affairs as yourself..."[24]

The letters of 1883 read like a death watch, and a pathetic plea for Hay to return. On January 30, 1883, Stone believed himself so ill that he would not be able to return to his office for "some time." He then beseeched, "I can only say that I wish you were here now."[25] Four days later, he complained of lack of sleep and again more than hinted that Hay should return. "It is commendable in my wife, not to distrust your plans, and ask you to come home, and I do not do it, but I think it my duty to inform you of the situation of things, which my wife would not really appreciate."[26] On February 8, Stone diagnosed his illness as the fragmentation of his business empire. It was "nervous prostration caused by anxiety about many matters. It is seemed to be everything combined to go wrong all at once. All that I anticipated in the way of depreciation in values seemed to be realized. This worry and sleepless nights brought me into an unpleasant condition in which forebodings of being entirely disabled occupied my mind."[27] Stone was in a financial and emotional meltdown, at least from his perspective. Part of his financial problem was Hay himself. At least four times Stone accompanied his pleas with a $2,000 check. On Christmas day, 1882, Stone plainly stated that he was pained to know Hay's expenses "exceeded his estimation," but that he would send another $2,000.[28]

All was not a tale of woe. In October of 1882, Stone wrote, "Say to the children that Jumbo has been here, and that I have seen him and twenty-one other elephants of all sizes. They are trained to maneuver like cavalry." He also felt vindicated in the founding of Adelbert College. "All eminent men from abroad agree that it is the best unverified field for

---

[24] Letter, Amasa Stone to John Hay, December 10, 1882. Courtesy of John Hay Library Archives, Brown University.

[25] Letter, Amasa Stone to John Hay, January 30, 1882. Courtesy of John Hay Library Archives, Brown University.

[26] Letter, Amasa Stone to John Hay, February 3, 1883. Courtesy of John Hay Library Archives, Brown University.

[27] Letter, Amasa Stone to John Hay, February 8, 1883. Courtesy of John Hay Library Archives, Brown University.

[28] Letter, Amasa Stone to John Hay, December 25, 1882. Courtesy of John Hay Library Archives, Brown University.

building up a great, incredible institution of any in the land."[29] He further stated, "I am conscious of having done a good thing in this last one (the College.) The buildings fully meet my expectations, but this would not have been the result had I not given their construction most of my time… "[30] He also took satisfaction in the other two charities (Home for Aged Women and Children's Aid Society) which he had founded, having given him "much pleasure and satisfaction. They are giving out a full measure of my expectations."[31]

On April 13, 1883, Stone informed Hay he had not been to his office for some four weeks, and that sleeplessness and indigestion still prevailed and "I am really miserable."[32] On April 21, he wrote his last letter to Hay, and as far as we know, the last letter that he ever wrote to anyone. "As to myself, you specially inquire about, I cannot give you encouragement – I have not been to my office for some time – nervous prostration seemed to be my first misfortune and sleeplessness has followed from then until now…."[33] One newspaper reported upon Stone's death that he also suffered from eczema, a disease of the skin. He turned to opium in order to sleep, but that only left him more depressed.[34]

Why John Hay did not cut his 12-month vacation short remains a mystery. Even his biographer admits he acted selfishly with the rationalization, "I came abroad, hoping to get some benefit to my own health." And, "As this is the last visit we shall make to Europe in many years, I want to see as much as convenient." The day after the Hay family departed for America, Stone committed suicide on May 11. Having boarded the steamship *Germanic*, they knew nothing of his death until they reached the New York harbor, a week later. Taliaferro writes, "Clearing customs with the bounty of clothing, jewelry, artwork, furniture, and antiques acquired dur-

---

[29] Letter, Amasa Stone to John Hay, October 19, 1882. Courtesy of John Hay Library Archives, Brown University.

[30] Letter, Amasa Stone to John Hay, November 13, 1882. Courtesy of John Hay Library Archives, Brown University.

[31] Ibid.

[32] Letter, Amasa Stone to John Hay, April 13, 1883. Courtesy of John Hay Library Archives, Brown University.

[33] Letter, Amasa Stone to John Hay, April 21, 1883. Courtesy of John Hay Library Archives, Brown University.

[34] Unidentified newspaper article courtesy Case Western Reserve University Archives.

ing eleven months of travel, was no small chore, but they got through as quickly as they could and hastened to Cleveland."[35]

To Hay's credit, he attempted to encourage his father-in-law as much as possible. On May 2, 1883, he wrote the following:

> I have been reading the life of Carlyle, and the other day I walked down to the little house where he lived and died, and near which his statue now stands in bronze. At your age, he suffered precisely as you do, deep nervous depression, persistent indigestion and loss of sleep – a general disorder and irritation of the entire nervous system. His misery seems to have been of the keenest character. Yet he lived to be eighty-six years of age, and the last twenty-five years of his life were comparatively healthy and free from pain. I met the other day at dinner an old gentleman named H., eighty-two years old. He told me that between sixty and sixty-five his digestion seemed hopelessly impaired. He could eat and drink nothing and slept very little. Now he dines out every night and is the gayest of the company wherever he is. I rely on your strong constitution, your sober and moral life, the reserve of vitality you have about you, to wear out all your present troubles and to bring you to a healthy and happy condition again. You have so much to live for, to enjoy the results of the good you have done and to continue your career of usefulness and honor. [36]

Stone's Death

Amasa Stone was as methodical and punctilious in death as he had been in life or as the *Chicago Daily Tribune* put it, his "suicide was in keeping with Mr. Stone's life – secret, dark, and thorough."[37] He would be in control of his demise as he had attempted to be in control of every event and person in his life. He would not leave his fate in the hands of physicians, not even in the hands of God. Not that he did not consult physicians. On December 25, 1882, he informed Hay that "Dr. Cushing visits me daily." How long the "daily" went on, we do not know. It is safe to say, he did not consult Cushing on how he should end his life.

On May 11, 1883, Stone was feeling so badly that he stayed in bed until 11:00 a.m. This was due to weakness, depression, and lack of sleep throughout the night. At 1:00 p.m., he informed his wife he would lie down for a nap. At 4 p.m., Julia knocked on his bedroom door but received no

---

[35] Taliaferro, 223.

[36] Thayer, Vol. I, 416-417.

[37] "Obituary," *Chicago Daily Tribune* (May 13, 1883).

answer. She then checked the bathroom door, finding it locked. After calling her husband's name, she assumed that he might have fainted and been unconscious. She called for the butler, who climbed through the transom above the door. He found Stone, shirt off, slumped over the bathtub with a hole in his chest, to the left of his sternum, in the approximate location of his heart. The instrument of death had been a .32 caliber Smith and Wesson revolver lying by the tub. The one empty chamber indicated that Stone had been efficient; he did not need another round.

The butler, along with the coachman, carried the body to the bed. "We didn't know at first what was the matter, but when we laid him in bed, we saw that his night shirt was stained with blood. Then we found the revolver, which was on the floor to the left of where he lay. It was a plain black handled weapon that he had in his room." A coroner named Isom examined the body and filed a report that "mental derangement led to self-destruction. The fatal bullet passed directly through the heart, and is still imbedded in the body."[38]

Stone left no suicide note. A newspaper reporter somehow learned that he had given some instructions to his secretary S. A. Raymond on that afternoon, but as to the content we have no knowledge. It may have been the location of his will, since no one else seemed to have known that he had left one. As much as possible he had put his house in order. He would control the end of his life, if he could not control the events in his life, the financial freefall of three rolling mills in which he was heavily invested. One reporter raised the existential question: "As one walked under the two maples of the gate of the dead man's residence on Euclid Av., the third door east of Muirson, this bright morning, he could not help exclaiming: 'Why did the possessor of all these attractions, desire to shorten his life?'"[39] The "attractions" in the grip of sickness and failure had become much less attractive. Failure was not in Amasa Stone's vocabulary. Stone would never offer an apology for the Ashtabula Bridge disaster.

No Cleveland newspaper account that I have read related his suicide to the worst incident in his life, but that was left for those further away. The *Chicago Daily Tribune* paralleled the Collins and Stone suicides: "The

---

[38] Unidentified newspaper article, courtesy of Case Western Reserve University Archives. There is a contradiction in that we do not know if Stone had his shirt on or off.

[39] "Startling Suicide: Amasa Stone the Millionaire Shoots Himself", unidentified newspaper courtesy of the Case Western Reserve University Archives.

Ashtabula Bridge which went down with horrible loss of life, December 29, 1876, was built from plans made by Mr. Stone who was President of the Road, and was constructed under the supervision of Mr. Collins, the Chief Engineer. Both men ended their lives by tragic suicide."[40] When Collins died, *The Nation* in New York City did not let Stone off the hook. "The Ashtabula Bridge was erected under the supervision of the president of the road, who was himself a bridge builder; it was an expensive structure, containing a large excess of material, but its design was peculiar, and one which no one familiar with iron bridge work would be willing to adopt."[41] A reporter, except for not noting Stone's bald pate, gave a fairly accurate description of his physical appearance in the later years of his life.

Amasa Stone was not a man to attract attention to himself. He was about six feet tall, of medium build, inclined rather toward slimness, of dark complexion, had a full iron gray beard and moustache and rather a heavy head of gray hair. His nose was of a strong type, but not overly large, and his forehead was expansive and indicative of natural abilities. He was somewhat stoop shouldered and always dressed in plain black broadcloth, cut with little regard to the latest style. He wore little or no jewelry.[42]

Funeral

Three days after his death, May 14, at 2:00 on Monday afternoon, a funeral was held at the Stone residence. The prediction that Stone would have the least people at his funeral did not prove to be true. The house was packed; everybody who was somebody in Cleveland was present: "Colonel W. H. Harris, George H. Ely, Colonel H. C. Parsons, General James Barnett, Mr. T. P. Handy, General J. H. Devereux, Hon. Horace L. Foote, Rev. Charles Terry Collins, ex-Postmaster Allen, Mr. James Clark, General H. H. Dodge, Judge Prentiss, Mr. J. H. Wade, Mr. J. H. Hoyt, Hon. Amos Townsend, Mr. R. K. Winslow, Judge Williamson, Hon. H. B. Payne, R. D. Noble, Esq., Judge R. P. Ranney, Colonel William Edwards, Ex-Mayor Charles Otis, D. W. Cross, Judge B. Burke, Dr. D. B. Smith, Judge B. O. Griswold and a host of others."[43]

---

[40] Ibid.

[41] Quoted in "The Late Charles Collins," *Cleveland Plain Dealer* (January 29, 1877).

[42] Description comes from an unidentified newspaper article, courtesy of the Case Western Reserve University Archives.

[43] "Simple Services at the Funeral of Amasa Stone Today," *Cleveland Plain Dealer* (May 14, 1883).

Stone was as unpretentious in death as he was in life. Perhaps he had indicated his desires to Samuel Raymond on the day of his death. "The body attired in a suit of black lay in a plain cloth casket, the lid of which is covered with a profusion of lilies and ivy leaves."[44] The service was simple and brief. A choir from Old Stone Church sang an appropriate hymn, "Art Thou Weary?" Rev. Arthur Mitchell, pastor of Old Stone, read from the book of Job, and attempted to provide a rationale for Stone's suicide. "Mr. Stone overtaxed the powers of his brain. There is no difference…between dying suddenly of a troubled heart and of falling a victim to disease of the brain. Both are dispensations from Above, and must be born in patience." The moralizing of the *Chicago Daily Tribune* may have been more accurate than the platitudes of Arthur Mitchell:

> The lesson of Mr. Stone's blighted life is one which young men might study with advantage. Something more than an indomitable will is needed to insure complete success. No man is so strong or so wise or so able that he can safely rely upon his own judgement in matters in which other men are stronger, wiser, abler than he. Mr. Stone's life would have been happier and in no sense less successful, had he had the perception to recognize, and the courage, to confess, his own limitations and deficiencies. He would have been a greater man had he been less conscious of his greatness – a stronger man had he distrusted more his own strength.[45]

The service ended with the choir singing, "When Our Hearts Are Bowed with Woe." It was a mournful affair. The pallbearers consisting of A. B. Stone of New York, a brother; H. P. Stone, of Springfield, MA., W. Thompson of New York; A. G. Stone, E.E. Stone, George W. Howe, S. A. Raymond of Cleveland and Samuel Mather, bore the body to a horse-drawn hearse, to be transported to Lake View Cemetery and be placed in the Brainard vault until the arrival of the Hays. The vault is still in existence in section 1, close to Lake View Avenue.

The Plain Dealer predicted more "pretentious services" at the interment, but that was not to be. Ten days after the funeral, burial took place at Lake View Cemetery. Again, Arthur Mitchell officiated, and the Adelbert College Glee Club sang. Twelve friends and business associates served as pallbearers.[46] In the preceding days, Adelbert Stone's body had been moved from Woodland Cemetery and placed in what would become the Amasa

---

[44] Ibid.

[45] "Amasa Stone," *Chicago Daily Tribune* (June 8, 1883).

[46] "Amasa Stone's Funeral." *New York Times* (May 25, 1883).

Stone plot, section 10 of Lake View Cemetery. In the end the *Cleveland Plain Dealer* attempted the best spin possible. "Near friends say that the depression of spirits which culminated in suicide was not caused by Mr. Stone brooding over his own losses, but by the keen chagrin he felt that many of his friends have suffered through investing by his advice in the iron works which failed."[47]

The family received many letters of condolence, but none more meaningful than the following from George Duffield, author of the famed hymn, "Stand Up for Jesus,"

Detroit, May 12, 1883

My Dear Friends,

The first and strongest impulse of my heart is at once to take the train for Cleveland, but this not being possible, the same impulse leads me to assure you in writing that this day I remember you where most of all you desire to be remembered.

So recently called into the Valley of the Shadow of Death myself, I feel near enough to reach out my hand, and thus express a sympathy beyond all power of words.

I knew your dear husband well. I loved him much, more every time we met, and I shall ever reverence his memory as that of one of the noblest of men with whom it has been my privilege to be acquainted.

I knew not, until long after, the sudden and overwhelming grief into which you both had been plunged by the loss of that beloved son. The last time I saw him he spoke so tenderly of your own health in this connection, and I could readily see that he was bearing it for both, for you and for himself. And now that overtaxed and overwearied brain and overburdened heart are at length at rest.

The ways of Providence are indeed "a mighty deep." Take the case of Mrs. Garfield - but under date of July 30, 1882, she wrote me as follows: "In the great future we hope we may find the answer to all our questionings. Until then I remain in search of what wisdom I may be able to gather out of time."

---

[47] "Simple Services," *Cleveland Plain Dealer* (May 14, 1883).

May you, my dear friend, be able eventually to say the same.

In heartfelt sympathy,

Sincerely yours,

George Duffield[48]

Drowning in a Sea of Red

The weight around Amasa Stone's neck was as heavy as the iron rails which his rolling mills had produced. On February 3, 1883, he had written his son-in-law: "The enclosed slip taken from yesterday's *Leader* will account to quite an extent for my nervousness for some time past." The title of the article read, "Heavy Failure – The Union Iron and Steel Company of Chicago Financially Embarrassed – Too Much Material, High Wages and A Decline in the Price of Rails, Amasa Stone of Cleveland, One of the Heaviest of Its Creditors." When Stone was approached by a reporter, he said he was too busy to answer questions and only made a brief statement: "Too Much Material, High Wages, and The Decline in the Price of Rails." The newspaper further stated, "The President of the Company, A. B. Stone, was said to be a Man of Large Experience," but the workmen indicted A. B. by accusing him of hiring a superintendent who had mismanaged the company due to the fact that his only qualification for the job was his being part of the Stone family.[49]

Throughout 1882-1883, Amasa Stone was drowning in a sea of red, while battered by labor strikes and lawsuits, some by him and some against him. On February 3, 1883, the Kansas City Rolling Mill shut down.[50] A. B. Stone was President, and Amasa was heavily invested. The Mill was "greatly behind in its payroll." As men gathered at the gates of the closed mill, some made "vague and impossible threats while others were broken down and cried like children." The Rolling Mill was perhaps the greatest

---

[48] John Hay (ed.), *Amasa Stone: Born April 27, 1818, Died May 14, 1883* (Cleveland: De Vinne Press, n.d.) 42-43.

[49] "Heavy Failure," *Cleveland Leader* (February 2, 1883).

[50] "The Mills Shut Down," *The Kansas City Evening Star*, Vol. 9, No.121 (February 3, 1883).

failure of a Kansas City industry in the latter half of the nineteenth century. The plant employed between five and six hundred personnel.[51]

The Kansas City Rolling Mill and The Chicago Union Iron and Steel Company were joint corporations: the failure of one guaranteed the failure of the other. Amasa and A. B. were the leading stockholders in both ventures. On February 3, 1883, the *Chicago Daily Tribune* reported the Chicago firm to be in debt $2,300,000.[52] The day before it had locked its gates, the *Tribune* reported Amasa Stone was one of the heaviest creditors. The closing of the mill came at a great loss to Chicago, putting 2,000 men out of work. The plant had partially shut down in March of 1882 because of strikes.

The company hired immigrant labor as scabs and employed Pinkerton detectives to protect them. A mob of some one thousand people marched down State Street, but there was hardly any violence beyond throwing stones, mud, bricks, and rotten vegetables.[53] The *Tribune* concluded that the failure of the police to protect substitute labor would dissuade companies from locating in Chicago, perhaps costing the city $10,000,000. When a *Tribune* reporter interviewed Stone on February 2, 1883, asking the reasons for shutting down the mill, he responded, "As I understand their orders were light, prices low and the efforts that were made to reduce wages were not successful by one half." The *Tribune* concluded that Stone had not been kept posted but we can be sure that Stone knew much more than he communicated.[54]

Finding a Diagnosis

In the same year Stone committed suicide, George Miller Beard died. At the time of his death, he was the leading authority on "neurasthenia," which was the scientific word for nervousness. The term had first been used by Edwin H. VanDeusen, Superintendent of the Michigan Asylum for the Insane at Kalamazoo from 1859-1878. VanDeusen described neurasthenia as a state of nervous exhaustion which might lead to insanity without proper intervention. But VanDeusen, because of his administra-

---

[51] "Scientific Miscellany: Kansas City Industries," *Kansas City Review of Science and Industry*, Vol. V, No. 3 (July 1881) 185-186.

[52] "The Iron 'Age,'" *Chicago Daily Tribune* (February 3, 1883).

[53] Mob Rule," *Chicago Daily Tribune* (February 28, 1882).

[54] "Mr. Amasa Stone Apparently Had Not Been Kept Posted," *Chicago Daily Tribune* (February 2, 1883).

tive duties, did not fully develop his thesis. Thus it was left to Beard, who died at the age of 41 to write several volumes on this term, and introduce the newly discovered malady to the general public, at least to those with enough time and money to learn about and treat such an ailment.

It was difficult for Beard and his disciples to describe exactly of what neurasthenia consisted. With 250 individuals writing 330 articles on neurasthenia between 1870 and 1910, there was a wide variance as to causes, symptoms, and cures. Almost every disease known to humankind was placed under the term. According to neurasthenia scholar Daniel Schuster, "Ultimately, the popularization of neurasthenia built upon its array of symptoms and lack of a clear psychology, which made the condition protean, and responsive to what varied Americans sought in a diagnosis. An advertising tool to sell products, medical flair to spruce up stories, and ultimately a way to understand suffering."[55]

The tendency to relate every blister, rash, hot flash, and every bout with tiredness, anxiety, anemia, irritability, melancholy, and you name it to neurasthenia enabled quackery to abound. Dr. Miles Medical Association advertised restorative tonic, restorative blood purifier, heart restorative, and restorative nervine. From our perspective, the most amusing was the "electric belt," which would take care of lethargy, sexual debility, and other limitations. The American Electronic Company sold their version for $5.00, and for an extra $2.00 could be added an optional "family attachment for the discreet parts of the body that needed recharging." Not even the margins of Stone's personality had room for such foolishness.

In spite of our snidely laughing at such remedies as the above from the perspective of our understanding of pathology and in particular psychology, the advocates of neurasthenia were on to something. The telegraph, the railroad, electric lights, and other sweeping technological changes were disrupting humanity's rhythms with nature, and neurasthenia was "Part of the compensation for our nation's progress and refinement."[56] We now refer to this cultural challenge as modernity, with the collapsing of time and space as one of its central characteristics.

Added to the above was a growing body of literature addressing the meaning of manliness. Schuster accurately assessed that during the Vic-

---

[55] David G. Schuster. *Neurasthenic Nation: America's Search for Health, Comfort, and Happiness, 1869-1920* (New Brunswick, NJ: Rutgers University Press, 2011) 38.

[56] Ibid., 20.

torian period, a real man was in control; in control of his wife, his children, his economics, while at the same time being in control of his inner world, which consisted of his thoughts and emotions. Narrating the "market revolution" that transformed America, historian John Larson states, "A man's fortune was supposed to bear witness to his character, industry, and frugality, and attention to hearth and home."[57] The poet W. H. Auden believed that "the most striking difference between Americans and Europeans was to be found in their attitudes toward money. No European associates wealth with personal merit or poverty with personal failure. But to the American what is important is not so much the possession of money but the power to earn it 'as a proof of one's manhood.'"[58]

The harsh realities of existence, which in Stone's case consisted of the shrinking of his fortune, the death of a son, and a catastrophic event which he may not have been able to do anything to prevent, all subtracted from his sense of self-worth. To confess his inability to cope with life's circumstances was an alternative worse than suicide. Nineteenth century bankruptcy scholar Edward J. Balleisen observes, "For tens of thousands of nineteenth-century Americans, bankruptcy represented dashed hopes and circumscribed prospects. Indeed, in a society that endlessly celebrated entrepreneurial success, business failure could exact a substantial psychological toll, not infrequently leading those who endured it to the whiskey bottle or the insane asylum."[59]

In 1882, Cornelius Vanderbilt's son, Cornelius Jeremiah, put a bullet through his brain. Being Amasa Stone or Cornelius Vanderbilt brought a lot of liabilities. The new age of technology was as deleterious for Stone as it was for the twelve hour day wage earner. Both were stuck on an industrial treadmill which would hardly allow for changing pace, much less getting off. In Veblen's words, "So long as the comparison is distinctly unfavorable to himself, the normal average individual will live in chronic dissatisfaction with his present lot; and when he has reached what may be called the normal pecuniary standard of the community or of his

---

[57] John Lauritz Larson. *The Market Revolution in America: Liberty, Ambition, and the Eclipse of the Common Good* (New York: Cambridge University Press, 2010) 99.

[58] Lewis. H. Lapham. *Money and Class in America* (New York: Ballantine Books, 1988) 266.

[59] Edward J. Balleisen. *Navigating Failure: Bankruptcy and Commercial Society in Antebellum America* (Chapel Hill, NC: The University of North Carolina Press, 2001) 14.

class within the community, this chronic dissatisfaction will give place to a restless straining to place a wider and ever widening pecuniary interval between himself and the average standard."[60]

At age 65 Amasa Stone had strained all he could strain.

---

[60] Thorstein Veblen. *Theory of the Leisure Class* (New York: Viking Penguin, 1994) 31.

Chapter 14

Assessing Responsibility for the Ashtabula Bridge Disaster

The Origin of Tragedy

In 1888, statistics showed that for ten years or more train bridges on American railroads had been failing at the rate of 25 per year, which averaged one bridge per year for every 5,000 miles of railroad. "Structural steel with a safe working strength 20% greater than wrought iron came into general use in about 1890. In less than five years, it was exclusively used for bridges, and wrought iron shapes were no longer rolled."[1] Steel gradually replaced iron as it was stronger and less expensive to make. Almost all changes in technology are created by the discovery that the present technology is inefficient, too expensive or a liability so great that neither seller nor buyer can rationalize the risk. Thus, for every accident, especially those sufficiently destructive to demand extensive analysis, the question is asked, "Was this event due to human error, or an inherent flaw in the system that can be remedied?" Of course, whatever the remedy or decision, the verdict does nothing to reverse the misfortunes of those who have experienced a horror staggering the imagination.

One false remedy is to participate in a collective defense mechanism, which assigns guilt to a particular person or problem and justifies or exonerates other individuals or corporations who may be equally guilty. It was a ploy put into play in the all-encompassing account of the beginning of human existence, at least as the Bible tells the story. In fact, almost every evil enacted by humanity is in the atavistic narrative found in Genesis 1-4.

---

[1] Henry G. Tyrell. *History of Bridge Engineering* (Chicago: Self-published, 1911) 184.

Whether one interprets the account of our beginnings as a comprehensive myth, or a historical event which actually took place, human beings are caught in life's ultimate plot, foreshadowing every sin and hardship plaguing human existence: greed, lust, murder, lying, condemnation, descent into evil, family conflict, nature's challenges to human welfare, the pain which is the common lot for all humanity, and yes, casting blame: "The woman made me do it." and, "The serpent made me do it." Whatever our literary, religious, or historical opinion on this archetypical incident recorded in the Bible, all other explanations of our coming into existence woefully fail to challenge our imagination and satisfy our existential longing to bring meaning out of simply being an infinitesimal individual with the other seven billion individuals residing with us on planet Earth.

While explaining some of our constant duress, what Henry David Thoreau diagnosed as "quiet desperation," the Genesis account created a problem which has never been resolved by Christianity or any other religion or philosophy. "Why did a good and gracious God create persons whom He knew would sin and fail, causing unimaginable suffering such as the Ashtabula Bridge disaster?" Without searching for a satisfying answer as to what really caused the deadliest train wreck up until that time into a full-blown theodicy, justification of the ways of God, I adopt an explanation championed by Nathaniel Taylor (whom we discussed in Chapter One) as well as other New England divines. Taylor, a contemporary of Stone, argued that the first sin committed by Adam was "certain" but that he and all future human beings had "power to the contrary."[2] Jonathan Edwards would have believed that everyone who died in the Ashtabula accident was predestined to die, and even Stone was predestined to build the bridge. But for Taylor, common sense states that both Stone could have decided not to build the bridge, and the passengers could have chosen not to get on the train.

Taylor, a teacher at Yale at the same time that Stone was growing up on his Charlton farm not many miles away, attempted an ameliorating position for the iron-clad determinism of Calvinism, a theological paradigm which would allow human beings some input into everyday events, as well as their eternal destiny. Taylor's writings were a valiant effort to retain the freedom of human will, while keeping intact the sovereignty of God. Taylor admitted along with Calvin, Wesley, and Edwards that all men sin. But

---

[2] Sidney Earl Mead. *Nathaniel William Taylor 1786-1858* (Chicago: University of Chicago Press, 1942) 189.

the sin is not because of some defect in human motives, but the circumstances in which we find ourselves. Outside of the grace of God, individuals "do freely and voluntarily set their hearts, their supreme affections on the world, rather than on God; they are conscious that this supreme love of the world is the fountain and source of all other sins."[3] Like many Gilded Age Americans, Amasa Stone was consumed by wealth, or more correctly, wealth had consumed him.

The Reductionism of Mere Survival

The Church was not the only institution providing energy for the belief in unmitigated progress. Herbert Spencer's "survival of the fittest" had invaded America, providing a strange blend of optimism and pessimism. As in animal and plant life, according to Spencer, the weak, not being able to acclimate to a changing environment, would, if not die, at least not be able to thrive. The strong would not only prevail, but be able to harness the rapidly changing discoveries, explorations, and inventions readily available to the person who could and would adapt and persevere. Of course, this formula provided an air tight rationalization for the expanding gap between the rich and poor, the inequities that defined the Gilded Age more than any other epoch in American history.

Not until Washington Gladden and Walter Rauschenbusch would anyone challenge this hell-bent philosophy which would eventuate into isogenics. If there were any "Jeremiads" to offer corrections or counter this prevailing theory on poverty, these voices were muted. To discover a sermon that undermined the prevailing zeitgeist requires diligent research and often comes up empty. Few turned for guidance to the wisest man who ever lived and certainly knew something about success: "The race is not to the swift, or the battle to the strong, nor does food come to the wise or wealth to the brilliant, or favor to the learned: but time and chance happen to them all" (Ecclesiastes 9:11).

In other words, there are circumstances beyond our control, and that was especially true for railroad employment; in short, other than white collar managerial jobs, they were dangerous. Any responsibility actually demanding physical contact with part of a railroad's operation was more hazardous than any other vocation. Gilded Age historian Rebecca Ed-

---

[3] Nathaniel Taylor. "Native Depravity and Free Will," in *American Christianity: An Historical Interpretation with Representative Documents* Vol. II, H. Shelton Smith, et. al. eds. (New York: Charles Scribner's Sons, 1963) 33.

wards claims that "From 1860 to 1900 counting employees, passengers, and bystanders, some 200,000 were killed by railroads."[4] "Brakemen" were especially vulnerable; they "often suffered crushed hands and arms, especially in windy, rainy, or icy weather, and many fell to their deaths. On the Illinois Central Railroad between 1874 and 1884, one out of every 20 trainmen suffered death or disabling injury; among the brakemen the rate was one in seven."[5]

Almost unbelievably, the Interstate Commerce Commission reported "for the year July 1, 1888 to June 30, 1889, 1,972 railway employees had been killed on the job, and 20,028 men were injured."[6] Or to put it another way, "In 1889 for every 117 trainmen employed, one was killed, for every 12, one was injured."[7] Working for the railroad in England seemed to be much safer in that, "One trainman in every 329 was killed, and one in 30 injured."[8] We can only speculate that the more deadly statistics of American trains were due to irregular terrain, longer working hours, and lack of safety precautions. There is also the possibility of good old fashioned American pride and competition entering into the equation, boasting efficiency, speed, and power. Walter Rauschenbusch, whom we discuss below, stated, "We do everything more strenuously and recklessly than others. Our machinery is speeded faster; our capital centralizes faster; we use up human life more carelessly; we are less hampered by custom and prejudice. If we are headed toward a social catastrophe, we shall get there ahead of scheduled time."[9]

Of course, the American populace quickly became aware that the railroads were a mixed blessing. In the 1840s as the railroads began to challenge the efficiency of canals, one luddite wrote,

> The railroad stems directly from Hell. It is the Devil's own invention, compounded of fire, smoke, soot, and dirt, spreading its infernal poison throughout the fair countryside. It will set fire to houses along its slimy tracks. It will throw burning brands into the ripe fields of the honest hus-

---

[4] Edwards, 4.

[5] Ibid., 60.

[6] Walter Licht. *Working for the Railroad: The Organization of Work in the Nineteenth Century* (Princeton, New Jersey: Princeton University Press, 1983) 190.

[7] Ibid., 191.

[8] Ibid.

[9] Walter Rauschenbusch. *Christianity and the Social Crisis* (New York: The MacMillan Company, 1907) 239.

bandman and destroy his crops. It will leave the land despoiled, ruined, a desert where only sable buzzards shall wing their loathsome way to feed upon the carrion accomplished by the iron monster of the locomotive engine. No, sir, let us hear no more of the railroad.[10]

As late as 1900, a comedian told his New York audience "When I finally reached Washington and a stationary bed, I had to hire two men to shake the bed all night and pour cinders down my neck."[11]

Walter Rauschenbusch Confronts the Gilded Age

Walter Rauschenbusch was the first American theologian who comprehensively described the fault lines quickly expanding between the rich and the poor and that there was a large part of the American population being shaken by a technological and industrial earthquake driving them to urban cores which only increased their problems. These problems did not originate in America, and no one had foreseen them by having researched the cotton mills of Manchester, England and the crowed slums of London. After reviewing the Church's escape into monasticism and eschatology, Rauschenbusch commented on England: "From 1760 to 1818 the population of England increased 70%. The poor relief increased 530%.... Here then we have the incredible paradox of modern life. The instrument by which all humanity could rise from want, and the fear of want, actually submerged a larger part of the people in want and fear."[12]

Rauschenbusch awakened to the plight of the poor after taking a pastorate in the Hell's Kitchen area of New York City. "Laboring there in a squalid tenement section for eleven years (1886-97) he came to see that the acute social problems with which he was surrounded demanded a rethinking of the Gospel as he had previously understood it."[13] He would later remember: "I saw how men toiled, all their lifelong, hard toilsome lives, and at the end had almost nothing to show for it; how strong men

---

[10] Stewart H. Holbrook. *The Story of American Railroads* (New York: American Legacy Press, 1947) 41.

[11] Sean Dennis Cashman. *Americans in the Gilded Age: From the Death of Lincoln to the Rise of Theodore Roosevelt* (New York: New York University Press, 1984) 29.

[12] Rauschenbusch, 217.

[13] H. Shelton Smith, et al., 202

begged for work and could not get it in hard times."[14] Rauschenbusch became the Social Gospel's greatest prophet, and in the span of his brief life, wrote its most defining literary works, *Christianity and the Social Crisis*, *Christianizing the Social Order* and *A Theology for the Social Gospel*. Unfortunately, the later individualistic personal salvationists who clung to "fundamentalism" and its myriad expressions of evangelicalism pigeonholed him as a "liberal" and ignored him, much to the American Church's impoverishment. This wholesale banishment was the result of theological ignorance on a grand scale which haunts American evangelicalism even today.

Though he did not explicitly state he was undermining Adam Smith, Rauschenbusch must have been aware of his antithesis to the ruling capitalist philosophy. No clergyman summed up more accurately the working man's plight. He wrote in 1907,

> No attempt is made to allot to each workman his share in profits in the joint work. Instead, he is paid a fixed wage. The upward movement of their wage is limited by the productiveness of his work; the downward movement of it is limited only by the willingness of the workman to work at so low a return. His willingness will be determined by his needs. If he is poor or has a large family, he can be induced to take less. If he is devoted to his family, he can be induced to take less. The less he needs, the more he can get; the more he needs, the less he can get.[15]

In particular, Rauschenbusch condemned Beecher and Russell Conwell and all other American preachers who claimed if a person was poor, it was their own fault. He was the first American Christian theologian to delineate the contradictions in capitalism.

> In the same way we shall have to see through the fiction of capitalism. We are assured that the poor are poor through their own fault; that rent and profits are the just duties of foresight and ability; that the insurgents are the cause of corruption in our city politics; that we cannot compete with foreign countries unless our working class will descend to the wages paid abroad. These are all very plausible assertions, but they are lies dressed up as truths.[16]

---

[14] Christopher H. Evans. *The Kingdom is Always but Coming: A Life of Walter Rauschenbusch* (Grand Rapids: William B. Eerdmans Publishing Company, 2004) 61.

[15] Rauschenbusch, 231.

[16] Ibid., 350.

Personal Ethics vs. Corporate Behavior

Theologian Reinhold Niebuhr argued that though it may be possible for an individual to be "moral," it is impossible for a group, corporation, or nation to be moral. The individual will act according to his own group's self-interest or advantage, rather than a competing group or foreign country's best interest. Thus, a soldier will kill and maim in wars, something that would not enter his thought processes back home in Middletown, USA. But in so doing, the soldier will scar himself and those closest to him for life.

Groups are held together by mutual self-interest at least implicitly accepted or necessitated, and maybe even explicitly stated, or elaborately drawn out. The peaceful and well-intended "Mayflower Compact" was a declaration of war on the Indians, though none of the "Pilgrims" would have interpreted the document in this manner. They would simply develop a theology and government which allowed them to steal land belonging to someone else and remove the occupants from the soil which had been inhabited for generations by their ancestors. The laws and interests of any group are always at the expense of other groups. Of course, there will always be both corporations and nations with economic and political goals which are perceived to be collaborative and advantageous, but these are the exceptions. Railroading could only thrive at the expense of other railroads.

Transcendence over the interest of one's group or nation is rare. A moral magnificence and magnanimity, which directed and defined the lives of both Roger Williams and Abraham Lincoln, have not for the most part defined the interests of either political or corporate America. Ethicist Lawrence Kohlberg depicted a person who is at the highest level of ethical development as a person who is autonomous, oblivious to what the group thinks, and transcends receiving affection or esteem from one's peers. Kohlberg scholar Doug Scholl wrote, "Those at the highest level of moral maturity are characterized by a major thrust towards autonomous moral principles, which have validity and application apart from authority of the group or persons who hold them, and apart from the individual's identification with those persons or groups."[17]

---

[17] Doug Scholl. "The Contribution of Lawrence Kohlberg to Religious and Moral Education" in *Religious Education,* Randolph Miller, ed. (New York: The Religious Education Association, 1971) 366.

The above "moral maturity" would be disastrous for a profit motive. In Stone's situation, his only route to self-worth was pleasing America's richest man, Cornelius Vanderbilt, and showing a profit each year to the shareholders of the New York Central. Thus, during the years surrounding the 1873 panic, Stone unbelievably showed a profit and paid dividends to the shareholders. None of the board of directors questioned his integrity, or asked, "How can we be making a profit with dilapidated equipment, cut freight rates, less freightage, lower passenger fares, and fewer passengers?" All of these questions were obliterated by the profit motive. Maury Klein and Harvey Kantor accurately describe the chasm between individual conviction and corporate actions for the railroads.

> If the corporation lacked a body to kick or a soul to damn, so, too, did it lack a conscience to prick. As an impersonal organization it was dedicated to the single purpose of making money. Individuals within the corporation might be sensitive to social needs or feel some twinge of social responsibility, and on occasion they might even attempt to formulate policy with these nobler ends in mind. More likely they discharged their consciences in private acts of charity outside the company. The business of the corporation was business. It had little room for sentiment, and whatever social benefits it conferred were byproducts of the relentless quest for profits.[18]

No railroad director tried to find the disconnect and contradictions, therefore suggesting that maybe more money should be spent on the Ashtabula Bridge, than used to fatten their bank accounts. But the lack of integrity, and neglect of details, was the singular reputation of the New York Central, owner of the Lake Shore. Railroad scholar Walter Licht wrote concerning the Central,

> A systematized plan of management did not evolve there. For a variety of reasons, including the manner which the company was formed, and the individuals who were involved, the senior executives of the railroad paid little attention to administrative problems. The railroad ran in an ad hoc fashion, with a general superintendent and a few immediate officers, supervising practically all aspects of the line's operation. A vague centralized, departmental structure, emerged from the situation, with top executives assuming responsibility over specific functions.[19]

---

[18] Klein and Kantor, 54.
[19] Licht, 18.

Niebuhr would not have interpreted the Central's fuzzy lines of communication, lack of holding employees (especially peers) accountable, and placing financial success above passenger safety as particularly unique to the Central, because both nations and corporations have little ability to critique themselves. In fact, the ability to objectively assess one's actions may be the salient characteristic of what it means to be human. In other words, the nation is a corporate entity held together much more by force and emotion than by mind. "Since there can be no ethical actions without self-criticism, and no self-criticism without the rational capacity of self-transcendence, it is natural that national attitudes can hardly approximate the ethical."[20] Niebuhr condemned John Hay for his stance against the 1877 strikers which we have already narrated: "When a man like John Hay regarded the labor riots of 1877, which arose from the injustices of buccaneer capitalism as evidence of the venality of labor, and took occasion to reaffirm his individualistic creed, the judgement can hardly be regarded as an honest one."[21] Some 25 years before Niebuhr's' *Moral Man and Immoral Society*, Water Rauschenbusch, in his *Christianity and the Social Crisis,* had picked up on this same theme of group loyalty. "There has never been a social class or group which has not punished to the best of its ability anyone who betrayed the interest of the class and which did not visit bitter condemnation on those actions which endangered its safety than any others."[22]

The Character of Amasa Stone

What kind of person was Amasa Stone? Was he honest, trustworthy, and faithful, contributing to the general weal? Did he tell the truth? Was he ethical in his business transactions? Was he a person of integrity, respected by those who knew him best, as opposed to those who only knew him by reputation? Stone lived in the acquisitive age, and possibly no one is more representative and even stereotypical of taking a hard-earned seat, and in his case many seats, in the post war industrial colossus with its many appendages and faces.

Stone had betrayed himself, or perhaps one could say that the American ethos of hard work, constant vigilance, and the rule of empire had betrayed him. There was no way to count the cost and calculate the risk

---

[20] Niebuhr, 88.

[21] Ibid., 124.

[22] Rauschenbusch, 326.

of juggling multiple enterprises that exponentially multiplied, becoming more entangling and more enslaving. When in his 20s Stone woke up in the middle of the night to solve the structural problem of the Howe Truss, he did not simply find a solution; he had created his own nightmare. He had traded a good night's sleep after a hard day of farm work for constantly being on the job. The new world of industrial output on a massive scale, presenting engineering challenges that had never been encountered, not only required labor during the day but a constant racking of the brain as to how to go about that work.

These entangling complications leading one where he had never been before exemplified the Gilded Age. Stone was a pioneer in the vanguard of building bridges and extending railroad lines, often in record time. "How To" manuals did not exist. Often when Stone initiated a project, it was a first in scope, size, and in particular obstacles, which were daunting and would have stymied anyone other than a man who believed anything could be done. Railroad buff Stewart Holbrook describes the pioneering challenge which Stone constantly assaulted: "The building of many a road in the first half of the last century was a great story. They were built at a time when men proceeded by trial and error; there were no rules, no sort of science. Neither the grading of a road bed, nor the laying down of iron rails had any precedent. Nor had the operation of trains."[23] Amazingly, in the year of the Ashtabula accident, 1876, the American Association of Passenger Traffic Officers had stated, "Up to this time the railway system is an accident. The mass of railway men are also by accident…"[24]

Such trial-and-error methods were bound to fail, and for Stone, one of his projects had failed in spectacular fashion. It was the kind of event on which tabloid newspapers could double their circulation over the next several days with graphic description of a holocaust that within a couple of hours had consumed almost 100 people. This was far more profitable than investigative reporting, which sorted through the killing and starvation of hundreds of Chinese workers laying the Transcontinental Railroad, the thousands of coal miners who were coughing their lungs out and slowly dying from "black lung" that had not yet been diagnosed, and the hundreds who would die in labor strikes that would not explode until after Stone's death. The corruption of Reconstruction is beyond our purview here, only

---

[23] Holbrook, 56.

[24] Quoted in Sarah H. Gordon. *Passage To Union: How the Railroads Transformed American Life, 1829-1929* (Chicago: Ivan R. Dee, 1996) 244.

to know that as many as 150 blacks were murdered by the Ku Klux Klan in the "Colfax Massacre" in Grant Parish Louisiana on April 13, 1873. Hardly anyone in the North took note of the black man's plight that had not been solved by emancipation. And newspapers gave little attention to the cost of life demanded by prosperity. "For years the Colorado coal fields had been scarred by labor warfare. This was raw competition such as Karl Marx had predicted: dangerous mines run by harsh bosses, and policed by armed guards in a desolate, hellish place. During 1913 alone, 464 men were killed or maimed in local mining accidents."[25]

It has been said that a person is a success if he is loved and respected by his children. Clara and Flora Stone adored their father. Two of the most prominent men of the nineteenth century moved into the family; they respected their father-in-law even to the fault of seeing and interpreting the world through his eyes. Though the following fits the criteria of hagiography, there is no reason to doubt the veracity of a short biographical sketch which appeared in the *Magazine of Western History* in 1886 written by John Hay.

> He remained to the end of his days one of the simplest and most unassuming of men. This does not mean that there was anything of diffidence or distrust in his nature; on the contrary he was perfectly aware of his own powers and confident in the exercise of them. But he never lost the inherent American Democracy of his character; the puddler from the rolling mill, the brakeman of the railroad, was always as sure of as courteous and considerate a hearing from him as a senator or millionaire. There was no man in the country great enough to daunt him, and none so simple as to receive from him the treatment of an inferior. He was a man extraordinarily clean in heart, in hand, and in lips. His closest intimates never heard a word from him which might not have been spoken in presence of ladies in a drawing room. He hated slander and scandal of every kind; he not only would never share in it, but would not tolerate it in his presence. His Puritan conscience, which afforded him so high and so rigid a standard of conduct in great affairs, was equally unbending in all the lesser matters of daily concern.[26]

In spite of the above, Stone had lost; he had lost his son in a freak drowning incident, his health in a freak carriage accident, his ability to escape through sleep by insomnia, a malady that had only become worse

---

[25] Chernow, 579.

[26] Hay, *Magazine of Western History*, 111-112.

through the years, his ability to enjoy anything because of excruciating ulcers, and now he had lost his reputation. The man who had once been the most commanding person in Cleveland now walked with a cane, hobbled and bent over, a captive in his own house, not wanting to hear the gossip and feel the stares of those who blamed him not only for the worst accident in the history of northern Ohio, but up until that time, the worst train disaster ever. Despair and depression overwhelmed him until he took his own life.

Stone was victimized as well as victimizer. Ironically what had made him so successful, the environment of the Gilded Age, had wrapped its choking tentacles around him. As Sean Cashman states, Stone's contemporaries were "consumed with ambition, many worked to excess, damaging their health in the process. Most suffered from chronic stomach complaints. Morgan's face was often a rash of spots and Rockefeller was a victim of premature hair loss."[27] Several railroad magnates died young: Jay Gould, Edward Harriman, and Henry Villard, all in their 60s. Steve Fraser accurately describes the vicious cycle in which Stone found himself when he was attempting to pay dividends to the New York Central's directors during the 1873 Panic.

> The very process of innovation was conjoined to the coercions of the competitive marketplace. Indeed, the ruthless efficiency of the market, expressed in the ceaseless outdating of existing forms, methods, and outputs of production, chronically devalued older, suddenly antiquated entrepreneurial property titles to the means of production. Fourteen of the twenty-five years between 1873 and 1897 were years of depression or recession. It was a "daimonic" economy, whipsawed by ferocious acts of creation and destruction. Ordinary businessmen saw themselves less as the titanic, Faustian movers and shakers of legend and more like the harassed and anxious strivers they really were, haunted by chaos and insecurity.[28]

The Leisure Class's Never-Ending Treadmill

When Stone moved into his mansion on Euclid Avenue, he officially joined Thorstein Veblen's "Leisure Class," but without leisure. The word which most defined the leisure class was "invidious," the desire and at-

---

[27] Cashman, 41.

[28] Steve Fraser. *Every Man A Speculator: A History of Wall Street in American Life* (New York: Harper Collins Publishers, 2005) 111.

tempt to prove one's self-worth by making others envious of attainments, possessions, money, clothes, etc. The problem is easy to perceive, at least for anyone who desires perception. Worthiness is a never-ending pursuit with no goal line, no arrival, and no retirement. Stone's curse was not that he could no longer play, which he hardly ever did, but he could no longer work. He could no longer prove himself, and whatever philanthropy he exhibited would always be interpreted as expiation for his guilt.

No street in America was a more graphic illustration of Veblen's leisure class than Euclid Avenue in Cleveland. But no mansion was large or elegant enough to house the gargantuan egos living within them. Restlessness and anxiety were the most frequent howling ghosts for the affluent. Veblen wrote

> The standard of expenditure which commonly guides our efforts is not the average, ordinary expenditure already achieved; it is an ideal of consumption that lies just beyond our reach, or a reach which requires some strain. The motive is emulation - the stimulus of an invidious comparison which prompts us to outdo those with whom we are in the habit of classing ourselves.[29]

Though Stone had built 23-inch walls for his house, they were not thick enough to ward off the accusing, ominous, ever nagging, pointing finger of "guilty." Though he had built the Ashtabula Bridge with the best of intentions, even according to Joseph Tomlinson not cutting costs, he could not sufficiently compartmentalize his conscience. Compartmentalization is a defense mechanism that attempts to block out with shut doors and fortified walls the unpleasantries of life. They are the proverbial skeletons in the closet. As we have argued, no American businessman during the latter half of the nineteenth century was more compartmentalized than John D. Rockefeller. He was the perfect legalist, hiding his business shenanigans behind his Baptist piety of no drinking, smoking, card playing, or dancing.

But even Rockefeller could not escape the bark of Ida Tarbell. When her muckraking series on Standard Oil ran in *McClure's* magazine, Chernow claims "Rockefeller, his wife, his son, and his three daughters were affected by serious medical problems, or nervous strain."[30] His daughter Bessie "Suffered a stroke or heart ailment that left her sadly demented," and John D's wife Cettie, in April 1904, suffered an affliction "that left her

---

[29] Veblen, 103.
[30] Chernow, 458.

severely paralyzed."[31] Chernow depicted Rockefeller's success as enabled by "His visionary leadership, his courageous persistence, his capacity to think in strategic terms, but also his lust for domination, his messianic self-righteousness, and his contempt for those short-sighted mortals who made the mistake of standing in his way."[32] After the Ludlow Massacre, even the polite Helen Keller said of Rockefeller, "He is the monster of capitalism, who gives to charity, and in the same breath he permits helpless workmen, their wives and children to be shot down."[33]

The developmental psychologist Erik Erikson argued that one comes down to the end of life with ego-integrity or despair. Ego-integrity is the assurance that life has had order and meaning, the acceptance of one's life as what ought to have been, and emotional integration which permits participation by fellowship, as well as acceptance of the responsibility of leadership.[34] Stone had accepted that leadership, but it had boomeranged on him in disgust with the memory of almost 100 people being incinerated. Despair led him to a point to which no reader of this book has ever been, suicide. He could not do life over, perhaps make a change in the bridge, which could have been corrected if he had only foreseen the defect and its consequences. Without medical remedies, prescriptive medicine and psychosomatic insights that are readily available today, death became a more favorable alternative than life.

The characters in the plot who were involved in the engineering tragedy on December 29, 1876, are almost too numerous to list: Stone, Vanderbilt, Rockefeller, Collins, Tomlinson, the board of directors, the managers, the engineers, conductors, inspectors, the persons who got on the train in spite of the weather that forbade any kind of travel, and even the residents of the small village of Ashtabula, who looked daily on an object that looked as if it defied all laws of physics but somehow never asked themselves either individually or collectively, "What are we going to do in the event of a disaster, something that will forever identify us with one event?" Even Dwight Moody, who believed himself to communicate with God, was on a different frequency than Philip Paul Bliss who was convinced he should

---

[31] Ibid.

[32] Quoted in Steve Weinberg. *Taking on the Trust: The Epic Battle of Ida Tarbell and John D. Rockefeller* (New York: W. W. Norton and Company) 78.

[33] Chernow, 579.

[34] Erik Erikson. *Childhood and Society* (New York: W. W. Norton and Company, 1978) 268-269.

stay in Rome, Pennsylvania with his family. Moody persuaded Bliss that God wanted him in evangelistic work, rather than fulfilling what he believed to be his God-given gifts in writing Gospel music, which demanded solitude and reflection. Against Bliss's better judgement, he yielded to the persuasive pressure put upon him by Moody to return to Chicago and caught a train which assured he would never write another hymn. But such historical judgement almost sounds sacrilegious in light of Moody's great success as a "soul winner."

No one individual could have carried out a catastrophic event unless he was an evil schemer bent on blowing up a bridge at 7:28 PM. Certainly, Amasa Stone was not that person, and quite possibly he was no more guilty than some of the others caught in a drama not of their own choosing. He was the most powerful, most visible, and in the minds of his detractors, the most responsible. "Come Hell or high water," Stone was going to see the Ashtabula Bridge erected, just as he had unbelievably built a bridge over the Connecticut River at Warehouse Point, Connecticut in record time some quarter of a century earlier. Stone had plenty of victories to validate his self-confidence, but the one failure would forever mar his legacy.

Stone was not the kind of person to process his motives, to make sure his ego boundaries were realistically defined and to calculate the risk between engineering achievement and safety precautions. He was a hard-nosed builder of tangible objects, many of them quite impressive. He was not a philosopher; he did not surround himself with a "think tank." As one engineer recently said, "Look, some of us were attracted to engineering because we're good at figures and because we're not good with people. We can do our work and solve most of the problems you give us, but please don't try to drag us into longwinded discussions about the meaning of life."[35]

A Confluence of Factors and Individuals

The factors which brought down the Ashtabula Bridge must be divided into indirect distant causes, and direct immediate causes. The former includes destructive forces set in motion by Rockefeller, who financially weakened the railroads because of demanding rebates, Vanderbilt, who

---

[35] Samuel C. Florman. *Blaming Technology: The Irrational Search for Scapegoats* (New York: St. Martin's Press, 1981) 149.

gobbled up much of whatever he could, rather than improving what he already owned, Jay Gould and Jay Cooke, who both did their part in bringing on the '73 Panic and the Directors who were not willing to jeopardize their earnings by asking questions about what really mattered, the safety of railroad passengers. And then there was the Church that often faced the almost hopeless dialectic of having to choose between enculturation and irrelevancy, between accommodation and bold truth. In many ways the failure of the Ashtabula Bridge was archetypical of the Gilded Age, a unique period of rapid industrial growth and unregulated capitalism. The residents of Ashtabula, Ohio had been "gifted with a bridge" that with its gothic appearance looked as if it should have been built in Sleepy Hollow with the headless horseman riding across it. Ashtabula citizens often were aghast at the dark forbidding framework that silhouetted the sky east of them.

As to direct immediate causes, my philosophical, historical and theological rambling should not obscure the reality that specific individuals were directly responsible for the bridge's failure. Tomlinson never informed Stone that he believe a Howe truss of any substance should not be used to traverse a large span.[36] Turning the braces to their vertical position called for all kinds of adjustments, but none of them adequate for attaching the braces as firmly as they would have been if the angle blocks had been cast otherwise. When the bottom lugs were sheared off, the compression braces were held only by friction supplied by the weight of the bridge and the tension rods holding the top and bottom chords together. No one suggested all of the angle blocks should be re-cast, which would have been anathema to Stone. Turning the braces necessitated shearing off lugs and chipping off corners of the I-beams, compromising the structural integrity of the bridge. Tomlinson's testimony could all be distilled into one statement, "I knew better, but did not do better."

Tomlinson was relieved that the Coroner's Jury seemed to exonerate him and condemn Stone. In a letter written March 18, 1877 to G. W. Dickinson for collecting his travel expenses he stated, "It is very hard on Mr. Stone, and I cannot help feeling sorry, but it seems like retribution as he was hard on others."[37] From Tomlinson's perspective, Stone had gotten what he deserved.

---

[36] *Joint Committee*, 138, 149.

[37] Letter, Joseph Tomlinson to G. W. Dickinson, March 18, 1877, courtesy of Leonard Brown in email to this author, 3/17/2022.

Albert Congdon manufactured parts and shipped them to A. L. Rogers, not stipulating their exact placement. This was all the more critical in that the exactitude and uniformity of I-beams for chords, braces, and counter-braces when compared to today's technology would have varied in weight and thickness. If anyone needed "instructions on the box," it was Rogers. Congdon, the master mechanist, failed to take responsibility for assembling the final product.

Rogers knew nothing about camber, and this caused all kinds of problems, all of them resolved by guesswork. Reducing camber by shaving off part of the top lugs weakened them. Placing shims and other makeshift devices to raise the camber robbed the bridge of the precision and frictionless interaction of parts, a professionalism which much of the bridge lacked. Rogers did not possess the temperament or expertise to superintend the erection of a 150-foot bridge made from iron, rather than wood. Stone ignored all of these disqualifications, and by default, appointed Rogers to erect the bridge.

During the Legislative hearing, Collins was thrown under the bus (in this case the train) more than anyone else. He was most blamed, and should have been least blamed. As we have already shown, being responsible for 1,800 miles of track with little to no help was impossible. But there was a caveat which does not completely exonerate Collins. Since he was not given ultimate authority for the bridge's construction, which he should have in light of his title and expertise, he seemed to be peeved and thus, would have nothing to do with the bridge, until Rogers beseeched him for help. Considering his emotional temperament, and his sensitivity to the rights and feelings of others, he may have not been the person to make the cold, calculated decisions often demanded by large construction projects. All achievements carry liabilities, and Collins' personality was more given to maintenance than high-risk ventures.

Engineer Gustavus "Pap" Folsom testified to hearing a loud pop or snap whenever he crossed the bridge. Did he ever inform anyone else of a noise which, no doubt, indicated some irregularity? Maybe it was the fissure in the angle blocks, but in the end, this crack was nothing more than the proverbial straw that broke the camel's back. A bridge properly constructed would have withstood this single flaw. The failure of the bridge was an accumulation of flaws, constantly becoming worse over its eleven years of existence.

Stone was set up for failure, not by God or any of the persons for or against him, but by his unbroken line of constant success. Such consistent accomplishment leads to a false sense of infallibility, which in this case eliminated several precautions: properly buttressing the chords, placing sufficient plates across them at frequent intervals, and fastening both the lateral and diagonal bracing more securely. Cutting corners by attaching the lateral bracing to every other angle block was fatal. (The sway bracing was also attached to only every other angle block.) The angle block which failed had no lateral brace attached to it. Welding was not a practical technology until the middle of the twentieth century, and spot-fusing whatever could be fastened would have eliminated much of the sway and vibration, and no doubt, spared the bridge and those who traveled over it that fateful night.

The only way to have ensured the safety of the bridge was to err on the side of caution with an abundance of brackets, bolts, and cross-ties providing stabilization whenever possible. In 1878 John Hay was overseeing construction of a building in Cleveland. Stone wrote him from Florida.

> You are doing right in increasing the strength of the box girders to a size that will make them safe beyond doubt. Their strength should be determined outside of the party that rolls the iron, as it is the interest of the party that rolls it to get the dimensions as heavy as possible - as to the I-beams I am certain that they will be strong enough - Do not fail to have the ends of the I-beams punched for the connecting bolts. This should be a large ¾ hole 3 inches from the end of every bar for a loop connecting bolt thus (At this place in the letter, Stone drew a wrought-iron U bolt upside down going through each end of the buttressed I-beams and fastened with a large nut on each end of the bottom of the U bolt) and for anchoring to the walls. As to the I-beams I am certain they have an abundance of strength.[38]

"Safe beyond doubt and an abundance of strength" were not true of the Ashtabula Bridge. Throughout life Stone had learned as he went and passed down his experiential knowledge to the people who were under his influence. The above instructions to Hay were too late for the 98 people who perished in Ashtabula, December 29, 1876. Also note Stone recommending that the "strength be determined outside the party that rolls it." In other words, there needed to be a quality control person outside of the rolling mill which fabricated the metal. Stone's relationship with and dependence on his brother A. B., President of the Cleveland Rolling Mill, would always be perceived in the eyes of many as a conflict of interest.

---

[38] Letter, Amasa Stone to John Hay, March 28, 1878, courtesy of John Hay Library Archives, Brown University.

Simmons and Johannesen

David Simmons, a bridge historian as well as being versed in many facets of Ohio history, speculates, not without validity, as to why Stone used iron instead of wood for the Ashtabula bridge.

> With his brother newly established as president of a Cleveland company capable of producing rolled wrought iron beams and with himself as a major stockholder in that firm, Amasa Stone may simply have been experimenting with this new building material in the hopes of expanding the market for their products. Coupled with this may have been a sense of pride and family solidarity that led Stone to insist on all-iron construction for the new Ashtabula bridge, to take full responsibility for the design, and to ignore the admonitions against an iron Howe truss made by Joseph Tomlinson, a railroad-employed bridge designer.[39]

Simmons condemned Stone for his headstrong obdurate temperament. "He certainly demonstrated the improper method for introducing new technological concepts into a corporation, assuming too much responsibility himself, dismissing the learned advice of his colleagues, and ignoring the need for training all levels of the organization in the use of the new and unfamiliar technology."[40] He likened Stone's fall from grace to a Greek tragedy, as the protagonists were often "brought low by *hubris*, the sin of overweening pride, could have so predicted."[41]

Eric Johannesen's (a specialist in the architectural history of northeast Ohio) assessment of Stone was similar to Simmons, stating that his house on Euclid Avenue "perfectly expressed the self-made ambitious arrogance of the nineteenth century empire builder."[42] According to Johannesen, Stone had the reputation of a man "who operated in an autocratic manner, and tolerated no opposition."[43] After noting Stone's engineering success, which included the Cleveland Union Passenger Depot over 2 football fields long, entirely open without supporting columns, the large edifice with mansard roof and domed belvedere for the Children's Aid Society, the large Adelbert Hall which was a college unto itself, Johannesen does

---

[39] David A. Simmons. "Fall From Grace: Amasa Stone and the Ashtabula Bridge Collapse," *Timeline* (June – July 1989) 37.

[40] Ibid., 43.

[41] Ibid.

[42] Eric Johannesen. "Stone's Trove: The Legacy of an American Oligarch," *Timeline* (June – July, 1989) 27.

[43] Ibid., 28.

not outright condemn Stone. The railroad shed demonstrated Stone's "understanding of the need between competing parties in the business world." Even though Stone's primary interest was the welfare of his own railroads, he understood the need for "general cooperation."[44] The author concluded by asking the question; "What is the true measure of enigmatic Amasa Stone – villain or hero?"[45] He holds off on a verdict. "Today perhaps the most visible index to Stone's life can be found in the important structural record he created, as actual engineer and contractor, or as client and patron, in northeastern Ohio."[46]

In the popular historical interpretation, Stone bears more responsibility for the Ashtabula Bridge failure than any one person. And this is largely true in that he built a structure in a way it had never been built before, out of a substance that had never been used for that large a bridge. Multiple I beams and diagonal bracing improperly attached was a faulty design. There is no example of chords made of multiple iron I-beams before the Ashtabula Bridge. But often the more we look at something, the more complex an item or issue becomes. Certitude fades, and if Stone was guilty, he had many accomplices. I do not claim to have set the record straight, but have furthered the conversation that in all likelihood, will never find a clear conclusion.

---

[44] Ibid., 31.
[45] Ibid., 33.
[46] Ibid.

Chapter 15

Epilogue and Fragmented Reflections

The Lake Shore and Michigan Southern Railroad never admitted responsibility for the bridge failure. We have no record that Amasa Stone ever apologized to anyone. The Railroad paid out $495,722.42 to the families of the victims who lost their lives and for the injuries of those who survived. Almost all of these were paid out in $3,000-$5,000 increments, which seems like a paltry sum today, even considering inflation.[1]

In contrast to Cornelius Vanderbilt, John D. Rockefeller and Andrew Carnegie, the name of Amasa Stone has mostly been forgotten. The iconic Rockefeller Plaza in New York, Vanderbilt University in Nashville, and Carnegie-Mellon University in Pittsburgh carry on the family names. Many Americans are faintly aware that the Biltmore in Asheville, N.C. by far the nation's largest mansion at 170,000 square feet, was residence to one of the Vanderbilt descendants, but would not know that it was built by George Vanderbilt, one of the lesser known of the family.

The Stone Chapel which sits on the opposite side of Euclid Avenue from the main campus of Case Western Reserve University is not a building which stands out. Unless one is specifically looking for it, the small chapel goes unnoticed by the thousands of commuters and students who drive past it each day on congested Euclid Avenue. According to campus facilities coordinator Karen Cohan, there was serious discussion about tearing it down, but due to her intervention, as well as others with respect for tradition and heritage, saner minds prevailed. The interior is rather subdued and far less awe-inspiring than Old Stone Church on Cleveland Center Square. Again, as we have noted, there is no relationship between

---

[1] *Reports*, 1877, 11. Also see Corts, 166-161.

the Stone name and the exterior of the church. As we have already hinted, hardly anyone would know that Adelbert Hall was the first building of Adelbert College of Western Reserve University. It is certainly not designated as a centerpiece of the University as was intended by Amasa Stone. In fact, I'm not sure that there is an iconic building or location which defines Case Western Reserve University as compared to say the Chapel of Duke University, or the majestic colonial building at the center of the University of Virginia, Charlottesville. Nonetheless, Case Western Reserve University stands out as one of the most respected educational institutions in the world.

Of the three men I attempted to weave together in this book, Stone, Rockefeller, and Vanderbilt, the last had by far the more gargantuan ego. On June 30, 1871 Grand Central Station on New York's Forty-second Street was completed, the largest railway depot in America. T. J. Stiles describes the building for which Vanderbilt paid $6,000,000 out of his own pocket as the "second largest in the world, a brick bastion with white iron trim, standing three stories high (160 feet to the top of the central tower), 240 feet wide, and 692 feet deep, extending north from Forty-second Street. A huge train shed, or 'car house,' stretched 650 feet long under an arched glass roof. The statistics of what went into the depot were staggering: eight million pounds of iron, ten million bricks, twenty thousand barrels of cement, plus eighty thousand feet of glass in the roof".[2]

On November 10, 1869 thousands showed up for the unveiling of a twelve-foot bronze statue of Vanderbilt at St. John's Depot in New York; on either side were icons of the Commodore's career. In a long tribute, Mayor Hall referred to Vanderbilt as the "richest man on the continent," and did not mention that Vanderbilt had paid for this idol to himself. When the new Grand Central Station was built in 1913 the statue was placed on a small second story roof facing Forty-second Street. In an endeavor in which only a historian would delight, I asked several policemen, security guards, custodians at Grand Central, and whomever else I thought might have some knowledge of the statue's whereabouts, but no one knew where it was in the facility. Rather than the homage of both residents and tourists, the statue is left outside to endure the heat of summer, the snow and sleet of winter, and the excrement of pigeons who find Vanderbilt's bald shining head a convenient stop. As to the Moravian Cemetery on Staten Island where Vanderbilt is buried, unless one is in a helicopter, or has the skills

---

[2] Stiles, 515.

of Ethan Hunt of *Mission Impossible* fame, the very impressive grave site is inaccessible.

Nothing of the splendor of Euclid Avenue remains today. The mansions have been replaced with apartment buildings, gas stations, store fronts, medical buildings, and the overall demise of the inner city. The vanishing of the symbols of wealth serves as a parable, testifying to the Apostle Paul's reminder that things seen are temporary and those unseen are eternal. The Stone house was razed in 1910, and one might assume that since that was the year after Flora Stone died there might be some relationship between the two events, but I know of no official document supporting the connection. According to Euclid Avenue historian Jan Cigliano, several of the families such as the Rockefellers and Brushes stipulated that their residences "were to be razed immediately after they died; they shuddered at the thought of multiple family's occupying their personal havens."[3] Julia Stone had died in 1900, so seemingly the Stones did not dictate the destruction of their house; it may have been more to do with the completion of the Mather mansion in 1911. One can imagine the pain of watching a wrecking ball swing through the walls of that architectural wonder, though it may have been in a depreciated condition at the time of its demise. As Cigliano points out while, "A New England austerity—thrift, hard work, and earnest intent—created the city and moved it forward to greater financial and industrial strength.... eventually, by the turn of the century, their way of life collided with the industrialism of their own making—the factory smog, the commercial traffic, and the immigrant and black labor crowding and surrounding neighborhoods."[4] Industrial wealth had built beautiful mansions and ultimately destroyed its creations, and even some of the men who built them.

The Stone family seems to have suffered from a syndrome which would later be known as the "Kennedy Curse." We have already noted the freakish swimming accident of Amasa's only son Adelbert. Even more dismaying was the death of his namesake Adelbert Hay, 26-year-old son of John and Clara, who fell from a windowsill at the "New Haven House" in Connecticut on June 23, 1901. The best interpretation of the event was that sometime around midnight he sat on the edge of the window in order to get some fresh air, went to sleep, and fell 60 feet to the pavement below. The worst spin was the newspaper caption from the *Washington Sunday*

---

[3] Cigliano, 322.
[4] Ibid., 328.

Globe, "Adelbert S. Hay—drunk when he fell from the Hotel Window."[5] We will never know since there were no toxicology reports. The body was immediately taken to New York, and there the father John Hay viewed the corpse. *The New York Times* reported, "The Secretary immediately entered a carriage and with his luggage, consisting of two large bags, was driven in great haste to the residence of Seth Moseley, 36 Wall Street. In the drawing room of the house reposed the body of the son. In the presence of the beloved dead Col. Hay utterly collapsed, and, prostrated by his great grief, took to his bed. Dr. Samuel D. Gilbert was summoned and administered to the patient."[6]

Adelbert had accomplished much, even at his young age. Shortly after his graduation from Yale in 1898, he was appointed as a United States Consul to South Africa during the Boer War. Upon arrival in Pretoria he was given the responsibility of looking after the needs of 6,000 British POWs. Before his death, he had been offered a position as assistant secretary to President William McKinley, similar to the job his father had held under Lincoln. Of course, the tabloids had their day claiming the handsome young man had been seen with his actress girlfriend that evening drinking champagne. Taliaferro does not buy into the alcohol diagnosis: "Neither the doctors, the coroner, the hotel staff, Dell's friends, nor certainly his family said anything about drinking or suicide, a subject that had entered the consciousness of his parents at least fleetingly."[7] Neither does Taliaferro say anything about his relationship with the actress Marguerite Cassini which seemingly had an element of truth, even in the tabloids. The six-foot-two strapping young man with exceedingly good looks could have escorted the woman of his choice. Cut down in the prime of life, he would never realize the diplomatic success of his father.

Amasa Stone Mather, son of Samuel and Flora Mather, fared little better than his first cousin. Born in 1884, he became a handsome young man, featured in a silent film, *The Perils of Society.* After attending Yale, he with three friends took a trip around the world hunting big game in Africa, and upon returning to the U.S. was invited by Theodore Roosevelt to give

---

[5] "Death By Defenestration: The Short Life of Adelbert 'Del' Stone Hay." theesotericcuriosa.blogspot.com/2010/07/from-shadows-shortlife-of-adelbert. htwl.

[6] "Secretary Hay Prostrated, Overwhelmed with Grief Over the Death of His Son," *The New York Times* (June 24, 1901).

[7] Taliaferro, 402.

a verbal report about his trip. Amasa soon became head of the Pickland-Mather Iron Ore Department and was headed in the same direction as his millionaire father. Ironically, just before a planned trip to an International Red Cross conference with his father, he died of the flu in 1920 which had taken the lives of millions around the world.[8] Today there are no descendants living in Cleveland who carry either the Stone or Mather name.

The Gilded Age represented the most unregulated capitalism in American history. The most critical issues remain with us: the inequities between labor and management or ownership, the question of unfair monopolies, and the mindlessness of factory labor whether it be making silicon chips, or placing a hinge on a car door as a unit passes in front of a plant worker. Many of these problems have been "solved" by shipping low-skilled jobs overseas, or hiring immigrant labor to do them here in America, which I am not confident is a solution. Migrant workers still die of heat strokes in California fields, and sweat shops still exist where women earn 50 cents an hour making a garment for 40 cents that will sell for $40 in the United States. If it was difficult to understand the machinations of Jay Gould, it is even more difficult to understand the operations of Facebook, Microsoft and Amazon. Controlling the thought processes of Americans which eliminates the ability to think critically is probably a greater problem than controlling the gold market or the transportation system.

Eminent historian Joseph Ellis argues that we are in a second Gilded Age. He points out that "between 1972 and 2012, after adjusting for inflation, the average income for most Americans declined by 13 percent, it rose by 153 percent for the top 1 percent."[9] Further, the top one percent consisting of bankers, managers, and CEOs make "300 times more than the workers they employ."[10] CNN reported today, July 20, 2022, American CEOs make 324 times as much as their workers. Or as it has been put in another way, an increasingly greater gap exists between those who have two houses and those who have two jobs.

Politically, the Gilded Age is a strange candidate for paradise, burdened as it is with images of Robber Barons feasting at Delmonico's in self-indulged splendor, the millionaires' club of wholly bought-and-sold congressmen, and a reigning ideology called Social Darwinism that depicted entrenched

---

[8] Mackley, 34.

[9] Joseph J. Ellis. *American Dialogue: The Founders and Us* (New York: Alfred A. Knoft, 2018) 105.

[10] Ibid., 108.

poverty and permanent economic inequality as conditions sanctioned by some combination of God's will and Nature's laws.[11]

The never-ending debates about government regulations on corporations is the very nature of a democracy. Most Americans are convinced that China has too much regulation, and more primitive countries have too little with inept government unable to regulate anything. Hardly any of us would choose to live in North Korea or the Democratic Republic of the Congo. As a Syrian refugee to the United States said, "We have to appreciate this country...If anyone is complaining about this country, they have to take an airplane there (Syria) for two days. They are going to come back and kiss the ground."[12]

We continue to question philanthropy at the cost of degrading and dehumanizing men and women under brutal conditions. Libraries, universities, and museums were built on the backs of the toiling masses. But there is an argument for the amalgamation of smaller entities into giant corporations as accomplished by the monopolistic titans. The technology of the ultimate super power enabled car factories and steel mills to be converted into defense plants, munitions depots, and all kinds of fighting and killing artillery. America won two World Wars not only with brave men and women willing to fight, but also through the ability to out manufacture any other nation in the world. This prowess circles back to Stone, Vanderbilt, Rockefeller and Carnegie, as well as many other inventors and industrialists. Some of them were not native to the United States such as Albert Einstein, Neils Bohr, and Werner von Braun, but fortuitously, because of our genius forefathers and providence, or however one interprets American fortune, we were able to provide a haven to which these geniuses escaped and contributed to the defeat of the Axis powers personified by Hitler.

Even with the Environmental Protection Agency cleaning up the air in Cleveland, Ohio; Detroit, Michigan; Gary, Indiana; Pittsburgh, Pennsylvania; and Fairfield, Alabama carbon emissions still plague us, eating through the ozone layer, allowing the sun to cook the earth while carbon dioxide traps in heat, raising temperatures precarious to the polar ice caps and upsetting the delicate ecological balance so critical for plant and animal life. Or are we simply in a climatic cycle which will reverse itself after several centuries or millennia? Whatever the answer, humankind cannot

---

[11] Ibid., 112.

[12] Hannan Adley. "Ten Years After Fleeing War: Syrians Reflect On Life in USA," *USA Today* (January 11, 2022).

afford to stand around doing nothing until we find out. The possibility that we are contributing to our demise is a salubrious assumption.

Engineering continues to be a problem, perhaps even more of a problem since we are pushing the envelope to exponentially greater degrees. I just happen to live in Kansas City where on July 17, 1981 an overhead walkway at the Crown Center Hyatt Regency Hotel collapsed and killed 114 people. The failure was due to design flaws and an insufficient number of tie rods from the floor to the ceiling; eerie how similar this was to the Ashtabula Bridge, given that its tension members were two-inch round rods. And, during the year of writing this book, a condominium collapsed in Miami killing 98 people (the same number as the Ashtabula disaster) due to design flaws, aging, and lack of inspections. As of today, I heard our President, Joe Biden, claim that there were 45,000 bridges which needed to be rebuilt or replaced in America.

We continue to live in a "progress paradox," begging the question, "Has technology reached a point of diminishing returns?" Clearly some discoveries have made us better such as the almost total eradication of tuberculosis, polio, and getting a grip on AIDS at least in the Western World. But while I write this, the Omicron virus is out of control which may be partially to blame on the global village and urbanization. It used to take a rat carrying disease six months to cross the ocean, then six weeks, then six hours. I trust the situation will be different by the time you read this, but we will probably be confronting some other new virus variant and will have run out of Greek letters.

A highlight of the Salter family Christmas is to visit Union Station in Kansas City. Kansas City in the late nineteenth and early twentieth century rivaled Chicago as the railroad crossroads of the United States, both for cattle and soldiers. I stare at the pictures on the wall of the passengers sitting in Union Station waiting to take the trip of a lifetime in the 1920s-30s and wonder what is going on in the mind of a 12-year-old boy. From my limited perspective, the colors and the designs from the 2000 Union Station restoration project are more vivid and appealing than those found in Grand Central Station in New York City.

The attraction at Christmas is a huge model train display, scores of trains and hundreds of cars passing through picturesque villages. Webster defines "nostalgia" as a "Sentimental yearning for the happiness of a former place or time."[13] Nostalgia requires a certain amount of amnesia,

---

[13] Webster, 1325.

blocking out bad memories, and retaining good ones. The highlight of the visit is the enthrallment and wonderment expressed by my three-year-old grandson. He doesn't need amnesia to discover a brand new world. But this year was a bit different for me, observing and reflecting on the speeding trains travelling through tunnels and over bridges. My reflections were refracted through the lens of what happened in a tiny town in northeast Ohio. In 1876, even the Pacific Express parlor cars were not very comfortable. No air-conditioning in the summer, and during the winter one was either too close or too far away from one of the coal stoves on each end of the car.

Yes, I really would like to return to Ashtabula in 1876, but this book is the best I can do. A book allows us to travel with imagination, without enduring the concerns of everyday bodily functions with primitive facilities, or no facilities at all, much less the fear of not making our destination in a blinding snowstorm. But I'm not sure that's any worse than sitting in the back of a metal tube with rocket boosters on the back of it while it sways back and forth as it is hit with lightning strikes in a thunderstorm. I am still fascinated by flight, but having spent much of my life in a profession which required travel by flight, I no longer have any desire to be strapped down in the middle seat of an airplane.

Trains are no longer a viable option for long distance travel in the United States. If you want to ride on a real train, go to Spain where you can float along at 200 mph and make it from Madrid to Barcelona in a couple of hours. Amtrak is probably going to completely go out of business as passenger routes are increasingly becoming non-existent. This particular demise is probably due to the affluence of almost every American owning an automobile. Notice on any given commute how many cars have only one occupant, the driver. The ecological awakening and the runaway price of fuel may continue to enhance the use of light rail service into major cities and faster trains between the metropolises on the eastern seaboard, but cross-country train travel in America will soon be gone forever. My dad often used the expression "I would not do that or trade that or go there," or whatever the situation "for a share in a railroad." I have never heard someone from my generation use that frame of reference. If a "day trader" is buying railroad stock, he knows more than I do, though lumbering freight trains still seem a viable option for moving mass quantities of raw materials and to a lesser extent manufactured goods.

After having taken a half dozen trips to Ashtabula, Ohio, my concluding observation is that it is no longer a town defined by an 1876 train disaster. Many of those who reside in Ashtabula know that an accident took place but remember few of the details. Two historical markers give a sketch of December 29, 1876, one near Ashtabula Hospital not far from the wreck site and another off West 24th Street, where there are parking places for those who desire to explore the location of the wreckage. A walking bridge allows a visitor to cross a mill pond, but upon coming to the Ashtabula River, there is nothing to indicate which way to turn. Thus, the one bit of help I can render is that upon arriving at the river, the visitor should turn left and walk about one hundred yards. The train went down south of where the present-day stone arch railroad bridge is located. An abutment of the old bridge is on the east side of the river and directly under the four-track present bridge. Nothing on the river banks indicates a disaster occurred on December 29, 1876.

Some say that poltergeists retaining the identities of those who lost their lives that fateful night still linger in the trees, bushes and tall grass along the river banks. I am a sceptic. I would like to imagine that the spirit of Marion Shepard still hovers over the scene where she enabled several persons to escape and afterwards, washed and treated their wounds. She married and lived out a quite ordinary existence in San Diego, California. Her life serves as a parable that ordinary people often do extraordinary things. She also reminds us that in the darkest of nights, "the better angels of our nature" shine the brightest.

Bibliography

Primary Sources

Ammidown, Holmes. *Historical Collections*, Vol. II (New York: published by author, 1874).

Bartlett, J. Gardner. *Simon Stone Genealogy: Ancestry and Descendants of Deacon Simon Stone Watertown, Mass., 1320-1926* (Boston: Pinkham Press, 1926).

Burt, Henry M., ed. *Memorial Tributes to Daniel L. Harris with Biography and Extracts from His Journal and Letters* (Springfield, Massachusetts: Printed for the Family for Private Presentation, 1880).

Chanute, Octave and George Morrison. *The Kansas City Bridge* (New York: De Van Nostrum, 1876).

Crooks, George R., ed. *Sermons by Bishop Simpson* (New York: Harper and Brothers, 1885).

Crooks, George. *Life and Letters of the Rev. John McClintock* (New York, New York: Nelson and Phillips, 1876).

The Dedication of the New Buildings and Inaugural of Adelbert College of Western Reserve University, Cleveland, O., October 26, 1882, (Lately Western Reserve College, Hudson, Ohio) (Cleveland: A. W. Fairbanks, 1883).

Gary, Ferdinand Elsworth *Lake Shore and Michigan Southern Railroad System and Representative Employees* (Chicago: Biographical Publishing Company, 1900).

Guilford, Linda Thayer. *The Story of a Cleveland School, from 1848 to 1881* (Cambridge: John Wilson and Son, University Press, 1890).

Hay, John (ed.) *Amasa Stone: Born April 27, 1818, Died May 11, 1883* (Cleveland: De Vinne Press, n.d.).

Hay, John. *The Bread-Winners: A Social Study* (Ridgewood, New Jersey: The Gregg Press, Inc., 1967).

Inglis, William O. *John D. Rockefeller Interview: 1917-1920* (Sleepy Hollow, New York: Meckler Publishing/The Rockefeller Archive Center).

Joblin, Maurice. *Cleveland Past and Present: Its Representative Men* (Cleveland: Maurice Joblin, 1869).

Johnson, Cristfield. *History of Cuyahoga County, Ohio. In three parts: Part First, - History of the County, Part Second, - History of Cleveland, Part Third, - History of the Townships. With Portrait and Biographical Sketches of Its Prominent Men and Pioneers* (Cleveland: D. W. Ensign & Co., 1879).

Keenan, Henry. *The Money-Makers: A Social Parable* (New York: D. Appleton, 1885).

Large, Moina W. *History of Ashtabula County* Vol. I (Topeka, Indiana: Historical Publishing Company, 1934).

Laws of the General Assembly of Pennsylvania 1870, No. 201, "An Act."

Ludlow, Arthur G. *The Old Stone Church: The Story of a Hundred Years, 1820-1920* (Cleveland: Privately printed, 1920).

Mather, Samuel. "The Amasa Stone Memorial Chapel," *Western Reserve University Bulletin* Vol. XIV, No. 6 (November 1911).

Peet, Stephen D. *The Ashtabula Disaster* (Chicago: J. S. Goodman-Louis Lloyd and Company, 1877).

Robison, W. Scott. *History of the City of Cleveland: Its Settlement, Rise, and Progress.* (Cleveland: Robison and Cockett, 1887).

Rockefeller, John D. *Random Reminiscences of Men and Events* (New York: Sleepy Hollow Press and Rockefeller Archive Center, 1984).

Rosenfield, Richard N. *American Aurora* (New York: Saint Martin's Griffen, 1997).

Sing to the Lord (Kansas City, MO: Lillenas Publishing Company, 1993).

Taylor, Nathaniel. "Native Depravity and Free Will," *American Christianity: An Historical Interpretation with Representative Documents* Vol. II, H. Shelton Smith, et. al. eds. (New York: Charles Scribner's Sons, 1963).

Thayer, William Roscoe. *The Life and Letters of John Hay* Vol. I and II (Boston: Houghton Mifflin, 1915).

Veblen, Thorstein. *Theory of the Leisure Class* (New York: Viking Penguin, 1994).

Weber, Max. *The Protestant Work Ethic and the "Spirit" of Capitalism and Other Writings* (New York: Penguin Books, 2002).

Wesley, John. "The Use of Money," *The Works of John Wesley*, Vol. VI (Kansas City: Beacon Hill Press of Kansas City, 1978).

White, John Bart R. *Genealogy of the Descendants of Thomas Gleason of Watertown Massachusetts 1607-1909* (Haverhill, MA: Press of the Nicolas Print, 1909).

Whittle, D. W., ed. *Memoirs of P.P. Bliss* (New York: A. S. Barnes & Company, 1877).

Secondary Sources

Abbott, Karen. *Sin in the Second City: Madams, Ministers, Playboys, and the Battle for America's Soul* (New York: Random House Trade Paperback Edition, 2007).

Ackerman, Kenneth D. *Dark Horse: The Surprise Election and Political Murder of President James A. Garfield* (New York: Carroll & Graf Publishers, 2003).

Baznik, Richard E. *Beyond the Fence: A Social History of Case Western Reserve University* (Cleveland: Case Western Reserve University, 2014).

Beatty, Jack. *Age of Betrayal: The Triumph of Money in America 1865-1900* (New York: Alfred Knopf, 2007).

Beckert, Swen. *The Money Metropolis: New York City and the Consolidation of the American Bourgeoise* (New York: Cambridge University Press, 2001).

Birmingham, Stephen. *America's Secret Aristocracy* (Boston: Little, Brown & Company, 1987).

Brands, H. W. *American Colossus: The Triumph of Capitalism 1865-1900* (New York: Doubleday, 2010).

Brockmann, R. John. *Twisted Rails, Sunken Ships: The Rhetoric of Nineteenth Century Steamboat and Railroad Accident Investigation Reports, 1833-1879* (Amityville, New York: Baywood Publishing Company, Inc., 2004).

Brown, David S. *The Last American Aristocrat: The Brilliant Life and Improbable Education of Henry Adams* (New York: Scribner, 2020).

Burnham, Charles A. "The Ashtabula Horror," Corts, 78-99.

Cashman, Sean Dennis. *America in the Gilded Age: From the Death of Lincoln to the Rise of Theodore Roosevelt* (New York: University Press, 1984).

Chandler, Alfred D., Jr. *The Railroads: The Nation's First Big Business* (New York: Harcourt, Brace & World, 1965).

Chandler, Alfred D., Jr. *The Visible Hand: The Managerial Revolution in American Business* (Cambridge: The Belknap Press of Harvard University Press, 1977).

Chandler, David Leon. *Henry Flagler: The Astonishing Life and Times of the Visionary Robber Baron Who Founded Florida* (New York: McMillian Publishing Company, 1986).

Chernow, Ron. *Titan: The Life of John D. Rockefeller, Sr.* (New York: Random House, 1998).

Cigliano, Jan. *Showplace of America: Cleveland's Euclid Avenue, 1850-1910* (Kent, Ohio: Kent State University Press, 1991).

Clark, Judith Freeman. *America's Gilded Age: An Eyewitness History* (New York: Facts on File, 1992) 97.

Cochran, Thomas C. *Railroad Leaders 1845-1890: The Business Mind in Action* (Cambridge, Mass: Harvard University Press, 1953).

Cohen, Charles Lloyd. *God's Caress: The Psychology of Puritan Religious Experience* (New York: Oxford University Press, 1986).

Cohen, Jared. *Accidental Presidents: Eight Men Who Changed America* (New York: Simon & Schuster, 2019).

Commager, Henry Steele. "Should The Historian Make Moral Judgements?" *A Sense of History: The Best Writing from the Pages of American Heritage* (New York: American Heritage, 1985) 461-472.

Corts, Thomas E., ed. *Bliss and Tragedy: The Ashtabula Railway-Bridge Accident of 1876 and the Loss of P. P. Bliss* (Birmingham, AL: Samford University, 2003).

Cramer, C. H. *Case Institute of Technology: A Centennial History 1880-1980* (Cleveland: Case Western Reserve University, 1980).

Cramer, C.H. *Case Western Reserve: History of the University, 1826-1976* (Boston: Little Brown and Company, 1976).

Cross, Whitney R. *The Burned-Over District: The Social and Intellectual History of Enthusiastic Religion in Western New York, 1800-1850* (New York: Cornell University Press, 1981).

Dennett, Tyler. *John Hay: From Poetry to Politics* (New York: Dodd, Mead & Company, 1933).

Dolson, Hildegarde. *The Great Oildorado -The Gaudy and Turbulent Years of the First Oil Rush: Pennsylvania 1859-1880* (New York: Random House, 1959).

Dutka, Alan F. *Misfortune on Cleveland's Millionaires' Row* (Charleston, South Carolina: The History Press, 2015).

Edwards, Rebecca. *New Spirits: Americans in the Gilded Age 1865-1905* (New York: Oxford University Press, 2006).

Ellis, Joseph J. *American Dialogue: The Founders and Us* (New York: Alfred A. Knoft, 2018).

Erikson, Erik H. *Childhood and Society* (New York: W. W. Norton and Company, 1963).

Evans, Christopher H. *The Kingdom is Always but Coming: A Life of Walter Rauschenbusch* (Grand Rapids: William B. Eerdmans Publishing Company, 2004).

Evensen, Bruce J. *God's Man for the Gilded Age: D. L. Moody and the Rise of Modern Mass Evangelism* (Oxford: Oxford University Press, 2003).

Florman, Samuel C. *Blaming Technology: The Irrational Search for Scapegoats* (New York: St. Martin's Press, 1981).

Flynn, John T. *God's Gold: The Story of Rockefeller and His Times* (Westport, Conn: Greenwood Press, 1932).

Foer, Franklin. *World Without Mind: The Existential Threat of Big Tech* (New York: Penguin Press, 2017).

Foner, Philip. *The Great Labor Uprising of 1877* (New York: Pathfinder, 1977).

Fraser, Steve. *Every Man A Speculator: A History of Wall Street in American Life* (New York: Harper Collins Publishers, 2005).

Gordon, Sarah H. *Passage To Union: How the Railroads Transformed American Life, 1829-1929* (Chicago: Ivan R. Dee, 1996).

Goulder, Grace. *John D. Rockefeller: The Cleveland Years* (Cleveland: The Western Reserve Historical Society, 1972).

Greenberger, Scott S. *The Unexpected President: The Life and Times of Chester A. Arthur* (New York: DaCapo Press, 2017).

Greven, Phillip. "The Self Shaped and Misshaped: The Protestant Temperament Reconsidered," *Through A Glass Darkly: Reflections on Personal Identity in Early America*, Ronald Hoffman, et. al. eds. (Chapel Hill: The University of North Carolina Press, 1997) 348-369.

Gutman, Herbert. *Work, Culture and Society* (New York: Vintage Books, 1977).

Haddad, Gladys. *Flora Stone Mather: Daughter of Cleveland's Euclid Avenue & Ohio Western Reserve* (Kent, Ohio: The Kent State University Press, 2007).

Hamilton, Barbara J. "Who's Who? Identifying Victims of the Disaster," Corts, 21-38.

Hamilton, Darrell E. "Almost the Perfect Disaster," Corts, 1-20.

Harlow, Alvin F. *The Road of the Century: The Story of the New York Central* (New York: Creative Age Press, Inc., 1947).

Haydn, Hiram Collins. *Western Reserve University from Hudson to Cleveland: 1878-1890* (Cleveland: Western Reserve University, 1905).

Hiltzik, Michael. *Iron Empires: Robber Barons, Railroads, and the Making of Modern America* (Boston: Houghton Mifflin Harcourt, 2020).

Hofstadter, Richard. *The American Political Tradition* (New York: Vintage Books, 1989).

Holbrook, Stewart H. *The Story of American Railroads* (New York: American Legacy Press, 1947).

Howe, Daniel Walker. *What Hath God Wrought: The Transformation of America, 1815-1848* (New York: Oxford University Press, 2007).

Hungerford, Edward. *Men and Iron: The History of the New York Central* (New York: Thomas Y. Crowell, 1938).

Innes, Stephen. *Creating the Commonwealth: The Economic Culture of Puritan New England* (New York: W. W. Norton & Company, 1995).

Kalil, Timothy. "P. P. Bliss and Late Nineteenth-Century Urban Revivalism," Corts, 100-106.

Klein, Maury and Harvey A. Kantor. *Prisoners of Progress: American Industrial Cities 1850-1920* (New York: Macmillan Publishing Co., Inc., 1976).

Klein, Maury. *The Life and Legend of Jay Gould* (Baltimore: Johns Hopkins University Press, 1986).

Klein, Maury. *The Power Makers: Steam, Electricity, and the Men Who Invented Modern America* (New York: Bloomsbury Press, 2008).

Lane, Wheaton J. *Commodore Vanderbilt: An Epic of the Steam Age* (New York: Alfred A Knopf, 1942).

Lapham, Lewis. H. *Money and Class in America* (New York: Ballantine Books, 1988).

Larson, John Lauritz. *The Market Revolution in America: Liberty, Ambition, and the Eclipse of the Common Good* (New York: Cambridge University Press, 2010).

Licht, Walter. *Working for the Railroad: The Organization of Work in the Nineteenth Century* (Princeton, New Jersey: Princeton University Press, 1983).

Lupetkin, John M. *Jay Cooke's Gamble: The Northern Pacific Railroad, the Sioux, and the Panic of 1873* (Norman, Oklahoma: University of Oklahoma Press, 2006).

Mackley, Kathryn L. *Samuel Mather: First Citizen of Cleveland* (Cleveland: Tasora Books, 2013).

May, Henry F. *Protestant Churches and Industrial America* (New York: Harper & Brothers Publishers, 1949).

McCullough, David. *The Great Bridge* (New York: Simon and Schuster, 1972).

McCullough, Robert. *Crossings: A History of Vermont Bridges* (Barre, Vermont: Vermont Historical Society, 2005).

McLellan, Dave and Bill Warrick. *The Lake Shore and Michigan Southern Railway* (Polo, Illinois: Transportation Trails, 1989).

Mead, Sidney Earl. *Nathaniel William Taylor 1786-1858* (Chicago: University of Chicago Press, 1942).

Millard, Candice. *Destiny of the Republic: A Tale of Madness, Medicine, and the Murder of a President* (New York: Doubleday, 2011).

Miller, Carol Poh and Robert A. Wheeler. *Cleveland: A Concise History, 1796-1996* (Cleveland: Case Western Reserve University, 1997).

Miller, Perry. *Orthodoxy in Massachusetts 1630-1650* (New York: Harper Torchbook, 1933).

Moody, William R. *D. L. Moody* (New York: Garland Publishing, 1988).

Morris, Ray. *Railroad Administration* (New York: D. Appleton and Company, 1920).

Nasaw, David. *Andrew Carnegie* (New York: The Penguin Press, 2006).

Nevins, Allan. *John D. Rockefeller: The Heroic Age of American Enterprise* Vol. I (New York: Charles Scribner Sons, 1940).

Niebuhr, Reinhold. *Moral Man and Immoral Society* (New York: Charles Scribner's Sons, 1932).

O'Connell, Robert L. *Fierce Patriot: The Tangled Lives of William Tecumseh Sherman* (New York: Random House, 2015).

O'Toole, Patricia. *The Five of Hearts: An Intimate Portrait of Henry Adams and His Friends 1880-1918* (New York: Ballantine Books, 1990).

Petroski, Henry. *To Engineer Is Human: The Role of Failure in Successful Design* (New York: Vintage Books, 1992).

Plowden, David. *Bridges: The Spans of North America* (New York: The Viking Press, 1974).

Pollock, J. C. *Moody: A Biographical Portrait of the Pacesetter in Modern Mass Evangelism* (New York: The MacMillan Company, 1963).

Randel, William. *Centennial: American Life in 1876* (New York: Chilton Book Company, 1969).

Rauschenbusch, Walter. *Christianity and the Social Crisis* (New York: The MacMillan Company, 1907).

Rose, Kenneth W. "Why a University for Chicago and Not Cleveland? Religion and John D. Rockefeller's Early Philanthropy, 1855-1900," *From All Sides: Philanthropy in the Western Reserve* (Cleveland: Case Western Reserve University, 1995) 30-41.

Ruminski, Dan and Alan Dutka. *Cleveland in the Gilded Age: A Stroll Down Millionaires' Row* (Charleston, South Carolina: History Press, 2012).

Salter, Darius L. *"God Cannot Do without America;" Matthew Simpson and the Apotheosis of Protestant Nationalism* (Wilmore, Kentucky: First Fruits-The Academic Press of Asbury Seminary, 2017).

Salter, Darius. *America's Bishop: The Life of Francis Asbury* (Wilmore, Kentucky: First Fruits-The Academic Press of Asbury Seminary, 2020).

Salter, Darius. *American Evangelism: Its Theology and Practice* (Grand Rapids: Baker Books, 1996).

Sanford, Charles. *Quest for Paradise* (Urbana: University of Illinois Press, 1961).

Schlesinger, Sr. Arthur M. "A Critical Period in American Religion 1875-1900," *Religion in American History: Interpretive Essays,* John M. Mulder and John F. Wilson, eds. (Englewood Cliffs, NJ: Prentice-Hall, Inc., 1978)302-317.

Scholl, Doug. "The Contribution of Lawrence Kohlberg to Religious and Moral Education" *Religious Education,* Randolph Miller, ed. (New York: The Religious Education Association, 1971) 364-372.

Schuster, David G. *Neurasthenic Nation: America's Search for Health, Comfort, and Happiness, 1869-1920* (New Brunswick, NJ: Rudgers University Press, 2011).

Smith, H. Shelton, et. al. *American Christianity: An Historical Interpretation with Representative Documents,* Vol. I, 1607-1820 (New York: Charles Scribner's Sons, 1960).

Smith, Henry Nash, ed. *Popular Culture and Industrialism: 1865-1890* (New York: University Press, 1967).

Smith, Jean Edward. *Grant* (New York: Simon and Schuster Paperbacks, 2001).

Standiford, Les. *Meet You in Hell: Andrew Carnegie, Henry Clay Frick, and the Bitter Partnership that Transformed America* (New York: Three Rivers Press, 2005).

Stiles, T. J. *The First Tycoon: The Epic Life of Cornelius Vanderbilt* (New York: Alfred A. Knopf, 2009).

Stover, John F. *American Railroads* (Chicago: The University of Chicago Press, 1997).

Sweet, William Warren. *Circuit Rider Days in Indiana* (Indianapolis: W. K. Stewart, 1916).

Taliaferro, John. *All The Great Prizes: The Life of John Hay from Lincoln to Roosevelt* (New York: Simon & Schuster Paperbacks, 2013).

Tarbell, Ida. *The History of the Standard Oil Company* Vol. I (New York: The MacMillan Company, 1925).

Tuve, Jeanette. *Old Stone Church: In the Heart of the City Since 1820* (Virginia Beach, VA: Donning Company, 1993).

Tyrell, Henry G. *History of Bridge Engineering* (Chicago: Self-published, 1911).

Vogel, Charity. *The Angola Horror: The 1867 Train Wreck That Shocked The Nation and Transformed American Railroads* (Ithaca, NY: Cornell University Press, 2013).

Waite, Frederick Clayton. *The First Forty Years of the Cleveland Era of Western Reserve University,1881-1921* (Cleveland: Western Reserve University Press, 1954).

Ward, George Otis. *The Worcester Academy: Its Locations and Its Principals: 1834-1882* (Worcester, Massachusetts: The Davis Press, 1948).

Weinberg, Steve. *Taking on the Trust: The Epic Battle of Ida Tarbell and John D. Rockefeller* (New York: W. W. Norton and Company).

Weisberger, Bernard. *They Gathered at the River* (Boston: Little Brown, 1958).

Wheeler, George. *Pierpont Morgan and Friends: The Anatomy of a Myth* (Englewood Cliffs, New Jersey: Prentice-Hall, Inc., 1973).

White, John H. *The American Railroad Passenger Car* (Baltimore: The Johns Hopkins University Press, 1978).

White, Richard. *Railroaded: The Transcontinentals and the Making of Modern America* (New York: W. W. Norton and Company, 2012).

Wolmar, Christian. *A Short History of the Railroad* (London: Kindersley Unlimited, 2019).

Newspapers

Adley, Hannan. "Ten Years After Fleeing War: Syrians Reflect On Life in USA," *USA Today* (January 11, 2022).

"Amasa Stone," *Chicago Daily Tribune* (June 8, 1883).

"Amasa Stone's Funeral," *New York Times* (May 25, 1883).

"Amasa Stone Visiting Schools," *New Ulm Review* (August 18, 1883).

"Ashtabula," *The Cleveland Herald* (January 15, 1877)

"The Ashtabula Accident," Unidentified newspaper courtesy of Carrie Wimer, Genealogy/Local History Archivist at the Ashtabula Public Library, Ashtabula, Ohio.

"The Ashtabula Bridge: A Letter from Mr. Amasa Stone Concerning It," *New York Times* (January 31, 1877).

"Ashtabula: Continuation of the Coroner's Inquest," *The Cleveland Herald* (January 13, 1877).

"Ashtabula: January 11," *New York Times* (January 12, 1877) Leverich album.

Ashtabula News Extra (Wednesday, March 14, 1877).

Barensfeld, Tom. "Cleveland's Union Railroad Station," Unidentified newspaper courtesy of Case Western Reserve University Archives.

"The Buell Deficits," *The Cleveland Herald* (November 16, 1869).

"Charles MacDonald Engineer, Is Dead," *Brooklyn Daily Eagle* (July 9, 1928).

The Cleveland Herald (January 11, 1877)

"Cleveland, Ohio: January 16," *New York Tribune* (January 17, 1877) Leverich album.

"Cleveland, Ohio January 21," *New York Tribune* (January 22, 1877)

"Cleveland, Ohio: January 25, 1877," *New York Tribune* (January 26, 1877)

"The Commodore's Roads," *New York Tribune* (November 23, 1878).

"Death of Stillman Witt," and "Stillman Witt Dead," *Cleveland Plain Dealer* (May 4, 1875).

"The Disaster! What Has Been Said and Done at the Coroner's Inquest," *Ashtabula Weekly Telegraph* (February 2, 1877).

"Engineer Folsom's Testimony," *Ashtabula Weekly Telegraph* (January 26, 1877).

"The Funeral of the Late Charles Collins," *Cleveland Plain Dealer.* (January 22, 1877).

Harper's Weekly (July 15, 1876).

"Heavy Failure," *The Cleveland Leader* (February 2, 1883).

"Home of a Great Bridge Builder," *Cedar Rapids Weekly Gazette* (March 11, 1963).

"The Iron 'Age,'" *Chicago Daily Tribune* (February 3, 1883).

"The Last Tribute of Respect: Funeral Services of the Late Charles Collins at Ashtabula," *Cleveland Plain Dealer* (January 25, 1877).

"The Late Charles Collins," *Cleveland Plain Dealer* (January 29, 1877).

Leek, Charles. "To the Editor of the Herald," Ashtabula, March 12, 1877. Bridge Disaster scrapbook, Ashtabula County Historical Society.

"Meeting a Terrible Fate – Nine Persons Crushed and Burned in a Collision – A Train Crashing into the Rear of the Atlantic Express – Nine, Perhaps Twelve Victims Caught in the Burning Cars – State Senator Wagner Among the Dead – Narrow Escape of Many – Others – Terrible Scene at the Wreck." *New York Times* (January 14, 1882).

"Melancholy Suicide," *The Worcester Palladium* (Worcester, Massachusetts: December 1863).

"The Mills Shut Down," *Kansas City Evening Star*, Vol. 9, No.121 (February 3, 1883).

"Mob Rule," *Chicago Daily Tribune* (February 28, 1882).

"Mr. Amasa Stone Apparently Had Not Been Kept Posted," *Chicago Daily Tribune* (February 2, 1883).

New York Times (January 12, 1877; August 27, 1922) Leverich album.

New York Tribune (January 31, 1877; February 7, 1877) Leverich album.

"Obituary," *Chicago Daily Tribune* (May 13, 1883).

"Painful Accident," *The Cleveland Leader* (October 17, 1867).

Rich, Bob. "Bank Rolling Higher Learning Philanthropists Feud Led to Founding of Two Schools," *Cleveland Plain Dealer* (February 4, 1996).

"Secretary Hay Prostrated, Overwhelmed with Grief Over the Death of His Son," *New York Times* (June 24, 1901).

"A Shocking Tragedy," *Cleveland Plain Dealer* (November 15, 1869).

"Smyth, Anson. "To the Memory of Amasa Stone," *New York Evangelist* (July 12, 1883).

"Simple Services at the Funeral of Amasa Stone Today," *Cleveland Plain Dealer* (May 14, 1883).

"SUICIDE! Charles Collins, Engineer of the Lake Shore Railway Dead!" *Cleveland Plain Dealer* (January 20, 1877).

"Warned in Dreams: People Who Were Saved From Ashtabula by Presentiment," *Chicago Daily Tribune,* n.d., Bridge Disaster scrapbook, Ashtabula County Historical Society.

"What Has Been Said and Done at the Coroner's Inquest," *Ashtabula Weekly Telegraph*   (January 26, 1877).

"What Has Been Said and Done at the Coroner's Inquest," *Ashtabula Weekly Telegraph* (February 9, 1877).

"What Has Been Said and Done at the Coroner's Inquest," *Ashtabula Weekly Telegraph*   (February 16, 1877).

Periodical Articles

Anderson, William P. and J. A. L. Waddell "Memoir of Joseph Tomlinson," *Transactions of the Canadian Society of Engineers.* Vol. 19, 321-325.

"The Death of Mr. Charles Collins," *Railway World.* (February 3, 1887) 99.

Gasparini, Dario and David Simmons, "American Truss Bridge Connections in the 19th Century, 1850-1900," *Journal of Performance of Constructed Facilities* (August 1997) 130-140.

Gasparini, Dario and Melissa Fields. "Collapse of the Ashtabula Bridge on December 29, 1876," *Journal of Performance of Constructed Facilities* (May 1993) 109-125.

Hicks, Granville. "The Conversion of John Hay," *The New Republic* (June 10, 1931)100-101.

"How the Lake Shore Railroad Became Great," *Conductor and Brakeman* (March 1898) 158-163.

"Albert H. Howland," *Journal of the Boston Society of Engineers* Vol. 2 (Boston: 1915) 131-132.

"In Memoriam, Charles Paine," *Journal of the Western Society of Engineers* Vol XI (July-August 1906) 472-473.

Jaher, Frederick Cople. "Industrialism and the American Aristocrat: A Social Study of John Hay and His Novel, *The Bread-Winners,*" *Journal of the Illinois State Historical Society,* Vol. 65 (Spring 1972) 69-93.

Johannesen, Eric. "Stone's Trove: The Legacy of an American Oligarch," *Timeline* (June – July 1989) 27-33.

Kansas City Review of Science and Industry. "Scientific Miscellany: Kansas City Industries," Vol. V, No. 3 (July 1881)185-186.

Kemp, Emory L. "The Introduction of Cast and Wrought Iron in Bridge Building," *The Journal of the Society for Industrial Archeology* Vol. 19, No. 2 (1993) 5-16.

Kennedy, James Harrison. "Bankers and Banks of Cleveland, Ohio," *Magazine of Western History* (July 1885) 272-290.

Leedy, Walter. "Henry Vaughan's Cleveland Commission: A Study of Patronage, Context, and Civic Responsibility." 1-12. http: //architronic. saed.kent.edu/v3n3/v3n3.04.html.

Lepore, Jill. "Historians Who Love Too Much: Reflections on Microhistory and Biography," *The Journal of American* History (June 2001)129-144.

Love, Jeannine. "John LaFarge and Cleveland's 1885 Amasa Stone Memorial Window: A Case for Re-evaluation?" Courtesy of Don Guenther, Old Stone Presbyterian Church, Cleveland, Ohio.

Petroski, Henry. "On 19th Century Perceptions of Iron Bridge Failures," *Technology and Culture* Vol. 24, Number 4 (October 1983) 655-659.

Prevost, Lewis M., Jr. "Description of Howe's Patent Truss Bridge Carrying The Western Railroad Over The Connecticut River at Springfield, Massachusetts," *Journal of the Franklin Institute of the State of Pennsylvania, and Mechanics' Register* (May 1842) *289-303.*

"R. F. Hawkins Iron Works: Bridge Building and Boiler Making for the World at Large." *Progressive Springfield: Successful Springfield Industries.* No. III. n.d.

Simmons, David A. "Fall From Grace: Amasa Stone and the Ashtabula Bridge Collapse," *Timeline* (June – July 1989) 34-43.

"Warehouse Point R.R. Covered Bridge," *Connecticut River Valley Covered Bridge Society Bulletin.* (Summer 1988) 6.

Archives

Amasa Stone Chapel, Case Western Reserve University, Cleveland, Ohio.

Ashtabula County Historical Society Archives, Geneva on the Lake, Ohio.

Ashtabula, Ohio Public Library Archives, Ashtabula, Ohio.

Bentley Library Archives, Ann Arbor, Michigan.

Case Western Reserve University Archives, Cleveland, Ohio.

Cedar Rapids Public Library, Cedar Rapids, Iowa.

Cleveland City Hall Archives, Cleveland, Ohio.

Cleveland Public Library Archives, Cleveland, Ohio.
Kelvin Smith Library Special Collections, Case Western Reserve University, Cleveland, Ohio.
John Hay Library Archives, Brown University, Providence, Rhode Island.
The Lincoln Presidential Library, Springfield, Illinois.
Missouri Valley Special Collections, The Kansas City Public Library.
Old Stone Presbyterian Church, Cleveland, Ohio.
University of Rochester Archives, Rochester, New York.
Western Reserve Historical Society, Cleveland, Ohio.

Reference Works

"Devereux, John H.," *Encyclopedia of Cleveland History.* (https://case.edu/ech/)
Webster's Encyclopedic Unabridged Dictionary of the English Language (San Diego: Thunder Bay Press, 2001).

Reports

Annual Report of the Directors of the Cleveland, Painesville and Ashtabula Railroad Company for the year ending December 31, 1864. (Cleveland: Fairbanks, Benedict, and Company, 1865).
Annual Reports: Lake Shore and Michigan Southern Railroad Company, 1875: https://quod.lib.umich.edu/r/railroad/0549714.1875.001?rgn=main;view=fulltext. 1876.
The Penn Central Transportation Company Records, 1835-1960, Oversize Vol. 24, Bentley Library Archives, Ann Arbor, Michigan.
Report of the Joint Committee Concerning the Ashtabula Bridge Disaster, Under Joint Resolution of the General Assembly (Columbus: Nevins & Myers, State Printers, 1877) 63.

Website Sources

"Amasa Stone, American Industrialist," Datahub platform for finance.
"Deadliest Train Accidents in American History," Enjuris.com.
"Death By Defenestration: The Short Life of Adelbert 'Del' Stone Hay." theesotericcuriosa.blogspot.com/2010/07/from-shadows-shortlife-of-adelbert.htwl.
library.timelesstruths.org.

libwww.library.phil.gov/cencol/exh-testimony.tem.

"Miss Marion Shepherd." Engineering Tragedy: The Ashtabula Train Disaster. Retrieved July 15, 2022. https://www.engineeringtragedy.com/marion-shepard

"Transportation Research Board." National Academies Sciences, Engineering, Medicine. Retrieved July 8, 2022. www.trb.org/Profile/AnnualMeeting/Registration.

Unpublished Works

Danko, George. *The Evolution of the Simple Truss Bridge, 1790-1850: From Empiricism to Scientific Construction* (unpublished Ph.D. dissertation, University of Pennsylvania, 1979).

Dow, Burton Smith III. "Amasa Stone, Jr.: His Triumph and Tragedy." (unpublished M.A. thesis, Western Reserve University, 1956).

Laning, Paul F. "The History of the Lake Shore and Michigan Southern Railway in Ohio," (unpublished M.A. thesis, Ohio State University, 1938).

Neil, Bob J. *Phillip P. Bliss (1838-1876): Gospel Hymn Composer and Compiler* (unpublished Ph.D. dissertation, New Orleans Baptist Theological Seminary, 1977).

Smucker, David. *Philip Paul Bliss and the Musical, Cultural, and Religious Sources of the Gospel Music Tradition in the United States, 1850-1876* (unpublished Ph.D. dissertation, Boston University, 1981).

Wilcox, Nancy T. *The Ashtabula Bridge Disaster.* Appendix B.

Presentations

Bradley, Bruce. "A Library of First Resort for Science, Engineering, and Technology: The Linda Hall Library." Linda Hall Library, United States of America.

Gasparini, Dario. "The Western Railroad's Bridge Over the Connecticut River at Springfield, MA (September 26, 2017) utube.com/watch?v=6eJTtOWXthc.

MacDonald, Charles. "The Failure of the Ashtabula Bridge: Transactions of the American Society of Civil Engineers" (New York: American Society of Civil Engineers, February 21, 1877).

Ressler, Stephen. "The Ashtabula Bridge Disaster and the Advent of Civil Engineering Professionalism." 2022 Kahn Distinguished Lecture Series. Lehigh University, February 18, 2022.

Index

CPSIA information can be obtained
at www.ICGtesting.com
Printed in the USA
JSHW012111151222
34971JS00010B/228